VCD
Spectroscopy for Organic Chemists

VCD
Spectroscopy
for Organic
Chemists

Philip J. Stephens
Frank J. Devlin
James R. Cheeseman

CRC Press
Taylor & Francis Group
Boca Raton London New York

CRC Press is an imprint of the
Taylor & Francis Group, an **informa** business

CRC Press
Taylor & Francis Group
6000 Broken Sound Parkway NW, Suite 300
Boca Raton, FL 33487-2742

First issued in paperback 2019

ISBN-13: 978-1-4398-2171-8 (hbk)
ISBN-13: 978-0-367-38120-2 (pbk)

Library of Congress Cataloging-in-Publication Data

Stephens, Philip J., 1940-
   VCD spectroscopy for organic chemists / Philip J. Stephens, Frank J. Devlin, James R. Cheeseman.
      p. cm.
   Includes bibliographical references and index.
   ISBN 978-1-4398-2171-8 (hardback)
   1. Vibrational spectra. 2. Vibrational circular dichroism. I. Devlin, Frank J., 1949- II. Cheeseman, James R., 1963- III. Title.

QD96.V53S74 2012
547--dc23   2012000294

Visit the Taylor & Francis Web site at
http://www.taylorandfrancis.com

and the CRC Press Web site at
http://www.crcpress.com

Contents

Preface

Chiral organic molecules are currently of widespread interest to organic chemists and pharmaceutical chemists. In addition to synthetic chiral molecules, naturally occurring molecules, which are invariably chiral and generally enantiomerically enriched, are of potential interest as leads for new drugs. The increasing importance of chiral molecules has stimulated the development of improved research techniques, especially chromatography, and of new asymmetric synthesis methods as well as spectroscopic methods for their structural characterization.

Circular dichroism (CD) is the differential absorption of left- and right-circularly polarized light. The vibrational circular dichroism (VCD) spectrum of a molecule, first observed in the mid 1970s, is the CD resulting from vibrational excitations of the molecule. The VCD spectra of the two enantiomers of a chiral molecule are of equal magnitude and opposite sign: mirror-image enantiomers give mirror-image VCD spectra. In principle, the absolute configuration (AC) of a chiral molecule can therefore be determined from its VCD spectrum. In practice, the determination of the AC of a chiral molecule from its experimental VCD spectrum requires a methodology that reliably predicts the VCD spectra of its enantiomers. The development of a rigorous quantum-mechanical theory of VCD and its implementation in quantum chemistry programs provides a reliable systematic technique for determining ACs from experimental VCD spectra.

Given the availability of commercial VCD instrumentation and quantum chemistry software, it became possible in the late 1990s for chemists to utilize VCD in elucidating the stereochemistries of chiral organic molecules. The purpose of this book is to increase the awareness of organic chemists of the utility of VCD spectroscopy and to provide them with sufficient knowledge to incorporate the technique into their own research.

PJS is profoundly grateful to all of his former graduate students and postdoctoral research associates. He especially thanks his former postdoctoral research associates Dr. Jack Cheng, Professor Larry Nafie (Syracuse University) and Professor Tim Keiderling (University of Illinois at Chicago) and his mentor, Professor A. David Buckingham (Cambridge University). FJD thanks his mentor Professor Hector E. Rubalcava (University College, Dublin) for teaching him the fundamentals of molecular spectroscopy. JRC thanks Dr. Michael Frisch and Dr. Gary Trucks at Gaussian, Inc. for their tremendous support over the years. We all thank the wonderful collaborators who have been involved in our research projects.

We would also like to thank our very supportive editor, Lance Wobus, the project coordinator, David Fausel, project editor, Marsha Hecht, as well as the production staff at Taylor & Francis.

Finally, we thank our wives, Anne-Marie Stephens, Ann Marie Devlin and Joanne Hiscocks, for their patience and loving support.

Philip J. Stephens
University of Southern California

Frank J. Devlin
University of Southern California

James R. Cheeseman
Gaussian, Inc.

1 Introduction to Vibrational Circular Dichroism

Molecules are not totally rigid. Even at absolute zero (0 K), the lengths of bonds between atoms oscillate, the angles between adjacent bonds oscillate, and the dihedral angles between bonds separated by a bond oscillate. These motions are termed molecular vibrations. According to quantum mechanics, the energies of the vibrational states of molecules are quantized, the lowest energy state being termed the ground vibrational state.

When a molecule is exposed to electromagnetic radiation (light), the interaction between the radiation and the molecule can cause light photons to be absorbed by the molecule, and the molecule to be excited from the ground vibrational state, g, to higher energy vibrational states, e. The excitation g $\rightarrow$ e, of energy $\Delta E = E_e - E_g$, is caused by photons of energy $h\nu = \Delta E$, where $h\nu = hc/\lambda = hc\bar{\nu}$ (ν, λ, c, and $\bar{\nu}$ are the light frequency, wavelength, velocity, and reciprocal wavelength, respectively, and h is Planck's constant).

The absorption of light, resulting from vibrational excitations of a molecule, as a function of the light frequency, is termed the vibrational absorption spectrum of the molecule. The vibrational absorption spectrum of a molecule is measured using an infrared (IR) absorption spectrometer, in which IR light is passed through a sample containing the molecule. The sample can be a pure solid, liquid, or gas, or a solid, liquid, or gaseous solution of the molecule in a solvent. When the molecules in the sample are selectively oriented, as in a crystalline solid sample, the absorption spectrum is dependent on the linear polarization of the light. When the molecules are randomly oriented, as is the case in pure liquid and gaseous samples, and in liquid and gaseous solutions, the absorption spectrum is linear polarization independent. Most commonly, vibrational absorption spectra are measured using unpolarized IR light and samples in which the molecules are randomly oriented. An example of a molecular vibrational absorption spectrum is shown in Figure 1.1. The molecule is camphor; the spectrum was measured using unpolarized IR light, in a cell of pathlength 236 microns (μ), and over the IR frequency range, of reciprocal wavelengths (wavenumbers) 1,530–825 cm^{-1}. Absorption is observed at many frequencies, demonstrating the existence of many vibrationally excited states.

All molecules belong to one of two classes: achiral and chiral. By definition, an achiral molecule is identical to its mirror image; i.e., if the molecule is reflected in a mirror and then rotated, it can be superimposed on the original, unreflected molecule. A chiral molecule is different: the molecule and its mirror image are not superimposable, and therefore constitute different molecules. The two forms of the molecule are termed enantiomers. Since human left and right hands are mirror

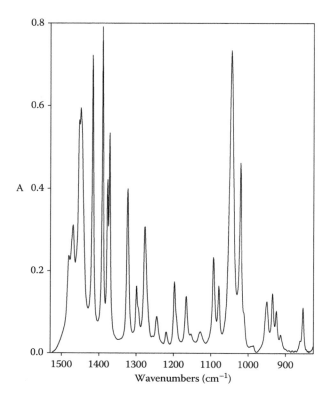

FIGURE 1.1 The mid-IR vibrational absorption spectrum of a 0.38 M CCl_4 solution of 1*R*,4*R* camphor. A is the absorbance (defined in Chapter 2).

images, and not superimposable, the two mirror image forms of a chiral molecule are sometimes also referred to as left-handed and right-handed.

A simple example of a chiral molecule is CHFClBr. The two enantiomers are:

Any molecule of the formula $CR_1R_2R_3R_4$ (R_1, R_2, R_3, and R_4 all being different) is also chiral. Of great biological significance is the chirality of amino acids, in which $R_1 = H$, $R_2 = NH_2$, $R_3 = COOH$, and R_4 depends on the specific amino acid (e.g., in alanine, $R_4 = CH_3$). A C atom bonded to four different groups is termed a stereogenic C atom. Many chiral organic molecules contain multiple stereogenic C atoms. For example, in the steroid natural product cholesterol, C atoms 1, 2, 3, 4, 5, 6, 7, and 8 are stereogenic:

cholesterol

In 1956, Cahn, Ingold and Prelog introduced a notation that specifies the chiralities of stereogenic C atoms: a C atom is either R or S [1]. The overall three-dimensional (3D) structure of an enantiomer of a chiral molecule can then be defined by listing which atoms are R or S. For example, naturally occurring cholesterol is $1S,2R,3S,4S,5S,6R,7R,8R$ [2]. This label is termed the absolute configuration (AC).

Although the 3D geometries of the two enantiomers of a chiral molecule are not identical, they do possess considerable similarity. In particular, all bond lengths, bond angles, and nonbonded interatomic distances are unchanged on reflection in a mirror. As a result, the vibrational excitation energies of the two enantiomers and the vibrational absorption spectra, measured using samples of randomly oriented molecules and unpolarized IR light, are identical.

The electric and magnetic fields of a linearly polarized light wave each oscillate sinusoidally in a plane containing the propagation direction, the electric field and magnetic field planes being perpendicular to each other. Passage of a linearly polarized light wave through an optical device called a quarter-wave plate [3] converts the light wave into a circularly polarized (CP) light wave. Two forms of CP light can be generated, termed right circularly polarized (RCP) and left circularly polarized (LCP). In both RCP and LCP light, the electric and magnetic fields rotate helically about the propagation direction of the wave. In RCP light the helix is right-handed and in LCP light the helix is left-handed. Thus, RCP and LCP light waves of the same frequency are mirror images.

The vibrational absorption spectrum of a molecule can also be measured using CP light. If the molecule is achiral and randomly oriented, the spectra obtained using RCP and LCP light are identical. However, if the molecule is chiral, this is not the case. The difference in absorption of RCP and LCP light is termed circular dichroism (CD). Conventionally, CD is defined as the absorbance (defined in Chapter 2) of LCP light (A_L) minus the absorbance of RCP light (A_R): $CD = \Delta A = A_L - A_R$. CD is therefore positive if $A_L > A_R$ and negative if $A_L < A_R$. For the two enantiomers of the chiral molecule, the CD at every light frequency is of equal magnitude, but is opposite in sign; their CD spectra are thus mirror images. The vibrational circular dichroism (VCD) spectrum of a molecule is the CD resulting from vibrational excitations of the molecule. Examples of the VCD spectra of the enantiomers of a chiral molecule are shown in Figure 1.2. The chiral molecule is camphor. The two enantiomers are:

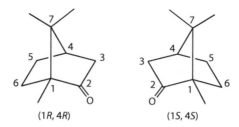

The VCD spectra of the two enantiomers were measured using 0.38 M solutions of camphor in the achiral solvent CCl_4 and a cell of pathlength 236 μ. The mirror image property of the VCD spectra of the two enantiomers is qualitatively obvious. Quantitatively, it is proven by addition of the two VCD spectra; as shown in Figure 1.2, the sum of the two VCD spectra is very close to zero at all frequencies.

The phenomenon of circular dichroism was first discovered by the French scientist Aimé Cotton in 1896 [4] and subsequently became known as the Cotton effect. The CD measured by Cotton was in the near-ultraviolet (UV) spectral region and originated in electronic excitations of molecules. As with the vibrational states of molecules, the electronic states of molecules are quantized. Light photons of the same energy as the energy of excitation from the lowest energy (ground) electronic state to a higher energy (excited) electronic state are absorbed by the molecule.

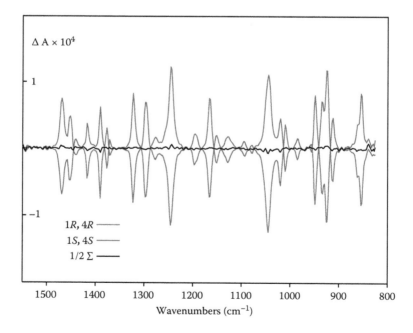

FIGURE 1.2 (SEE COLOR INSERT.)   The mid-IR VCD spectra of 0.38 M CCl_4 solutions of 1R,4R and 1S,4S camphor, using a cell of pathlength 236 μ. Σ is the sum of the spectra. The measurement of the spectra is discussed in Chapter 2.

Electronic absorption and CD typically occur in the visible-ultraviolet (VIS-UV) spectral region (200–1,000 nm, 10,000–50,000 cm^{-1}), where light frequencies are much higher than in the IR spectral region.

Electronic CD (ECD) spectra of chiral molecules were not widely studied until after the Second World War, when new instrumentation for VIS-UV CD measurement was developed, using modulation techniques and electro-optic modulators named Pockels cells, permitting ECD spectra to be more efficiently measured [5]. This led rapidly to a much higher level of interest in the application of ECD spectra to the elucidation of the stereochemistries of organic molecules. An important development, which facilitated such applications, occurred in 1961 when Moffitt, Woodward, Moscowitz, Klyne, and Djerassi proposed the octant rule, which predicts the sign of the ECD of the lowest energy electronic excitation of a carbonyl (C=O) group in a chiral molecule [6]. The octant rule enabled the ACs of chiral molecules containing carbonyl groups to be determined. In addition, it led to the development of similar rules for the electronic excitations of other functional groups, which further widened the application of ECD spectra to the determination of ACs [7,8].

The reason for the interest in the determination of the ACs of chiral molecules using ECD spectroscopy was that it provided a less laborious procedure than other available methods. Two approaches were predominant in determining ACs prior to the introduction of the ECD approach: (1) x-ray crystallography and (2) chemical synthesis. X-ray crystallography was used in two ways. One procedure was developed by Bijvoet et al. [9] and used the anomalous x-ray scattering dispersion of a high atomic number atom (a "heavy atom") in the molecule. For example, the AC of camphor was determined by replacing one of its H atoms by a Br atom and determining the AC of the resulting 3-Br-camphor [10]. Since the bromination of camphor does not change its AC, the AC of the 3-Br-camphor is identical to that of camphor. A second x-ray crystallography procedure used a derivatization reaction of the chiral molecule with a second chiral molecule, of known AC. Determination of the relative stereochemistry of the product molecule via x-ray crystallography then determines the AC of the underivatized chiral molecule. Since x-ray crystallography is widely used, and the most definitive method for determining the geometry of a molecule, these two procedures are highly reliable ways to determine the ACs of chiral molecules. However, there are disadvantages: (1) in the first procedure, if the molecule does not possess a heavy atom, a chemical reaction must be carried out; (2) in the second procedure, a chemical reaction must always be carried out; and (3) in both procedures, single crystals of sufficient size to permit x-ray crystallography must be obtainable. Since, sometimes, neither the reactions chosen nor the crystallization of the products are practical, x-ray crystallography is not always easily used in determining ACs. The principal alternative approach to x-ray crystallography was to synthesize the chiral molecule of interest from a precursor chiral molecule of known AC, using reactions whose mechanisms are understood and whose impacts on the molecular stereochemistry are predictable. This procedure is useful if such a synthetic procedure is practical, which is often, but not always, the case.

An additional application of ECD spectroscopy was also of interest to organic chemists after the development of ECD instrumentation: the conformational analysis of conformationally flexible chiral molecules. In the 1950s, it became clear, especially due to the work of Barton [11], that some organic molecules can have more than one structure: the multiple structures are termed conformations, and such molecules are termed conformationally flexible. An early example of a conformationally flexible molecule was cyclohexane, whose C6 ring can have two structures, termed chair and twist-boat conformations, discussed in Chapter 5:

Chair   Twist-boat

When the energy barriers between the conformations of a conformationally flexible molecule are not very high, the conformations can interconvert rapidly at room temperature, and therefore exist in equilibrium. The percentage populations of the conformations are determined by their relative free energies and the temperature, according to Boltzmann statistics [12]. Since the ECD of a molecule is sensitive to its geometry, different conformations of a given enantiomer of a chiral molecule exhibit different ECD spectra. Consequently, ECD spectroscopy provides a technique for elucidating the conformations populated in a chiral molecule [13].

In addition to CD, chiral molecules exhibit other properties, which are different for the two enantiomers. The earliest such property to be discovered was optical rotation (OR) [14]. When linearly polarized light is passed through a sample containing randomly oriented chiral molecules, the plane of the polarization is rotated by an angle α. The OR α is equal in magnitude, but opposite in sign, for the two enantiomers. Historically, OR was most often measured using light emitted by a sodium lamp at a wavelength referred to as the sodium D line (589 nm), and converted to the specific rotation, $[\alpha]_D$, defined by $[\alpha]_D = \alpha/\ell c$, where ℓ is the cell pathlength in dm and c is the concentration of the chiral molecule in the sample in g/100 ml. The two enantiomers with positive and negative $[\alpha]_D$ values were then termed (+) and (–), respectively. The AC of a chiral molecule is determined for either the (+) or the (–) enantiomer. In reporting the conclusion, both the AC and OR sign are listed. Thus, for example, the AC of (+)-camphor is (1R,4R)-(+) and the AC of (–)-camphor is (1S,4S)-(–).

Following the development of efficient instrumentation for the measurement of VIS-UV ECD spectra and the widespread application of ECD spectra to the elucidation of the ACs and/or conformational structures of chiral organic molecules, the obvious questions arose: Can CD due to vibrational excitations, vibrational circular dichroism (VCD), be measured in the IR spectral region, and if so, can VCD also be used to determine the ACs and/or conformational structures of chiral organic molecules? As a result, in the early 1970s, instruments capable of measuring CD in the IR spectral region were designed and built in two laboratories: the Stephens laboratory at the University of Southern California (USC) [15] and the Holzwarth laboratory at the University of Chicago (UC) [16]. In the 1970s experimental VCD spectra of chiral organic and organometallic molecules were measured and published: one molecule, 2,2,2-trifluoro-1-phenylethanol, at UC [17] and the 23 molecules listed in Table 1.1

TABLE 1.1
Chiral Molecules Whose VCD was Measured at USC in the 1970s

1. 2,2,2-Trifluoro-1-phenylethanol
2. α-Methylbenzylamine
3. N,N-α-Trimethylbenzylamine
4. 3-Methyl-cyclopentanone
5. 3-Methyl-cyclohexanone
6. Menthol
7. α-Pinene
8. β-Pinene
9. Camphor
10. 3-Br-camphor
11. Borneol
12. Tris (3-trifluoromethylhydroxymethylene-d-camphorato) praseodymium
13. Tris (3-trifluoromethylhydroxymethylene-d-camphorato) europium
14. Poly-1-methyl-propyl-vinyl-ether
15. Poly-4-methyl-1-hexene
16. Dimethyl tartrate
17. Alanine
18. Camphoric anhydride
19. 1,6-Spiro [4.4] nonadiene
20. Exo-3-deutero-isoborneol
21. Exo-3-deutero-camphor
22. α-Deutero-propylbenzene
23. $Fe(C_5H_5) (P(C_6H_5)_3) (CO) (Et)$

at USC [18]. This work proved that VCD spectra could indeed become a practical technique for determining the stereochemistries of chiral organic molecules. In order to realize this promise, two developments remained to be accomplished. First, the frequency range of the existing VCD instrumentation, which was limited to frequencies of >1,600 cm^{-1}, had to be extended, to permit a wider fraction of the IR spectral region to be accessed, and the sensitivity (i.e., the signal-to-noise ratio) of the existing VCD instrumentation had to be increased, to permit VCD to be measured reliably for a larger number of molecules. Second, a methodology by which molecular stereochemistries could be reliably deduced from experimental VCD spectra had to be developed; otherwise, the spectra would be of no practical value. By the mid-1980s the frequency range of the USC VCD instrument had been greatly expanded by Devlin and Stephens [19], the lower frequency limit having been extended to ~650 cm^{-1}, and the sensitivity substantially increased. At the same time, a rigorous quantum mechanical theory of VCD had been developed by Stephens [20], which was implemented for a number of chiral molecules using the *ab initio* Hartree-Fock (HF) molecular orbital theory [21]. Comparison of *ab initio* HF calculations of VCD spectra using the Stephens theory to experimental VCD spectra led to great optimism that VCD spectroscopy could soon become a widely used technique. Two further developments added to this optimism. First, the explosion in the late 1980s of *ab initio* density functional theory (DFT) and the documentation of its much greater accuracy than HF theory in predicting molecular properties made it desirable to implement the Stephens theory of VCD using DFT. This was carried out in the 1990s by Cheeseman and Frisch at GAUSSIAN, Inc. [22], using the GAUSSIAN program, which was originally developed in John Pople's laboratory and subsequently has become a widely distributed program, frequently used by both quantum chemists and organic chemists for predicting molecular properties. Comparison of the VCD spectra of chiral organic molecules, calculated using DFT in the Gaussian programs G92, G98, G03, and G09 [23], to experimental VCD spectra proved the superior accuracy of DFT VCD spectra [24]. Second, the extension of the methodology used by Stephens and Holzwarth for measuring VCD using dispersive IR spectrometers to Fourier transform IR (FTIR) spectrometers demonstrated that VCD spectra could also be obtained using FT instrumentation [25]. Following the DFT implementation of the Stephens theory of VCD, several companies manufacturing and marketing FTIR spectrometers realized that a market for commercial FT VCD instrumentation could exist, and began the manufacturing and marketing of VCD instruments. As a result, potential users of VCD spectroscopy no longer had to build their own instrumentation.

Given the availability of commercial software, permitting the prediction of VCD spectra using DFT, and of commercial VCD instrumentation, it became possible in the late 1990s for chemists to utilize VCD in elucidating the stereochemistries of chiral organic molecules. As a result, the number of publications per year reporting VCD studies of chiral organic molecules substantially increased. Despite this boom, many organic chemists remain unfamiliar with VCD spectroscopy. The purpose of this book is to increase the awareness of organic chemists of the utility of VCD spectroscopy. To achieve this purpose, we discuss in detail the experimental measurement of VCD spectra and their analysis using the Stephens theory of VCD, implemented using *ab initio* DFT. In Chapter 2, we discuss the experimental measurement of vibrational absorption and VCD spectra. In Chapter 3, we discuss the fundamental quantum mechanical theory of the vibrational states of molecules and of their vibrational absorption and VCD spectra. In Chapter 4, we discuss the application of the *ab initio* HF and DFT methods of quantum chemistry to the prediction of the molecular structures and vibrational states of organic molecules. In Chapter 5, we discuss the conformational analysis of conformationally flexible molecules. In Chapter 6, we discuss the analysis of the vibrational absorption and VCD spectra of a number of conformationally rigid chiral organic molecules, in order to define the optimum basis sets and DFT functionals for calculations of vibrational absorption and VCD spectra, and to define the methodology by which ACs are deduced from VCD spectra. Finally, in Chapter 7, we present studies of a set of chiral organic molecules that further document the power of VCD spectroscopy and make clear how wide is the applicability of this technique.

REFERENCES

1. (a) R.S. Cahn, C.K. Ingold, V. Prelog, The Specification of Asymmetric Configuration in Organic Chemistry, *Experientia*, 12, 81–94, 1956; (b) R.S. Cahn, C.K. Ingold, V. Prelog, Specification of Molecular Chirality, *Angew. Chem. Int. Ed. Engl.*, 5, 385–415, 1966.

2. (a) E.J. Westover, D.F. Covey, The Enantiomer of Cholesterol, *J. Membrane Biol.*, 202, 61–72, 2004; (b) J.W. Cornforth, I. Youhotsky, G. Popjak, Absolute Configuration of Cholesterol, *Nature*, 173, 536, 1954.
3. E. Hecht, A. Zajac, *Optics*, Addison-Wesley, 1975.
4. A. Cotton, Absorption and Dispersion of Light, *Ann. Chim. Physique*, 8, 347, 1896.
5. L. Velluz, M. Legrand, M. Grosjean, *Optical Circular Dichroism, Principles, Measurements and Applications*, Academic Press, 1965.
6. W. Moffitt, R.B. Woodward, A. Moscowitz, W. Klyne, C. Djerassi, Structure and the Optical Rotatory Dispersion of Saturated Ketones, *J. Am. Chem. Soc.*, 83, 4013–4018, 1961.
7. (a) P. Crabbé, *Optical Rotatory Dispersion and Circular Dichroism in Organic Chemistry*, Holden-Day, 1965; (b) P. Crabbé, *ORD and CD in Chemistry and Biochemistry*, Academic Press, 1972.
8. D.A. Lightner, J.E. Gurst, *Organic Conformational Analysis and Stereochemistry from Circular Dichroism Spectroscopy*, Wiley, 2000.
9. (a) J.M. Bijvoet, A.F. Peerdeman, A.J. van Bommel, Determination of the Absolute Configuration of Optically Active Compounds by Means of X-Rays, *Nature*, 168, 271–272, 1951; (b) J.M. Bijvoet, Determination of the Absolute Configuration of Optical Antipodes, *Endeavour*, 14, 71–77, 1955; (c) J.M. Bijvoet, Structure of Optically Active Compounds in the Solid State, *Nature*, 173, 888–891, 1954; (d) C. Giacovazzo, H.L. Monaco, G. Artioli, D. Viterbo, G. Ferraris, G. Gilli, G. Zanotti, M. Catti, *Fundamentals of Crystallography*, 2nd ed., Oxford University Press, Oxford, 2002.
10. (a) F.H. Allen, D. Rogers, A Redetermination of the Crystal Structure of (+)-3-Bromocamphor and Its Absolute Configuration, *Chem. Comm.*, 837–838, 1966; (b) M.G. Northolt, J.H. Palm, Determination of the Absolute Configuration of 3-Bromocamphor, *Recueil des Travaux Chimiques des Pays-Bas*, 85, 143–146, 1966.
11. (a) D.H.R. Barton, The Conformation of the Steroid Nucleus, *Experientia*, 6, 316–329, 1950; (b) D.H.R. Barton, The Principles of Conformational Analysis, *Science*, 169, 539–544, 1970.
12. A.H. Carter, *Classical and Statistical Thermodynamics*, Prentice-Hall, 2001.
13. Chapter 5 of reference 8: Conformational Analysis of substituted cyclohexanones.
14. T.M. Lowry, *Optical Rotatory Power*, Longmans, Green and Co., 1935.
15. G.A. Osborne, J.C. Cheng, P.J. Stephens, A Near-Infrared Circular Dichroism and Magnetic Circular Dichroism Instrument, *Rev. Sci. Instrum.*, 44, 10–15, 1973.
16. I. Chabay, G. Holzwarth, Infrared Circular Dichroism and Linear Dichroism Spectrometer, *Appl. Opt.*, 14, 454–459 1975.
17. G. Holzwarth, E.C. Hsu, H.S. Mosher, T.R. Faulkner, A. Moscowitz, Infrared Circular Dichroism of Carbon-Hydrogen and Carbon-Deuterium Stretching Modes. Observations, *J. Am. Chem. Soc.*, 96, 251–252, 1974.
18. (a) L.A. Nafie, J.C. Cheng, P.J. Stephens, Vibrational Circular Dichroism of 2,2,2-Trifluoro-1-Phenylethanol, *J. Am. Chem. Soc.*, 97, 3842, 1975; (b) L.A. Nafie, T.A. Keiderling, P.J. Stephens, Vibrational Circular Dichroism, *J. Am. Chem. Soc.*, 98, 2715–2723, 1976; (c) T.A. Keiderling, P.J. Stephens, Vibrational Circular Dichroism of Overtone and Combination Bands, *Chem. Phys. Lett.*, 41, 46–48, 1976; (d) T.A. Keiderling, P.J. Stephens, Vibrational Circular Dichroism of Dimethyl Tartrate: A Coupled Oscillator, *J. Am. Chem. Soc.*, 99, 8061–8062, 1977; (e) R. Clark, P.J. Stephens, Vibrational Optical Activity, *Proc. Soc. Photo-Opt. Inst. Eng.*, 112, 127–131, 1977; (f) P.J. Stephens, R. Clark, Vibrational Circular Dichroism: The Experimental Viewpoint, in *Optical Activity and Chiral Discrimination*, ed. S.F. Mason, D. Reidel Publishing, Dordrect, The Netherlands, 1979, pp. 263–287; (g) T.A. Keiderling and P.J. Stephens, Vibrational Circular Dichroism of Spirononadiene. Fixed Partial Charge Calculations, *J. Am. Chem. Soc.*, 101, 1396–1400, 1979.

19. F. Devlin, P.J. Stephens, Vibrational Circular Dichroism Measurement in the Frequency Range of 800 to 650 cm^{-1}, *Appl. Spectrosc.*, 41, 1142–1144, 1987.

20. (a) P.J. Stephens, Theory of Vibrational Circular Dichroism, *J. Phys. Chem.*, 89, 748–752, 1985; (b) P.J. Stephens, M.A. Lowe, Vibrational Circular Dichroism, *Ann. Rev. Phys. Chem.*, 36, 213–241, 1985; (c) P.J. Stephens, Gauge Dependence of Vibrational Magnetic Dipole Transition Moments and Rotational Strengths, *J. Phys. Chem.*, 91, 1712–1715, 1987; (d) P.J. Stephens, The Theory of Vibrational Optical Activity, in *Understanding Molecular Properties*, ed. J. Avery, J.P. Dahl, A.E. Hansen, D. Reidel, Dordrect, The Netherlands, 1987, pp. 333–342.

21. (a) P.J. Stephens, M.A. Lowe, Vibrational Circular Dichroism, *Ann. Rev. Phys. Chem.*, 36, 213–241, 1985; (b) M.A. Lowe, P.J. Stephens, G.A. Segal, The Theory of Vibrational Circular Dichroism: Trans 1,2-Dideuteriocyclobutane and Propylene Oxide, *Chem. Phys. Lett.*, 123, 108–116, 1986; (c) M.A. Lowe, G.A. Segal, P.J. Stephens, The Theory of Vibrational Circular Dichroism: Trans-1,2-Dideuteriocyclopropane, *J. Am. Chem. Soc.*, 108, 248–256, 1986; (d) P. Lazzeretti, R. Zanasi, P.J. Stephens, Magnetic Dipole Transition Moments and Rotational Strengths of Vibrational Transitions: An Alternative Formalism, *J. Phys. Chem.*, 90, 6761–6763, 1986; (e) P.J. Stephens, The Theory of Vibrational Optical Activity, in *Understanding Molecular Properties*, ed. J. Avery, J.P. Dahl, A.E. Hansen, D. Reidel, Dordrect, The Netherlands, 1987, pp. 333–342; (f) R.D. Amos, N.C. Handy, K.J. Jalkanen, P.J. Stephens, Efficient Calculation of Vibrational Magnetic Dipole Transition Moments and Rotational Strengths, *Chem. Phys. Lett.*, 133, 21–26, 1987; (g) K.J. Jalkanen, P.J. Stephens, R.D. Amos, N.C. Handy, Theory of Vibrational Circular Dichroism: Trans-1(S), 2(S)-Dicyanocyclopropane, *J. Am. Chem. Soc.*, 109, 7193–7194, 1987; (h) K.J. Jalkanen, P.J. Stephens, R.D. Amos, N.C. Handy, Basis Set Dependence of *Ab Initio* Predictions of Vibrational Rotational Strengths: NHDT, *Chem. Phys. Lett.*, 142, 153–158, 1987; (i) K.J. Jalkanen, P.J. Stephens, R.D. Amos, N.C. Handy, Gauge Dependence of Vibrational Rotational Strengths: NHDT, *J. Phys. Chem.* 92, 1781–1785, 1988; (j) K.J. Jalkanen, P.J. Stephens, R.D. Amos, N.C. Handy, Theory of Vibrational Circular Dichroism: Trans-2,3-Dideuterio-Oxirane, *J. Am. Chem. Soc.*, 110, 2012–2013, 1988; (k) R.W. Kawiecki, F. Devlin, P.J. Stephens, R.D. Amos, N.C. Handy, Vibrational Circular Dichroism of Propylene Oxide, *Chem. Phys. Lett.*, 145, 411–417, 1988; (l) R.D. Amos, K.J. Jalkanen, P.J. Stephens, Alternative Formalism for the Calculation of Atomic Polar Tensors and Atomic Axial Tensors, *J. Phys. Chem.*, 92, 5571–5575, 1988; (m) K.J. Jalkanen, P.J. Stephens, P. Lazzeretti, R. Zanasi, Nuclear Shielding Tensors, Atomic Polar and Axial Tensors and Vibrational Dipole and Rotational Strengths of NHDT, *J. Chem. Phys.*, 90, 3204–3213, 1989; (n) P.J. Stephens, The *A Priori* Prediction of Vibrational Circular Dichroism Spectra: A New Approach to the Study of the Stereochemistry of Chiral Molecules, *Croat. Chem. Acta*, 62, 429–440, 1989; (o) P.J. Stephens, Vibronic Interactions in the Electronic Ground State: Vibrational Circular Dichroism Spectroscopy, in *Vibronic Processes in Inorganic Chemistry*, ed. C.D. Flint, Kluwer, 1989, pp. 371–384; (p) K.J. Jalkanen, P.J. Stephens, P. Lazzeretti, R. Zanasi, Random Phase Approximation Calculations of Vibrational Circular Dichroism: Trans-2,3-Dideuteriooxirane, *J. Phys. Chem.*, 93, 6583–6584, 1989; (q) P.J. Stephens, K.J. Jalkanen, R.D. Amos, P. Lazzeretti, R. Zanasi, *Ab Initio* Calculations of Atomic Polar and Axial Tensors for HF, H_2O, NH_3 and CH_4, *J. Phys. Chem.*, 94, 1811–1830, 1990; (r) K.J Jalkanen, R.W. Kawiecki, P.J. Stephens, R.D. Amos, Basis Set and Gauge Dependence of *Ab Initio* Calculations of Vibrational Rotational Strengths, *J. Phys. Chem.*, 94, 7040–7055, 1990; (s) R. Bursi, F.J. Devlin, P.J. Stephens, Vibrationally Induced Ring Currents? The Vibrational Circular Dichroism of Methyl Lactate, *J. Am. Chem. Soc.*, 112, 9430–9432, 1990; (t) R. Bursi, P.J. Stephens, Ring Current Contributions to Vibrational Circular Dichroism? *Ab Initio* Calculations

for Methyl Glycolate-d$_1$ and -d$_4$, *J. Phys. Chem.*, 95, 6447–6454, 1991; (u) R.W. Kawiecki, F.J. Devlin, P.J. Stephens, R.D. Amos, Vibrational Circular Dichroism of Propylene Oxide, *J. Phys. Chem.*, 95, 9817–9831, 1991.

22. Gaussian, www.gaussian.com.
23. M.J. Frisch, G.W. Trucks, H.B. Schlegel, G.E. Scuseria, M.A. Robb, J.R. Cheeseman, G. Scalmani, V. Barone, B. Mennucci, G.A. Petersson, H. Nakatsuji, M. Caricato, X. Li, H.P. Hratchian, A.F. Izmaylov, J. Bloino, G. Zheng, J.L. Sonnenberg, M. Hada, M. Ehara, K. Toyota, R. Fukuda, J. Hasegawa, M. Ishida, T. Nakajima, Y. Honda, O. Kitao, H. Nakai, T. Vreven, J.A. Montgomery Jr., J.E. Peralta, F. Ogliaro, M.J. Bearpark, J. Heyd, E.N. Brothers, K.N. Kudin, V.N. Staroverov, R. Kobayashi, J. Normand, K. Raghavachari, A.P. Rendell, J.C. Burant, S.S. Iyengar, J. Tomasi, M. Cossi, N. Rega, N.J. Millam, M. Klene, J.E. Knox, J.B. Cross, V. Bakken, C. Adamo, J. Jaramillo, R. Gomperts, R.E. Stratmann, O. Yazyev, A.J. Austin, R. Cammi, C. Pomelli, J.W. Ochterski, R.L. Martin, K. Morokuma, V.G. Zakrzewski, G.A. Voth, P. Salvador, J.J. Dannenberg, S. Dapprich, A.D. Daniels, Ö. Farkas, J.B. Foresman, J.V. Ortiz, J. Cioslowski, D.J. Fox, *Gaussian 09*, Gaussian, Wallingford, CT, 2009. For previous versions, see www.gaussian.com.
24. (a) P.J. Stephens, F.J. Devlin, C.F. Chabalowski, M.J. Frisch, *Ab Initio* Calculation of Vibrational Absorption and Circular Dichroism Spectra Using Density Functional Force Fields, *J. Phys. Chem.*, 98, 11623–11627, 1994; (b) P.J. Stephens, F.J. Devlin, C.S. Ashvar, C.F. Chabalowski, M.J. Frisch, Theoretical Calculation of Vibrational Circular Dichroism Spectra, *Faraday Discuss.*, 99, 103–119, 1994; (c) K.L. Bak, F.J. Devlin, C.S. Ashvar, P.R. Taylor, M.J. Frisch, P.J. Stephens, *Ab Initio* Calculation of Vibrational Circular Dichroism Spectra Using Gauge-Invariant Atomic Orbitals, *J. Phys. Chem.*, 99, 14918–14922, 1995; (d) F.J. Devlin, J.W. Finley, P.J. Stephens, M.J. Frisch, *Ab Initio* Calculation of Vibrational Absorption and Circular Dichroism Spectra Using Density Functional Force Fields: A Comparison of Local, Non-Local and Hybrid Density Functionals, *J. Phys. Chem.*, 99, 16883–16902, 1995; (e) P.J. Stephens, F.J. Devlin, C.S. Ashvar, K.L. Bak, P.R. Taylor, M.J. Frisch, Comparison of Local, Non-Local and Hybrid Density Functionals Using Vibrational Absorption and Circular Dichroism Spectroscopy, in *Chemical Applications of Density-Functional Theory*, ed. B.B. Laird, R.B. Ross, T. Ziegler, ACS Symposium Series 629, 1996, chap. 7, pp. 105–113; (f) J.R. Cheeseman, M.J. Frisch, F.J. Devlin, P.J. Stephens, *Ab Initio* Calculation of Atomic Axial Tensors and Vibrational Rotational Strengths Using Density Functional Theory, *Chem. Phys. Lett.*, 252, 211–220, 1996; (g) C.S. Ashvar, F.J. Devlin, K.L. Bak, P.R. Taylor, P.J. Stephens, *Ab Initio* Calculation of Vibrational Absorption and Circular Dichroism Spectra: 6,8-Dioxabicyclo[3.2.1]Octane, *J. Phys. Chem.*, 100, 9262–9270, 1996; (h) P.J. Stephens, C.S. Ashvar, F.J. Devlin, J.R. Cheeseman, M.J. Frisch, *Ab Initio* Calculation of Atomic Axial Tensors and Vibrational Rotational Strengths Using Density Functional Theory, *Mol. Phys.*, 89, 579–594, 1996; (i) F.J. Devlin, P.J. Stephens, J.R. Cheeseman, M.J. Frisch, Prediction of Vibrational Circular Dichroism Spectra Using Density Functional Theory: Camphor and Fenchone, *J. Am. Chem. Soc.*, 118, 6327–6328, 1996; (j) F.J. Devlin, P.J. Stephens, J.R. Cheeseman, M.J. Frisch, *Ab Initio* Prediction of Vibrational Absorption and Circular Dichroism Spectra of Chiral Natural Products Using Density Functional Theory: Camphor and Fenchone, *J. Phys. Chem. A*, 101, 6322–6333, 1997; (k) F.J. Devlin, P.J. Stephens, J.R. Cheeseman, M.J. Frisch, *Ab Initio* Prediction of Vibrational Absorption and Circular Dichroism Spectra of Chiral Natural Products Using Density Functional Theory: α-Pinene, *J. Phys. Chem. A*, 101, 9912–9924, 1997; (l) C.S. Ashvar, P.J. Stephens, T. Eggimann, H. Wieser, Vibrational Circular Dichroism Spectroscopy of Chiral Pheromones: Frontalin (1,5-Dimethyl-6,8-Dioxabicyclo [3.2.1] Octane), *Tetrahedron Asymmetry*, 9, 1107–1110, 1998; (m) C.S. Ashvar, F.J. Devlin, P.J. Stephens, K.L. Bak, T. Eggimann, H. Wieser,

Vibrational Absorption and Circular Dichroism of Mono- and Di-Methyl Derivatives of 6,8-Dioxabicyclo [3.2.1] Octane, *J. Phys. Chem. A*, 102, 6842–6857, 1998; (n) F.J. Devlin, P.J. Stephens, *Ab Initio* Density Functional Theory Study of the Structure and Vibrational Spectra of Cyclohexanone and Its Isotopomers, *J. Phys. Chem. A*, 103, 527–538, 1999; (o) C.S. Ashvar, F.J. Devlin, P.J. Stephens, Molecular Structure in Solution: An *Ab Initio* Vibrational Spectroscopy Study of Phenyloxirane, *J. Am. Chem. Soc.*, 121, 2836–2849, 1999; (p) A. Aamouche, F.J. Devlin, P.J. Stephens, Determination of Absolute Configuration Using Circular Dichroism: Tröger's Base Revisited Using Vibrational Circular Dichroism, *J. Chem. Soc., Chem. Comm.*, 361–362, 1999; (q) F.J. Devlin, P.J. Stephens, Conformational Analysis Using *Ab Initio* Vibrational Spectroscopy: 3-Methyl-Cyclohexanone, *J. Am. Chem. Soc.*, 121, 7413–7414, 1999; (r) P.J. Stephens, F.J. Devlin, Determination of the Structure of Chiral Molecules Using *Ab Initio* Vibrational Circular Dichroism Spectroscopy, *Chirality*, 12, 172–179, 2000; (s) A. Aamouche, F.J. Devlin, P.J. Stephens, Structure, Vibrational Absorption and Circular Dichroism Spectra and Absolute Configuration of Tröger's Base, *J. Am. Chem. Soc.*, 122, 2346–2354, 2000; (t) A. Aamouche, F.J. Devlin, P.J. Stephens, Conformations of Chiral Molecules in Solution: *Ab Initio* Vibrational Absorption and Circular Dichroism Studies of 4, 4a, 5, 6, 7, 8-Hexa Hydro-4a-Methyl-2(3H)Naphthalenone, and 3,4,8,8a,-Tetra Hydro-8a-Methyl-1,6(2H,7H)-Naphthalenedione, *J. Am. Chem. Soc.*, 122, 7358–7367, 2000; (u) A. Aamouche, F.J. Devlin, P.J. Stephens, J. Drabowicz, B. Bujnicki, M. Mikolajczyk, Vibrational Circular Dichroism and Absolute Configuration of Chiral Sulfoxides: tert-Butyl Methyl Sulfoxide, *Chem. Eur. J.*, 6, 4479–4486, 2000; (v) P.J. Stephens, F.J. Devlin, A. Aamouche, Determination of the Structures of Chiral Molecules Using Vibrational Circular Dichroism Spectroscopy, in *Chirality: Physical Chemistry*, ed. J.M. Hicks, Vol. 810, ACS Symposium Series, 2002, chap. 2, pp. 18–33.

25. (a) L.A. Nafie, M. Diem, D.W. Vidrine, Fourier Transform Infrared Vibrational Circular Dichroism, *J. Am. Chem. Soc.*, 101, 496–498, 1979; (b) E.D. Lipp, C.G. Zimba, L.A. Nafie, Vibrational Circular Dichroism in the Mid-Infrared Using Fourier Transform Spectroscopy, *Chem. Phys. Lett.*, 90, 1–5, 1982; (c) E.D. Lipp, L.A. Nafie, Fourier Transform Infrared Vibrational Circular Dichroism: Improvements in Methodology and Mid-Infrared Spectral Results, *Appl. Spectrosc.*, 38, 20–26, 1984.

2 The Experimental Measurement of Vibrational Absorption and Vibrational Circular Dichroism Spectra

In this chapter we discuss the measurement of the vibrational absorption spectrum and the vibrational circular dichroism (VCD) spectrum of a solution of a chiral solute in an achiral solvent.

VIBRATIONAL ABSORPTION MEASUREMENT

The vibrational absorption spectrum of a solution is measured as follows. A cell of pathlength ℓ (cm) is filled with the solution and placed in an IR absorption spectrometer between the IR light source, S, and the IR detector, D:

$$
\text{S} \xrightarrow{\quad I_0(\nu) \quad} \boxed{\text{Cell}} \xrightarrow{\quad I(\nu) \quad} \text{D}
$$

$$
\longleftarrow \ell \longrightarrow
$$

The absorption of the solution causes the intensity of the light of frequency ν entering the cell, $I_0(\nu)$, to be reduced to $I(\nu)$. The transmittance T and the absorbance A of the solution at frequency ν are defined by:

$$
T(\nu) = \frac{I(\nu)}{I_0(\nu)} \qquad\qquad A(\nu) = -\log_{10} T(\nu)
$$

If $T = 10^{-1}$, i.e., 90% of the light is absorbed, $A = 1$. The intensities I and I_0 are measured by the detector with the cell inside and outside the spectrometer, respectively. The IR absorption spectrometer measures T and A over a range of frequencies, generating the vibrational absorption spectrum of the solution, now termed the IR absorption spectrum.

The IR absorbance of the solution is the sum of the absorbances of the solute molecules and solvent molecules:

$$
A(\text{solution}) = A(\text{solute}) + A(\text{solvent})
$$

In order to determine the IR absorbance of the solute, the solvent absorbance must be subtracted. To do this, the solution is replaced in the cell by the pure solvent, the IR absorption spectrum of the solvent measured and subtracted from the solution spectrum, to give the absorbance spectrum of the solute.

For a dilute solution, the absorbance of the solute obeys Beer's law:

$$A(\text{solute}) = \varepsilon\, c\, \ell$$

where c is the concentration in moles/L of the solution and ε is the molar extinction coefficient. Determination of the frequency-dependent molar extinction coefficient spectrum of the solute is the ultimate goal in measuring the IR absorption of the solution.

Historically, commercial IR absorption spectrometers were dispersive spectrometers, in which the light from the light source was passed through a monochromator to make the light passing through the solution and measured by the detector monochromatic (i.e., of single frequency) [1]. Subsequently, Fourier transform IR (FTIR) spectrometers, in which an interferometer replaces the monochromator of a dispersive spectrometer, have become the dominant instrumentation for the measurement of IR absorption spectra [2]. Currently, many FTIR spectrometers are commercially available. Throughout the book, the IR vibrational absorption spectra presented have been measured using a Thermo Nicolet Nexus 670 FTIR spectrometer in the Stephens laboratories at USC. This spectrometer uses a silicon carbide globar light source (Ever-Glo), a CsI beamsplitter, and a deuterated triglycine sulfate (DTGS) room temperature pyroelectric bolometer detector with a CsI window, and measures spectra over the frequency range 200–6,400 cm^{-1}.

In measuring the vibrational absorption spectrum of an organic solute, the choices of cell and solvent are important. The IR absorption of both the cell windows and the solvent needs to be as small as possible to optimize the signal-to-noise ratio (S:N) of the solute absorption spectrum. Cells with windows of highly IR-transparent solids, such as crystalline KBr, are widely available. The greater problem is that organic solvents always exhibit IR absorption. For a given solute molecule, it is important to examine the IR absorption of the solvents, in which it is soluble and chemically stable, and select the solvent with the minimum absorption. The IR spectra of the organic solvents, $CHCl_3$, CCl_4, CS_2, C_6H_6, CH_3CN, and C_6H_{12}, over the frequency range 800–2,000 cm^{-1}, are shown in Figure 2.1.

If a solvent contains H atoms, the possibility exists to measure the solute IR spectrum in the deuterated solvent as well, since the substitution of H atoms by D atoms substantially shifts the vibrational frequencies of the solvent molecule. For example, the IR absorption spectra of $CHCl_3$ and $CDCl_3$ over the range 800–2,000 cm^{-1}, shown in Figure 2.2, are substantially different, so the use of both $CHCl_3$ and $CDCl_3$ solutions permits the IR absorption of a solute to be measured reliably over most of the frequency range 800–2,000 cm^{-1}.

A further issue of importance in selecting the solvent to be used is the magnitude of the solute-solvent interaction. In using the IR absorption of the solute molecule to analyze its structure, it is important that the solute be as little perturbed by the solvent as possible, so solute-solvent intermolecular interactions should be minimized as much as possible. For example, solute-solvent intermolecular hydrogen

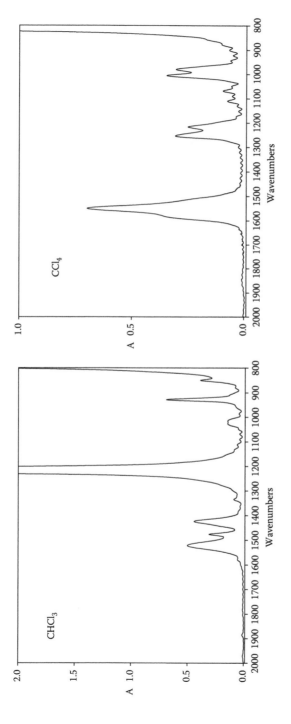

FIGURE 2.1   The IR absorption spectra of six organic solvents; pathlength 236 μ.

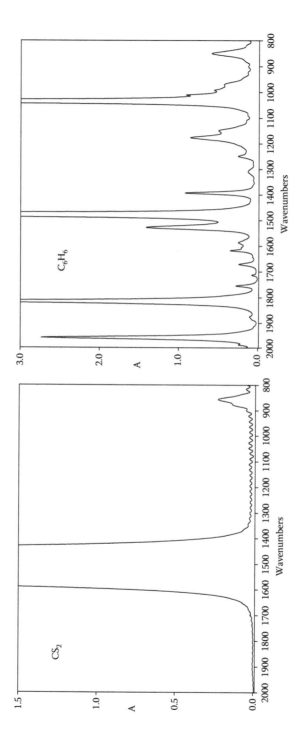

FIGURE 2.1 (CONTINUED)

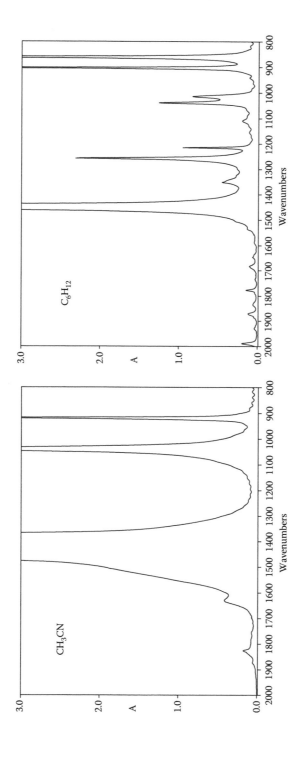

FIGURE 2.1 (CONTINUED)

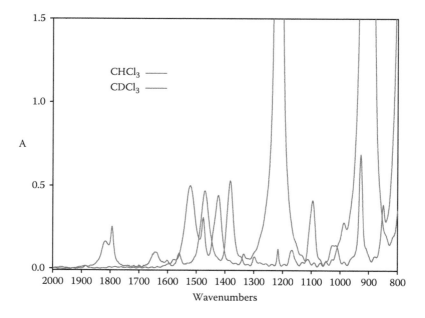

FIGURE 2.2 (SEE COLOR INSERT.)   The IR absorption spectra of CHCl$_3$ and CDCl$_3$; pathlength 236 μ.

bonding should be avoided. For this reason, nonpolar solvents such as CS$_2$ and CCl$_4$ are ideal.

Given the choice of solvent, it remains to choose the solution concentration, c, and the cell pathlength, ℓ. The concentration must be sufficiently low that Beer's law is obeyed. The pathlength must be sufficiently low that the absorption of the solvent is small. Beer's law is obeyed only when solute-solute intermolecular interactions are zero. For solute molecules that do not have a significant tendency to aggregate, it is generally the case that Beer's law is obeyed up to concentrations of ~1 M. For example, the concentration dependences of the IR absorption spectra of camphor and of α-pinene in CCl$_4$ solutions are shown in Figure 2.3 for the concentration ranges 0.981–0.033 and 2.965–0.099 M. For both molecules the molar extinction coefficient spectra are concentration independent over the concentration ranges studied. In the case of α-pinene, Beer's law is obeyed up to 3 M.

In contrast, alcohols and carboxylic acids are well known to aggregate at concentrations much lower than 1 M, due to intermolecular hydrogen bonding. For example the concentration dependences of the IR absorption of the alcohol *endo*-borneol in CCl$_4$ solution [3a] and the acid benzoic acid in CHCl$_3$ solution [3b] are shown in Figures 2.4 and 2.5 for the concentration ranges 0.306–0.008 M and 0.4–0.005 M, respectively. In both cases the O–H stretching IR spectra are very concentration dependent, demonstrating that Beer's law is not obeyed over these concentration ranges, due to intermolecular hydrogen bonding. In the case of benzoic acid, the C=O stretching IR spectrum at ~1,700 cm^{-1} is also very concentration dependent, since the C=O group participates in intermolecular hydrogen bonding. For alcohols and carboxylic acids, very low concentrations must be used in measuring their IR absorption spectra.

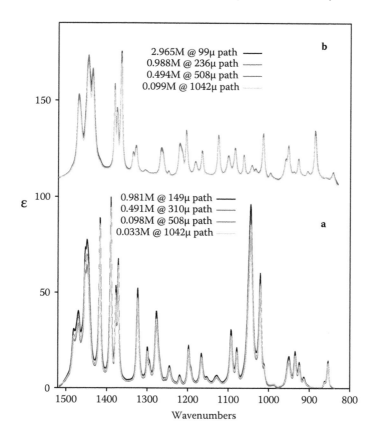

FIGURE 2.3   The concentration dependences of the IR absorption spectra of (1R,4R)-(+) camphor (a) and (1R,5R)-(+) α-pinene (b) in CCl₄ solutions.

Given the choice of solvent, solution concentration, and cell pathlength, there remain instrumental parameters that affect the solute IR absorption spectrum. These include the spectrometer resolution and the scan time. The resolution determines the degree to which bands close in frequency are resolved in the absorption spectrum. For solutions in which solute-solute intermolecular interactions are minimal, and the widths of vibrational absorption bands are consequently also minimal, 1 cm^{-1} resolution is generally sufficient to ensure a fully resolved absorption spectrum. The scan time determines the signal-to-noise ratio of the spectrum. By increasing the scan time, the noise level is decreased and the S:N ratio is increased. A typical solute IR absorption spectrum is the average of 32 scans and requires a collection time of about 1 min. However, if the sample solution is very dilute or if the solute has a particularly small molar extinction coefficient, then longer collection times are required in order to obtain a spectrum with an acceptable S:N ratio.

We have assumed so far that the solute is 100% pure. Since the IR absorption spectrum of a molecule is a sensitive function of both its formula and its geometrical structure, it follows that the presence of an impurity in the solute sample can be

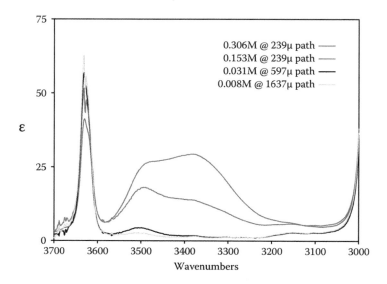

FIGURE 2.4 (SEE COLOR INSERT.)   The concentration dependence of the IR absorption in the O–H stretching region of *endo*-borneol in CCl$_4$ solution.

expected to give rise to additional bands in the absorption spectrum. As a result, the IR absorption spectrum is sensitive to the presence of impurities. Ideally, the solute is purified to 100% before its IR absorption spectrum is measured. However, this is not always the case. When only a single sample is available, there is no easy way to identify which bands in the absorption spectrum might be due to impurities. However, in the case of chiral solutes, one often has available samples of both enantiomers and the racemate. In such cases, it often occurs that their chemical purities vary, and as a result, their IR absorption spectra are not exactly identical. Absorption bands that are of different intensity in different samples can be assigned to impurities, and the sample for which these bands are the least intense is that of highest purity. For example, the IR absorption spectra of both enantiomers and the racemate of an oxadiazol-3-one derivative in CDCl$_3$ solution are shown in Figure 2.6. Both enantiomers show additional bands in their absorption spectra when compared to the racemate spectrum. Since the racemate has the least number of bands, it is therefore the compound of highest purity.

VCD MEASUREMENT

The vibrational circular dichroism (VCD) of a solution is

$$\Delta A = A_L - A_R$$

where A_L and A_R are the absorbances of left circularly polarized (LCP) and right circularly polarized (RCP) light. ΔA is termed the differential absorbance. When Beer's law is obeyed,

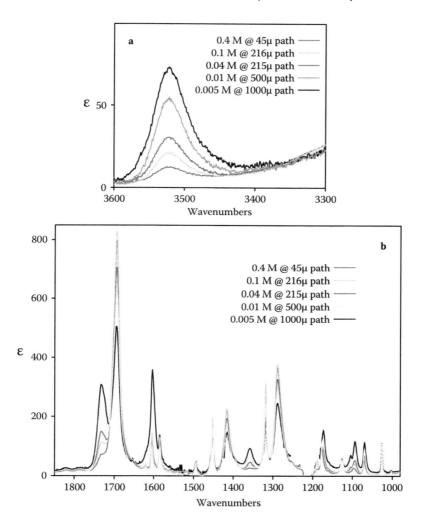

FIGURE 2.5 (SEE COLOR INSERT.) The concentration dependence of the IR absorption spectrum of benzoic acid in $CHCl_3$ solution, in the O–H stretching region (a) and in the mid-IR (b).

$$\Delta A = (\Delta \varepsilon) \, c\ell$$

where $\Delta\varepsilon$ is the differential molar extinction coefficient,

$$\Delta\varepsilon = \varepsilon_L - \varepsilon_R$$

The ratio of $\Delta\varepsilon$ to ε, $\Delta\varepsilon/\varepsilon$, is termed the anisotropy ratio. VCD anisotropy ratios are typically $<10^{-4}$.

All measurements of VCD have been made using a technique termed modulation spectroscopy. This technique was developed after World War II and applied to the measurement of ECD in the visible ultraviolet spectral region [4]. In a CD

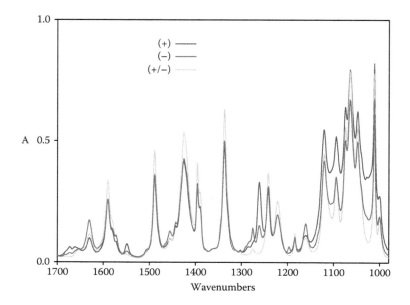

FIGURE 2.6 (SEE COLOR INSERT.)   The IR absorption spectra of 0.04 M $CDCl_3$ solutions of (+), (−), and (±) (8-(4-bromophenyl)-8-ethoxy-5-methyl-8H-[1,4]thiazino[3,4-c][1,2,4]-oxa-diazol-3-one; pathlength 597 μ.

instrument using modulation spectroscopy, the light from the source, S, is linearly polarized, using a linear polarizer, P, and then passed through a phase modulator, M, which generates light oscillating between right and left circular polarizations at a frequency v_M. The RCP and LCP light waves have the same intensities, $I_0(v)$. Passage of this light through a cell containing a solution of a chiral molecule with CD at the light frequency causes the intensities of the RCP and LCP light waves to be differently reduced, due to the difference in the absorbances, A_R and A_L. As a result, the intensity of light at the detector, D, oscillates between unequal intensities I_R and I_L at the frequency v_M. The magnitude of the oscillation is proportional to the CD, ΔA.

$$S \longrightarrow \boxed{}^{P} \longrightarrow \boxed{}^{M} \rightsquigarrow \boxed{\begin{array}{c} \text{SAMPLE} \\ A_L \neq A_R \end{array}} \longrightarrow D$$

The earliest CD instruments based on modulation spectroscopy used electro-optic modulators, in which an electric voltage, oscillating at frequency v_M, is applied to a crystal of potassium di-deuteriumphosphate (KD_2PO_4); these modulators were also termed Pockels cells. Subsequently, in the 1960s, a new type of phase modulator, termed photoelastic modulator (PEM), was invented, in which oscillating stress is applied to a solid optical element [5]. The earliest PEMs used such solid optical elements as fused quartz and crystalline CaF_2, which are transparent in the visible ultraviolet spectral region, and therefore can be used to measure ECD spectra.

To measure CD in the IR spectral region, a modulator transparent in the IR is required. The instruments that yielded the first measurements of VCD both used PEMs; the UC instrument used a Ge PEM [6] and the USC instrument used a ZnSe PEM constructed at USC [7]; both were new PEMs. Subsequently, ZnSe PEMs were commercialized by Hinds International [8], and thereafter almost all VCD instruments constructed have used commercial ZnSe PEMs. ZnSe is IR transparent down to ~650 cm^{-1}. The VCD of vibrational transitions at frequencies <650 cm^{-1} cannot be measured using a ZnSe PEM.

For a PEM to turn linearly polarized light of frequency ν into light oscillating between right and left circular polarizations, a specific magnitude of stress must be applied to the optical element. This magnitude is a function of ν. To measure VCD over a range of frequencies, the PEM stress magnitude has to be tuned. In order for a given solid to be usable as the optical element of a PEM, the stress magnitudes required to create oscillating circularly polarized light must be below the level of stress that causes the solid to fracture.

The magnitude of the oscillation of the light intensity at the detector, of frequency ν_M, is measured by the detector electronics, using a lock-in amplifier tuned to ν_M. In order to obtain the magnitude of the CD, ΔA, responsible for the oscillation, calibration is required. The calibration method used in the earliest VCD measurements at USC involved the substitution of the sample and cell by a linear polarizer and a crystalline window possessing linear birefringence, which together cause the intensity at the detector to oscillate at frequency ν_M with a predictable magnitude [9]. Comparison of the magnitudes of the oscillation of the light intensity due to the CD of the sample and the polarizer/birefringent window device permits the magnitude of the CD, ΔA, to be determined. This calibration method remains the standard used in VCD instruments.

The earliest VCD instruments were dispersive. Eventually, the modulation spectroscopic techniques used by these instruments were extended to Fourier transform IR (FTIR) instruments [10]. Currently, FTVCD instruments are commercially available [11]. Throughout the book, VCD spectra presented have been measured using a Bomem/BioTools FTVCD spectrometer in the Stephens laboratories at USC.

The accurate measurement of VCD requires that the optical elements of the instrument perform perfectly. In addition to the linear polarizer and the PEM performing perfectly, the windows of the cell and detector must be without linear birefringence and the detector must be insensitive to the polarization of the light. The use of a mirror to focus the light beam on the detector can cause distortion of the polarization; a lens without linear birefringence should be used instead. If these requirements are not satisfied, VCD spectra exhibit errors termed artefacts. To determine whether artefacts are present or not, the VCD spectra of the solvent and of the racemate of the chiral molecule of interest should be measured. In both cases, the VCD should be zero at all frequencies. In practice, when artefacts do occur, they are generally of larger magnitude at frequencies corresponding to strong absorption of the solvent or the racemate. For this reason, choosing the solvent to have minimal absorption over the frequency range being studied is important. In addition, the absorbance of the racemate should be limited to <1.0. When artefacts do occur, and are caused by the absorbance of the solution, the VCD spectra of the (+) and (–) enantiomers and the racemate of the chiral molecule should be measured using identical concentrations and cell pathlengths, so that their absorbances are identical. The VCD spectrum of the racemate is then used as the baseline for the spectra of the enantiomers, to cancel their artefact contributions.

When the (+) and (–) enantiomers have identical ee's the resulting racemate-subtracted VCD spectra, $\Delta\epsilon(+)$ and $\Delta\epsilon(-)$, should then be mirror images if their artefacts have been successfully removed by subtraction of the racemate VCD. To determine whether or not this is the case, the half-difference and half-sum spectra are plotted:

$$\tfrac{1}{2}\,\Delta = [\Delta\epsilon(+) - \Delta\epsilon(-)] \times 0.5$$

$$\tfrac{1}{2}\,\Sigma = [\Delta\epsilon(+) + \Delta\epsilon(-)] \times 0.5$$

In the absence of artefacts the half-sum spectrum is zero at all frequencies; if this is the case, the half-difference spectrum is the VCD spectrum of the (+) enantiomer.

Unfortunately, in some cases, for example, natural products, only a single enantiomer is available. In this case, the half-sum spectrum is not measurable and the magnitude of the artefact contributions cannot be determined. In such cases, the VCD spectrum of the solvent has to be used as the baseline for the VCD spectrum of the solution. The magnitudes of artefacts to be expected in the solvent-baseline-subtracted VCD spectrum of the solution can be estimated by measurement of the difference between the solvent-baseline-subtracted spectrum of another chiral molecule in the same solvent and its racemate-subtracted VCD spectrum. In 1975, Dr. Cheng at USC developed an instrumental technique for reducing artefact contributions to VCD spectra [12], in which a second PEM was inserted prior to the detector. This technique has been incorporated in the Bomem FTVCD instrument at USC. All of the VCD spectra in the book, including the spectra of camphor and α-pinene in the next sections of this chapter, have been measured using the Cheng artefact reduction technique.

IR ABSORPTION AND VCD SPECTRA OF SPECIFIC MOLECULES

The performance of a VCD instrument is optimally evaluated using measurements of the VCD spectra of conformationally rigid chiral molecules, whose enantiomers are both available, and both optically pure and chemically pure, and whose racemates are also available and chemically pure. Here, we present measurements of the VCD spectra of the two chiral molecules, camphor and α-pinene:

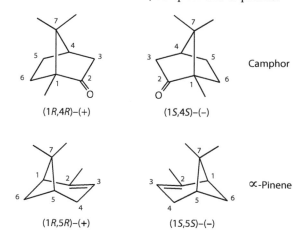

(1R,4R)–(+)   (1S,4S)–(–)   Camphor

(1R,5R)–(+)   (1S,5S)–(–)   α-Pinene

whose optically and chemically pure enantiomers and chemically pure racemates are commercially available. As discussed in Chapter 5, both molecules are conformationally rigid. The IR absorption spectra of camphor and α-pinene are simultaneously measured.

VCD spectra of camphor and α-pinene were first measured at USC in the 1970s [13]. Subsequently, these molecules were frequently used to test the performance of VCD instruments and the reliability of VCD theories [14].

CAMPHOR

The IR and VCD spectra of camphor were measured using samples of (1R,4R)-(+), (1S,4S)-(–), and (±) camphor obtained from Aldrich [15]. According to Aldrich, the purities of the (+), (–), and (±) samples were 98, 99, and 96% respectively. In order to assess the optical purities of the (+) and (–) samples, their $[\alpha]_D$ values were measured in ethanol solutions. According to Ramachandran et al. [16], for 100% ee (+) camphor $[\alpha]_D$ = + 43.6 (c5, EtOH). The Aldrich (+) and (–) samples at c5 concentration in EtOH had $[\alpha]_D$ values of +44.0 and −43.5, respectively, and therefore are both optically pure (ee 100%). The solvent used in measuring the IR and VCD spectra of camphor was CCl_4, to minimize the solvent IR absorption and the solute-solvent intermolecular interaction. Solutions of concentrations 0.38 M were used and Harrick cells [17] with KBr windows and pathlengths 236 and 546 μ.

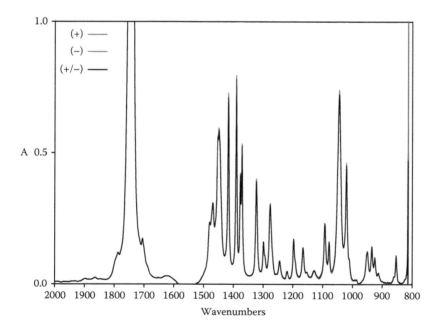

FIGURE 2.7   The IR absorption spectra of 0.38 M CCl_4 solutions of (+), (–), and (±) camphor; pathlength 236 μ.

The IR spectra obtained using the 236 and 546 μ cells are shown in Figures 2.7 and 2.8, respectively, over the frequency range 800–2,000 cm^{-1}. The excellent super-position of the spectra of the (+), (–), and (±) solutions confirms that their purities are identical, as specified by Aldrich. At the pathlength 236 μ, the absorbances of all bands are <1.0, except for the C=O stretching band at 1,745 cm^{-1}, whose absorbance is >>1.0. At the pathlength 546 μ, the absorbances of several bands in the range 1,000–1,500 cm^{-1} are >1.0. The IR spectra in Figures 2.7 and 2.8 are obtained using the IR spectra of neat CCl$_4$ as the baselines. The IR spectra of neat CCl$_4$ at 236 and 546 μ pathlengths are shown in Figure 2.9. The strongest absorption of CCl$_4$ is at the frequencies 790 cm^{-1} and 1,550 cm^{-1}. The impact of the absorption of CCl$_4$ on the IR spectra of camphor at these frequencies can be seen in Figures 2.7 and 2.8. In the region 800–825 cm^{-1}, the spectra are very noisy. At frequencies close to 1,550 cm^{-1}, the absorbances are negative.

The VCD spectra of the CCl$_4$ solutions of the (+), (–), and (±) samples, using the 236 and 546 μ cells, are shown in Figures 2.10 and 2.11, over the range 800–1,550 cm^{-1}. The VCD spectra of the (±) solution are compared to the VCD spectra of neat CCl$_4$ in Figure 2.12. At each pathlength the spectra of the (±) solution and neat CCl$_4$ differ. The largest difference is at the 546 μ pathlength at 1,045 cm^{-1}, coincident with the IR band at 1,045 cm^{-1}, whose absorbance is 1.6. The difference at this frequency at the 236 μ pathlength is smaller, demonstrating that the VCD artefact is absorbance dependent. To reduce the artefacts in the VCD spectra of the (+) and (–) solutions, the (±) VCD spectrum is subtracted, leading to the (±)-baseline-corrected VCD spectra

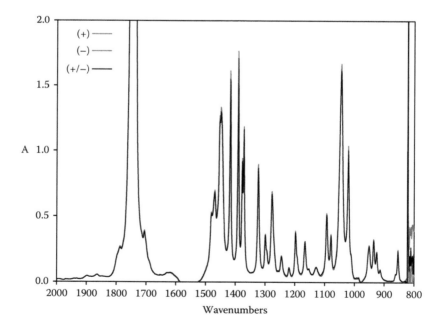

FIGURE 2.8 The IR absorption spectra of 0.38 M CCl$_4$ solutions of (+), (–), and (±) camphor; pathlength 546 μ.

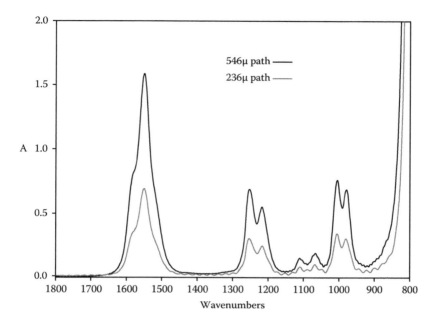

FIGURE 2.9   The IR absorption spectra of neat CCl_4; pathlengths 236 and 546 μ.

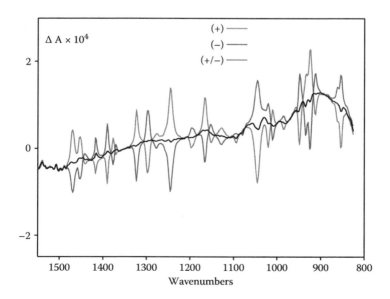

FIGURE 2.10 (SEE COLOR INSERT.)   The VCD spectra of 0.38 M CCl_4 solutions of (+), (−), and (±) camphor; pathlength 236 μ.

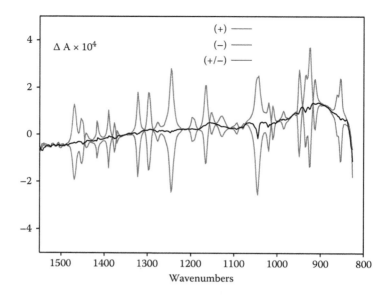

FIGURE 2.11 (SEE COLOR INSERT.)   The VCD spectra of 0.38 M CCl$_4$ solutions of (+), (−), and (±) camphor; pathlength 546 μ.

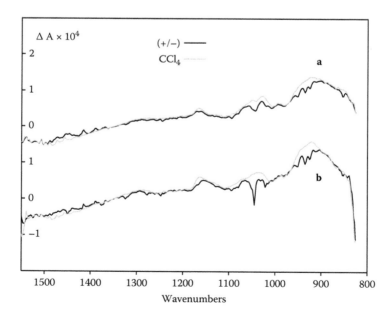

FIGURE 2.12   The VCD spectra of 0.38 M CCl$_4$ solutions of (±) camphor and neat CCl$_4$; pathlengths 236 μ (a) and 546 μ (b).

of the (+) and (–) samples shown in Figures 2.13 and 2.14. The half-difference and half-sum spectra, $\frac{1}{2}\Delta = \frac{1}{2}[\Delta A(+) - \Delta A(-)]$ and $\frac{1}{2}\Sigma = \frac{1}{2}[\Delta A(+) + \Delta A(-)]$, obtained thence are shown in Figures 2.15 and 2.16. The half-sum spectra are very close to zero at all frequencies, and at both pathlengths, although as a result of the greater absorbances at the 546 μ pathlength the deviations from zero of the 546 μ half-sum spectrum are greater than for the 236 μ spectrum.

To evaluate the relative reliability of the VCD spectra of (+) and (–) camphor, obtained using the VCD spectrum of neat CCl_4 as the baseline, the half-difference and half-sum VCD spectra have been obtained from the CCl_4-baseline-subtracted spectra, with the results shown in Figures 2.17 and 2.18. For both pathlengths, the half-sum spectra deviate from zero far more than when the (±)-baseline-subtracted spectra are used, confirming that the artefacts in the VCD spectra of the (+) and (–) samples are principally due to the camphor absorption bands.

The absorbances of the C=O stretching bands in Figures 2.7 and 2.8 are enormously large. As a result, the VCD of the C=O stretching band cannot be reliably measured using the 0.38 M solutions and the 236 and 546 μ pathlengths. The (±)-baseline-subtracted VCD spectra of (+) and (–) camphor are shown in Figures 2.19 and 2.20, together with the half-difference and half-sum spectra obtained thence. The (±)-base-line-subtracted spectra of (+) and (–) camphor are complex in structure and far from being mirror images. The half-sum spectra deviate much more from zero than do the half-difference spectra. In order to obtain a reliable VCD spectrum of the C=O stretching band, the IR and VCD spectra of 0.04 M solutions of (+), (–), and (±) camphor were measured at the 236 μ pathlength. The peak absorbances of the C=O stretching bands were 0.80. The (±)-baseline-subtracted VCD spectra of (+) and (–) camphor and the half-difference and half-sum spectra obtained thence are shown in Figure 2.21. The (±)-baseline-subtracted spectra of (+) and (–) camphor are similar in magnitude and opposite in sign. As a result, the half-sum spectrum is much closer to zero than the half-difference spectrum.

The above analysis shows that the most reliable VCD spectrum of (+) camphor in the range 800–1,550 cm^{-1} is obtained using the 0.38 M concentration and the 236 μ pathlength. In the region of the C=O stretching absorption, the most reliable VCD spectrum is obtained using the 0.04 M concentration and the 236 μ pathlength. The final VCD spectrum is therefore constructed using these two experiments. Conversion of the (±)-baseline-subtracted VCD spectra of (+) and (–) camphor from ΔA units to $\Delta\varepsilon$ units, and calculation of the half-difference spectra, $\frac{1}{2}[\Delta\varepsilon(+) - \Delta\varepsilon(-)]$, leads to the VCD spectrum of (+) camphor shown in Figure 2.22. The corresponding IR spectrum of (+) camphor in ε units is also shown in Figure 2.22. These IR and VCD spectra are the basis for the theoretical analysis discussed in Chapter 6.

α-PINENE

The IR and VCD spectra of α-pinene were measured using samples of $(1R,5R)$-(+), $(1S,5S)$-(–), and (±) α-pinene obtained from Aldrich [15]. According to Aldrich, the purities of the (+), (–), and (±) samples were >99, 99, and 98%, respectively; the enantiomeric excesses of the (+) and (–) samples, determined using gas chromatography,

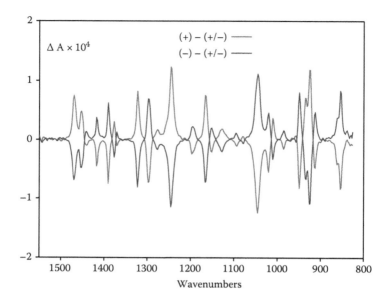

FIGURE 2.13 (SEE COLOR INSERT.)   The (±)-baseline-subtracted VCD spectra of 0.38 M CCl_4 solutions of (+) and (−) camphor; pathlength 236 μ.

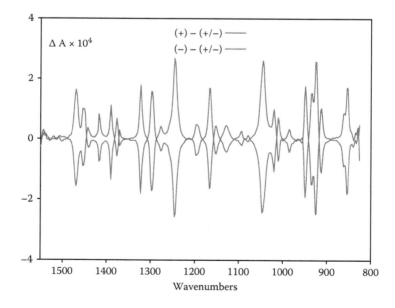

FIGURE 2.14 (SEE COLOR INSERT.)   The (±)-baseline-subtracted VCD spectra of 0.38 M CCl_4 solutions of (+) and (−) camphor; pathlength 546 μ.

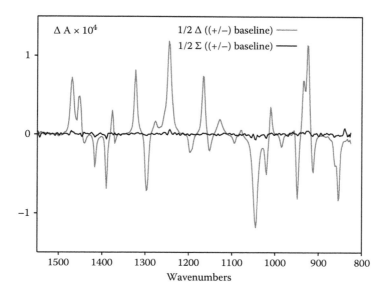

FIGURE 2.15 The half-difference and half-sum VCD spectra, obtained from the VCD spectra of (+) and (−) camphor in Figure 2.13; pathlength 236 μ.

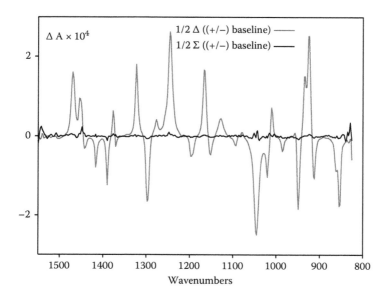

FIGURE 2.16 The half-difference and half-sum VCD spectra, obtained from the VCD spectra of (+) and (−) camphor in Figure 2.14; pathlength 546 μ.

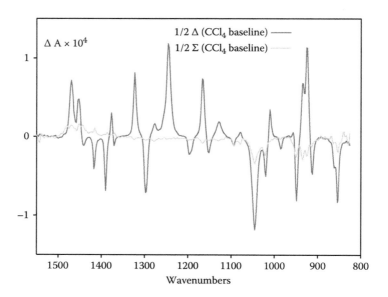

FIGURE 2.17 The half-difference and half-sum VCD spectra, obtained from the CCl$_4$-baseline-subtracted VCD spectra of (+) and (−) camphor; pathlength 236 μ.

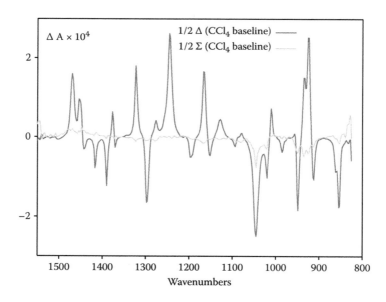

FIGURE 2.18 The half-difference and half-sum VCD spectra, obtained from the CCl$_4$-baseline-subtracted VCD spectra of (+) and (−) camphor; pathlength 546 μ.

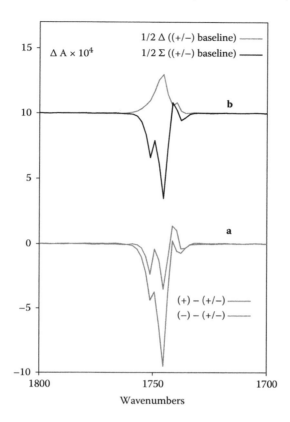

FIGURE 2.19   (a) The (±)-baseline-subtracted VCD spectra and (b) the half-difference and half-sum VCD spectra of (+) and (−) camphor; 0.38 M and 236 μ pathlength.

were 97 and 97%, respectively. The enantiomers of α-pinene are thus essentially optically pure. As with camphor, the solvent used in measuring the IR and VCD spectra of α-pinene was CCl_4. Solutions of concentrations 1.87 M were used, together with the KBr cells of pathlengths 236 and 546 μ. In addition, solutions of concentrations 0.93 M were used together with a KBr cell of pathlength 113 μ.

The IR spectra obtained over the frequency range 800–2,000 cm⁻¹ are shown in Figures 2.23–2.25. The excellent superposition of the spectra of the (+), (−), and (±) solutions of the same concentrations and at the same pathlength confirms that their purities are identical, as specified by Aldrich. At the 236 μ pathlength, the 1.87 M solutions exhibit bands with absorbances of <1.0 in the ranges 800–1,350 cm⁻¹ and 1,500–2,000 cm⁻¹. In the range 1,350–1,500 cm⁻¹ all bands have absorbances that are substantially >1.0. At the 546 μ pathlength many fewer bands have absorbances of <1.0. Diluting the solutions to 0.93 M and reducing the pathlength to 113 μ reduces all absorbances to <1.0.

The VCD spectra of the 1.87 M solutions at the 236 μ pathlength are shown in Figure 2.26, over the frequency range 800–1,700 cm⁻¹. In Figure 2.27, the (±)-baseline-subtracted VCD spectra of (+) and (−) α-pinene are shown. In Figure 2.28, the

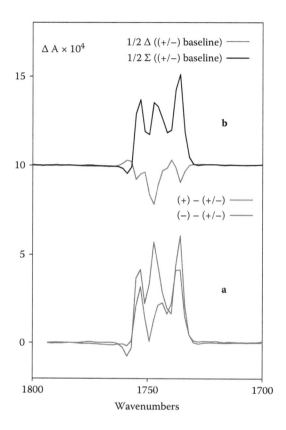

FIGURE 2.20   (a) The (±)-baseline-subtracted VCD spectra and (b) the half-difference and half-sum VCD spectra of (+) and (−) camphor; 0.38 M and 546 μ pathlength.

half-difference and half-sum spectra thence obtained are shown. In Figure 2.29, the half-difference and half-sum spectra obtained from the CCl$_4$-baseline-subtracted VCD spectra are also shown. The half-sum (±)-baseline-subtracted spectrum is very close to zero, except in the range 1,350–1,500 cm^{-1}, where the α-pinene absorbances are >>1.0. The deviation in the range 825–800 cm^{-1} is due to the CCl$_4$ absorption. The half-sum CCl$_4$-baseline-subtracted spectrum deviates much more from zero than the (±)-baseline-subtracted half-sum spectrum, showing that the VCD artefacts in the VCD spectra of (+) and (−) α-pinene originate predominantly from the α-pinene absorbances.

The corresponding spectra obtained using the 1.87 M solutions at the 546 μ pathlength are shown in Figures 2.30–2.33. The deviations from zero in the (±)-baseline-subtracted half-sum spectrum are enormously greater in the ranges 1,350–1,500 cm^{-1} and 800–825 cm^{-1} than in the 236 μ spectrum in Figure 2.28, due to the much greater absorbances of α-pinene and CCl$_4$, respectively. Even in the range 900–1,350 cm^{-1}, there are greater deviations, due to the greater absorbances of α-pinene. The deviations from zero in the CCl$_4$-baseline-subtracted half-sum spectrum are also greater than in the 236 μ spectrum in Figure 2.29.

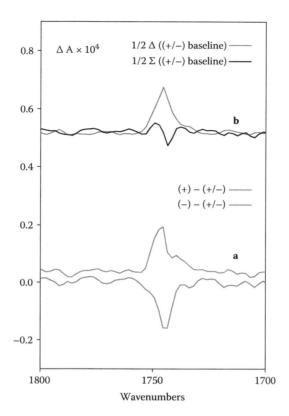

FIGURE 2.21   (a) The (±)–baseline-subtracted VCD spectra and (b) the half-difference and half-sum VCD spectra of (+) and (–) camphor; 0.04 M and 236 μ pathlength.

The corresponding spectra obtained using the 0.93 M solutions at the 113 μ pathlength are shown in Figures 2.34–2.37. Since the α-pinene absorbances in the region 1,350–1,500 cm⁻¹ are <1.0, the (±)-baseline-subtracted half-sum spectrum is close to zero in this region. The CCl₄-baseline-subtracted half-sum spectrum over the range 800–1,700 cm⁻¹ deviates much more from zero than the (±)-baseline-subtracted half-sum spectrum, again demonstrating that the artefacts in the VCD spectra of (+) and (–) α-pinene are due to the α-pinene absorption.

The VCD spectrum in Δε units of (+) α-pinene of optimum reliability, obtained from these experiments, is shown in Figure 2.38, together with the IR spectrum in ε units of (+) α-pinene. These IR and VCD spectra are the basis for the theoretical analysis discussed in Chapter 6.

CONCLUSIONS

The measurements of the IR absorption and VCD spectra of camphor and α-pinene, presented above, demonstrate the importance of the choices of the solvent, the solute solution concentration, and the cell pathlength in obtaining optimally reliable VCD

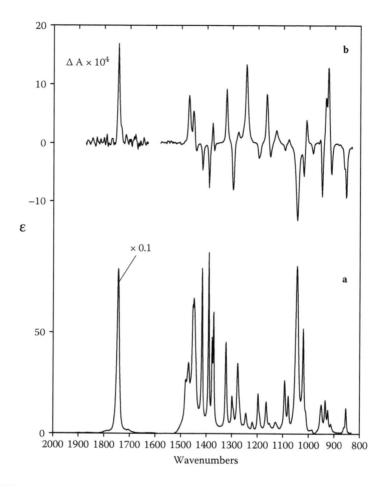

FIGURE 2.22 (a) The IR absorption spectrum of (+) camphor in CCl$_4$; 0.38 M, 2,000–1,769 cm^{-1} and 1,721–831 cm^{-1}; 0.04 M, 1,769–1,721 cm^{-1}; pathlength 236 μ. (b) The half-difference VCD spectrum of (+) camphor in CCl$_4$; 0.04 M, 1,873–1,630 cm^{-1}; 0.38 M, 1,582–831 cm^{-1}; pathlength 236 μ.

spectra. A solvent exhibiting minimal IR absorption should be chosen in order to optimize the reliability of the VCD spectrum of the solute. The solute IR absorbance is determined by the solution concentration and the cell pathlength. In order to reliably measure VCD over a given frequency range, the IR absorbance should be <1.0 over this range. In order to reliably measure VCD over the entire mid-IR frequency range, several combinations of solution concentration and cell pathlength can be required.

To measure the optimally reliable VCD spectrum of a chiral molecule, both enantiomers, of known ee's, and the racemate are required. The VCD spectra of solutions of the two enantiomers and the racemate of the same concentrations in the same solvent and using the same cell pathlength are measured, the racemate spectrum being

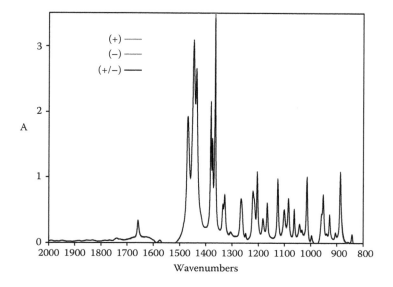

FIGURE 2.23 The IR absorption spectra of 1.87 M CCl_4 solutions of (+), (−), and (±) α-pinene; pathlength 236 μ.

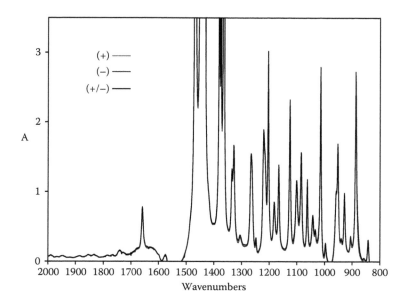

FIGURE 2.24 The IR absorption spectra of 1.87 M CCl_4 solutions of (+), (−), and (±) α-pinene; pathlength 546 μ.

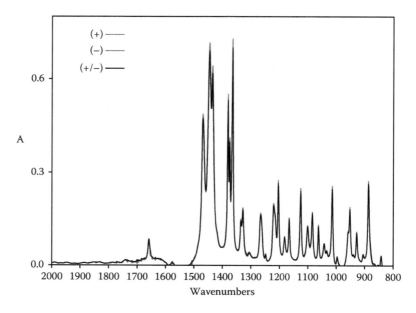

FIGURE 2.25 The IR absorption spectra of 0.93 M CCl_4 solutions of (+), (−), and (±) α-pinene; pathlength 113 μ.

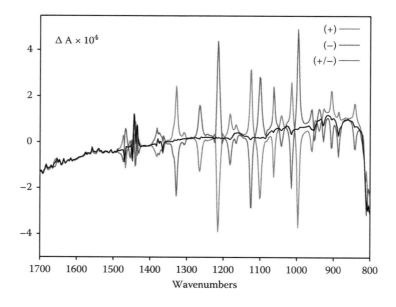

FIGURE 2.26 (SEE COLOR INSERT.) The VCD spectra of 1.87 M CCl_4 solutions of (+), (−), and (±) α-pinene; pathlength 236 μ.

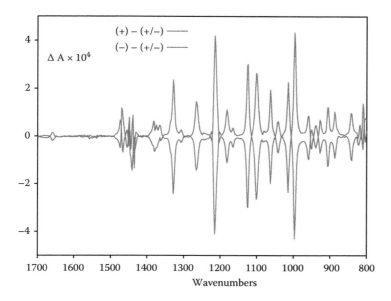

FIGURE 2.27 (SEE COLOR INSERT.)   The (±)-baseline-subtracted VCD spectra of 1.87 M CCl_4 solutions of (+) and (–) α-pinene; pathlength 236 μ.

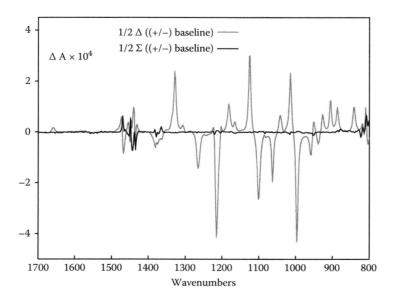

FIGURE 2.28   The half-difference and half-sum VCD spectra, obtained from the VCD spectra of (+) and (–) α-pinene in Figure 2.27; pathlength 236 μ.

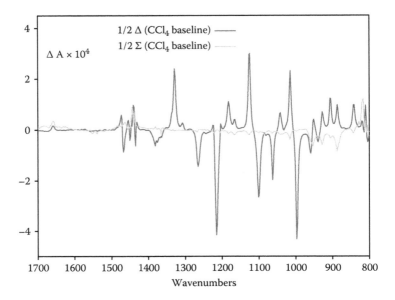

FIGURE 2.29 The half-difference and half-sum VCD spectra, obtained from the CCl_4-baseline-subtracted VCD spectra of (+) and (−) α-pinene; pathlength 236 μ.

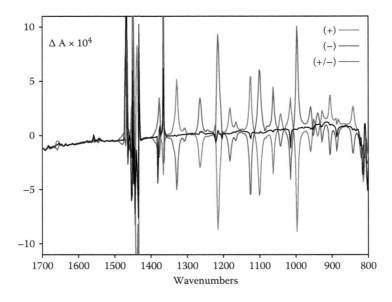

FIGURE 2.30 (SEE COLOR INSERT.)   The VCD spectra of 1.87 M CCl_4 solutions of (+), (−), and (±) α-pinene; pathlength 546 μ.

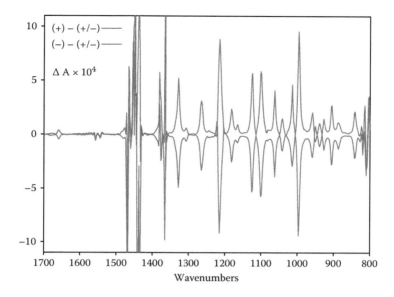

FIGURE 2.31 (SEE COLOR INSERT.)   The (±)-baseline-subtracted VCD spectra of 1.87 M CCl$_4$ solutions of (+) and (−) α-pinene; pathlength 546 μ.

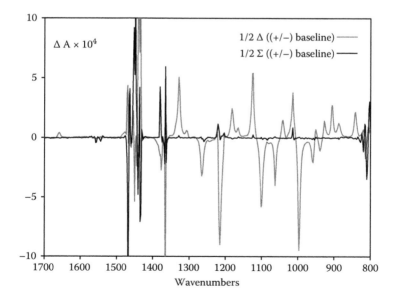

FIGURE 2.32   The half-difference and half-sum VCD spectra, obtained from the VCD spectra of (+) and (−) α-pinene in Figure 2.31; pathlength 546 μ.

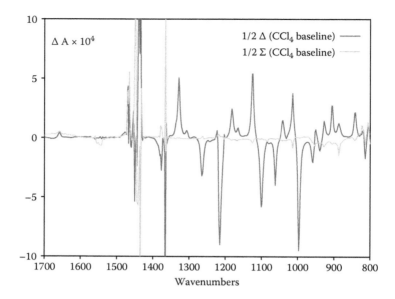

FIGURE 2.33 (SEE COLOR INSERT.) The half-difference and half-sum VCD spectra, obtained from the CCl$_4$-baseline-subtracted VCD spectra of (+) and (−) α-pinene; pathlength 546 μ.

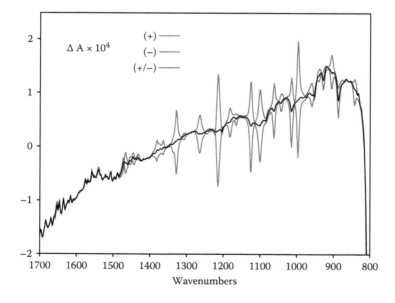

FIGURE 2.34 The VCD spectra of 0.93 M CCl$_4$ solutions of (+), (−), and (±) α-pinene; pathlength 113 μ.

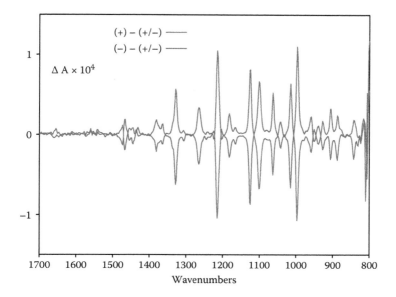

FIGURE 2.35 (SEE COLOR INSERT.)   The (±)-baseline-subtracted VCD spectra of 0.93 M CCl$_4$ solutions of (+) and (−) α-pinene; pathlength 113 μ.

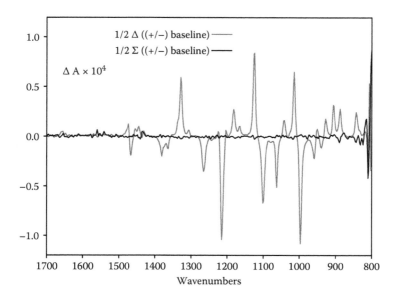

FIGURE 2.36   The half-difference and half-sum VCD spectra, obtained from the VCD spectra of (+) and (−) α-pinene in Figure 2.35; pathlength 113 μ.

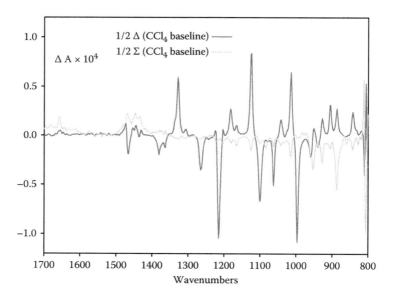

FIGURE 2.37 The half-difference and half-sum VCD spectra, obtained from the CCl_4-baseline-subtracted VCD spectra of (+) and (–) α-pinene; pathlength 113 μ.

used as the baseline for the spectra of the enantiomers. The resulting VCD spectra of the (+) and (–) enantiomers are $\Delta\varepsilon(+)$ and $\Delta\varepsilon(-)$. If the ee's of the enantiomers, ee(+) and ee(–), are <100%, the experimental VCD spectra have to be normalized to 100% ee's:

$$\Delta\varepsilon(+)[100\%ee] = \Delta\varepsilon(+) \times 100/ee(+)$$

$$\Delta\varepsilon(-)[100\%ee] = \Delta\varepsilon(-) \times 100/ee(-)$$

If the ee's of the enantiomers are in fact both 100%, normalization is not necessary. From $\Delta\varepsilon(+)[100\%ee]$ and $\Delta\varepsilon(-)[100\%ee]$ the half-difference and half-sum spectra are calculated:

$$\tfrac{1}{2}\Delta = \tfrac{1}{2}[\Delta\varepsilon(+)[100\%ee] - \Delta\varepsilon(-)[100\%ee]]$$

$$\tfrac{1}{2}\Sigma = \tfrac{1}{2}[\Delta\varepsilon(+)[100\%ee] + \Delta\varepsilon(-)[100\%ee]]$$

The reliability of the measurements of the VCD spectra of the enantiomers is assessed by the half-sum spectrum; if the half-sum spectrum is zero over the entire frequency range, the VCD spectra are 100% reliable. In frequency ranges where the half-sum spectrum deviates significantly from zero, the VCD spectra are less reliable. If the VCD spectra are optimally reliable, the half-difference spectrum is the reliable VCD spectrum of the (+) enantiomer.

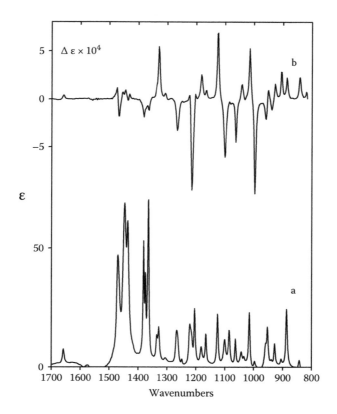

FIGURE 2.38   (a) The IR absorption spectrum of (+) α-pinene in CCl$_4$; 1,700–1,500 cm^{-1} and 1,358–837 cm^{-1}, 1.87 M and pathlength 236 μ; 1,500–1,358 cm^{-1}, 0.93 M and pathlength 113 μ. (b) The half-difference VCD spectrum of (+) α-pinene in CCl$_4$; 1,700–1,496 cm^{-1}, 1.87 M and pathlength 546 μ; 1,354–816 cm^{-1}, 1.87 M and pathlength 236 μ; 1,496–1,354 cm^{-1}, 0.93 M and pathlength 113 μ.

If the ee's of the two enantiomers are not known, the ratio of their ee's can be determined by measurements of their specific rotations at wavelength λ, using solutions of the same concentration in the same solvent:

$$\frac{ee(+)}{ee(-)} = \frac{[\alpha]_\lambda(+)}{[\alpha]_\lambda(-)}$$

If ee(+) is >ee(–) as a result, the VCD spectrum of the (–) enantiomer should be normalized to the ee of the (+) enantiomer:

$$\Delta\varepsilon(-)[ee(+)] = \Delta\varepsilon(-) \times \frac{ee(+)}{ee(-)}$$

Calculation of the half-difference and half-sum spectra, using $\Delta\varepsilon(+)[ee(+)]$ and $\Delta\varepsilon(-)[ee(+)]$, then enables the reliability of the measurements of the VCD spectra of the enantiomers to be assessed by the half-sum spectrum. If the VCD spectra are optimally reliable, the half-difference spectrum is the reliable VCD spectrum of the ee(+) (+) enantiomer.

If the racemate is not available, the baselines of the VCD spectra of the enantiomers are the VCD spectra of the solvent in the same cells. As demonstrated by the solvent-baseline-subtracted spectra of camphor and α-pinene, solvent-baseline-subtracted VCD spectra are less reliable than racemate-baseline-subtracted VCD spectra.

If only one enantiomer and the racemate are available, the VCD spectrum of the enantiomer is racemate baseline subtracted. In this case, the half-sum spectrum cannot be obtained and the level of reliability of the enantiomer VCD spectrum cannot be determined.

Clearly, the measurement of the VCD spectrum of a chiral molecule is optimal when the enantiomers of the molecule are obtained by chiral chromatography resolution of the racemate, and their ee's are determined by chromatography.

VCD spectra are only reliably measured when the samples are chemically pure. The relative purities of the enantiomers and the racemate are assessed by comparison of their IR spectra. If the samples are substantially impure, they should be purified before measurement of their VCD spectra.

REFERENCES

1. A.E. Martin, Instrumentation and General Experimental Methods, in *Infra-red Spectroscopy and Molecular Structure*, ed. M. Davies, Elsevier, 1963, chap. 2.
2. (a) P.R. Griffiths, *Chemical Infrared Fourier Transform Spectroscopy*, Wiley, 1975; (b) P.R. Griffiths and J.A. DeHaseth, *Fourier Transform Infrared Spectrometry*, Wiley, 1986; 2nd ed., 2007.
3. (a) F.J. Devlin, P.J. Stephens, and P. Besse, Conformational Rigidification via Derivatization Facilitates the Determination of Absolute Configuration Using Chiroptical Spectroscopy: A Case Study of the Chiral Alcohol endo-Borneol, *J. Org. Chem.*, 70, 2980–2993, 2005; (b) M. Urbanová, V. Setnička, F.J. Devlin, and P.J. Stephens, Determination of Molecular Structure in Solution Using Vibrational Circular Dichroism Spectroscopy: The Supramolecular Tetramer of *S*-2,2′-Dimethyl-Biphenyl-6,6′-Dicarboxylic Acid, *J. Am. Chem. Soc.*, 127, 6700–6711, 2005.
4. L. Velluz, M. Legrand, and M. Grosjean, *Optical Circular Dichroism, Principles, Measurements, and Applications*, Academic Press, 1965.
5. (a) M. Billardon and J. Badoz, Modulateur de Biréfringence, *Comptes Rendus*, 262B, 1672, 1966; (b) J.C. Kemp, *J. Opt. Soc. Am.*, 59, 950, 1969; (c) S.N. Jasperson and S.E. Schnatterly, An Improved Method for High Reflectivity Ellipsometry Based on a New Polarization Modulation Technique, *Rev. Sci. Inst.*, 40, 761, 1969; (d) L.F. Mollenauer, D. Downie, H. Engstrom, and W.B. Grant, Stress Plate Optical Modulator for Circular Dichroism Measurements, *Appl. Opt.*, 8, 661, 1969.
6. I. Chabay and G. Holzwarth, Infrared Circular Dichroism and Linear Dichroism Spectrophotometer, *Appl. Opt.* 14, 454–459, 1975.
7. (a) L.A. Nafie, J.C. Cheng, and P.J. Stephens, Vibrational Circular Dichroism of 2,2,2-Trifluoro-1-Phenylethanol, *J. Am. Chem. Soc.*, 97, 3842, 1975; (b) L.A. Nafie, T.A. Keiderling, and P.J. Stephens, Vibrational Circular Dichroism, *J. Am.*

Chem. Soc., 98, 2715–2723, 1976; (c) J.C. Cheng, L.A. Nafie, S.D. Allen, and A.I. Braunstein, Photoelastic Modulator for the 0.55–13 μm Range, *Appl. Opt.*, 15, 1960–1965, 1976.

8. Hinds International, www.hindspem.com.

9. (a) G.A. Osborne, J.C. Cheng, and P.J. Stephens, A Near-Infrared Circular Dichroism and Magnetic Circular Dichroism Instrument, *Rev. Sci. Inst.*, 44, 10–15, 1973; (b) L.A. Nafie, T.A. Keiderling, and P.J. Stephens, Vibrational Circular Dichroism, *J. Am. Chem. Soc.*, 98, 2715–2723, 1976.

10. (a) L.A. Nafie, M. Diem, and D.W. Vidrine, Fourier Transform Infrared Vibrational Circular Dichroism, *J. Am. Chem. Soc.*, 101, 496–498, 1979; (b) E.D. Lipp, C.G. Zimba, and L.A. Nafie, Vibrational Circular Dichroism in the Mid-Infrared Using Fourier Transform Spectroscopy, *Chem. Phys. Lett.*, 90, 1–5, 1982; (c) E.D. Lipp and L.A. Nafie, Fourier Transform Infrared Vibrational Circular Dichroism: Improvements in Methodology and Mid-Infrared Spectral Results, *Appl. Spectrosc.*, 38, 20–26, 1984.

11. Commercial FTVCD. BOMEM/BIOTOOLS: chiral-ir, http://www.btools.com; BRUKER OPTICS: PMA 50, http://www.brukeroptics.com/vcd.html; JASCO: FVS-4000, www.jascoinc.com/chiral; VARIAN: 7000 FT-IR, www.varian.com.

12. (a) J.C. Cheng, L.A. Nafie, and P.J. Stephens, Polarization Scrambling Using a Photoelastic Modulator: Application to Circular Dichroism Measurement, *J. Opt. Soc. Am.*, 65, 1031–1035, 1975; (b) P.J. Stephens and R. Clark, Vibrational Circular Dichroism: The Experimental Viewpoint, in *Optical Activity and Chiral Discrimination*, ed. S.F. Mason, D. Reidel, 1979, pp. 263–287.

13. (a) L.A. Nafie, T.A. Keiderling, and P.J. Stephens, Vibrational Circular Dichroism, *J. Am. Chem. Soc.*, 98, 2715–2723, 1976; (b) T.A. Keiderling and P.J. Stephens, Vibrational Circular Dichroism of Overtone and Combination Bands, *Chem. Phys. Lett.*, 41, 46–48, 1976; (c) P.J. Stephens and R. Clark, Vibrational Circular Dichroism: The Experimental Viewpoint, in *Optical Activity and Chiral Discrimination*, ed. S.F. Mason, D. Reidel, 1979, pp. 263–287.

14. (a) F.J. Devlin, P.J. Stephens, J.R. Cheeseman, and M.J. Frisch, Prediction of Vibrational Circular Dichroism Spectra Using Density Functional Theory: Camphor and Fenchone, *J. Am. Chem. Soc.*, 118, 6327–6328, 1996; (b) F.J. Devlin, P.J. Stephens, J.R. Cheeseman, and M.J. Frisch, *Ab Initio* Prediction of Vibrational Absorption and Circular Dichroism Spectra of Chiral Natural Products Using Density Functional Theory: Camphor and Fenchone, *J. Phys. Chem. A*, 101, 6322–6333, 1997; (c) F.J. Devlin, P.J. Stephens, J.R. Cheeseman, and M.J. Frisch, *Ab Initio* Prediction of Vibrational Absorption and Circular Dichroism Spectra of Chiral Natural Products Using Density Functional Theory: α-Pinene, *J. Phys. Chem. A*, 101, 9912–9924, 1997.

15. Aldrich Chemistry, www.sigmaaldrich.com.

16. P.V. Ramachandran, G.M. Chen, and H.C. Brown, *J. Org. Chem.*, 61, 88, 1996.

17. Harrick Scientific Products, www.harricksci.com.

3 The Theory of IR and VCD Spectra

In order to predict the vibrational absorption spectrum and the VCD spectrum of a molecule, we need to know the wavefunctions and energies of the states of the molecule, predicted by the Schrödinger equation:

$$H\Psi_i = E_i\Psi_i \tag{3.1}$$

The molecular Hamiltonian, H, is

$$H = V_{ee} + V_{en} + V_{nn} + T_e + T_n \tag{3.2}$$

where V_{ee} is the potential energy of electron-electron repulsion, V_{en} is the potential energy of electron-nucleus attraction, V_{nn} is the potential energy of nucleus-nucleus repulsion, T_e is the electronic kinetic energy, and T_n is the nuclear kinetic energy. E_i and Ψ_i are the energy and wavefunction of the ith state. The wavefunctions are functions of the electronic coordinates, $\mathbf{r}$, and nuclear coordinates, $\mathbf{R}$.

While the Schrödinger equation can be solved analytically for a H atom, with one electron and one nucleus, the analytical solution of the Schrödinger equation for a molecule with multiple electrons and nuclei is mathematically impossible. Consequently, the wavefunctions and energies of a molecule are obtained using the Born-Oppenheimer (BO) approximation [1], as follows. Initially, the Schrödinger equation is solved at the nuclear geometries, $\mathbf{R}$, with the nuclei fixed (i.e., $T_n = 0$):

$$H_{el}(\mathbf{r}, \mathbf{R})\Psi_i(\mathbf{r}, \mathbf{R}) = W_i(\mathbf{R})\Psi_i(\mathbf{r}, \mathbf{R}) \tag{3.3}$$

where

$$H_{el}(\mathbf{r},\mathbf{R}) = H - T_n = V_{ee} + V_{en} + V_{nn} + T_e \tag{3.4}$$

Equation 3.3 is termed the electronic Schrödinger equation. The wavefunction $\Psi_i(\mathbf{r}, \mathbf{R})$ defines the electron distribution of the ith electronic state at the nuclear geometry $\mathbf{R}$. The wavefunction and energy of the ith electronic state are both a function of $\mathbf{R}$; the energy function, $W_i(\mathbf{R})$, is termed the potential energy surface (PES). Next, the nuclear Schrödinger equation is solved:

$$H_n(i,\mathbf{R})\chi_{ij}(\mathbf{R}) = E_{ij}\chi_{ij}(\mathbf{R}) \tag{3.5}$$

where

$$H_n(i, \mathbf{R}) = W_i(\mathbf{R}) + T_n \tag{3.6}$$

and χ_{ij} is the nuclear wavefunction of the jth nuclear state. Finally, the BO molecular wavefunction is obtained by combining $\Psi_i(\mathbf{r}, \mathbf{R})$ and $\chi_{ij}(\mathbf{R})$:

$$\Psi_{ij}(\mathbf{r}, \mathbf{R}) = \Psi_i(\mathbf{r}, \mathbf{R}) \chi_{ij}(\mathbf{R}) \tag{3.7}$$

The energy of this wavefunction is E_{ij}.

We consider now the vibrational states of the lowest-energy (ground) electronic state, G. The PES of G is $W_G(\mathbf{R})$. For a molecule with N nuclei there are 3N nuclear coordinates (X_λ, Y_λ, Z_λ for each nucleus λ). In order to solve the nuclear Schrödinger equation, $W_G(\mathbf{R})$ must be known. We assume that for one nuclear geometry the potential energy W_G is a minimum on the PES, and that nuclear vibration is limited to geometries close to this minimum. This geometry, $\mathbf{R}^0$, is termed the equilibrium molecular geometry. As the nuclei move away from this geometry, the energy W_G increases. The change in energy with nuclear displacements from $\mathbf{R}^0$ is formulated by Taylor expansion of W_G with respect to the 3N Cartesian displacement coordinates:

$$X_{\lambda\alpha} = \mathbf{R}_{\lambda\alpha} - \mathbf{R}^0_{\lambda\alpha} \tag{3.8}$$

λ defines the nucleus and α defines the Cartesian axes, X, Y, and Z:

$$W_G(\mathbf{R}) = W_G(\mathbf{R}^0) + \sum_{\lambda,\alpha} \left(\frac{\partial W_G(\mathbf{R})}{\partial X_{\lambda\alpha}} \right)_0 X_{\lambda\alpha} + \frac{1}{2} \sum_{\substack{\lambda,\alpha \\ \lambda',\alpha'}} \left(\frac{\partial^2 W_G(\mathbf{R})}{\partial X_{\lambda\alpha} \partial X_{\lambda'\alpha'}} \right)_0 X_{\lambda\alpha} X_{\lambda'\alpha'} \tag{3.9}$$

The derivatives of W_G are evaluated at the equilibrium geometry, $\mathbf{R}^0$. Since $\mathbf{R}^0$ is the minimum on the PES, $W_G(\mathbf{R})$, the first derivatives, $(\partial W_G(\mathbf{R})/\partial X_{\lambda\alpha})_0$, termed the gradients, are zero. As a result,

$$W_G(\mathbf{R}) = W_G(\mathbf{R}^0) + \frac{1}{2} \sum_{\substack{\lambda,\alpha \\ \lambda',\alpha'}} \left(\frac{\partial^2 W_G(\mathbf{R})}{\partial X_{\lambda\alpha} \partial X_{\lambda'\alpha'}} \right)_0 X_{\lambda\alpha} X_{\lambda'\alpha'} \tag{3.10}$$

The second derivatives, $(\partial^2 W_G(\mathbf{R})/(\partial X_{\lambda\alpha} \partial X_{\lambda'\alpha'}))_0$, are termed the Hessian matrix of the PES.

In order to facilitate the solution of the nuclear Schrödinger equation, a further transformation of $W_G(\mathbf{R})$ is carried out, by converting the coordinates $X_{\lambda\alpha}$ to 3N new nuclear coordinates Q_i, termed normal coordinates, which are linear combinations of the Cartesian displacement coordinates:

$$Q_i = \sum_{\lambda,\alpha} S_{\lambda\alpha,i} X_{\lambda\alpha} \tag{3.11}$$

and which transforms Equation 3.10 into

$$W_G(\mathbf{R}) = W_G(\mathbf{R}^0) + \frac{1}{2}\sum_{i=1}^{3N} k_i Q_i^2 \tag{3.12}$$

where $k_i = \partial^2 W_G(\mathbf{R})/\partial Q_i^2$. Simultaneously, the nuclear kinetic energy T_n is transformed to

$$T_n = \frac{1}{2}\sum_i \dot{Q}_i^2 \tag{3.13}$$

where $\dot{Q}_i = \partial Q_i/\partial t$. The nuclear Schrödinger equation, in terms of the normal coordinates, is then

$$\left[W_G(\mathbf{R}^0) + \frac{1}{2}\sum_{i=1}^{3N} k_i Q_i^2 + \frac{1}{2}\sum_{i=1}^{3N} \dot{Q}_i^2 \right] \chi_{Gj}(Q_i) = E_{Gj}\chi_{Gj}(Q_i) \tag{3.14}$$

If there were only one coordinate Q_i, Q, this equation would be the Schrödinger equation for a one-dimensional (1D) harmonic oscillator of force constant k:

$$\left[W(Q=0) + \frac{1}{2}kQ^2 + \frac{1}{2}\dot{Q}^2 \right]\chi(Q) = E\chi(Q) \tag{3.15}$$

Solution of the 1D harmonic oscillator Schrödinger equation yields

$$E = W(Q=0) + \left(v + \frac{1}{2}\right)h\upsilon \tag{3.16}$$

where v is the vibrational quantum number, $v = 0, 1, 2, ...,$ and υ is the harmonic frequency, $\upsilon = (1/2\pi)\sqrt{k}$.

Since the Hamiltonian in Equation 3.14 is the sum of the Hamiltonians of 3N 1D harmonic oscillators, the wavefunction χ_{Gj} and energy E_{Gj} are given by

$$\chi_{Gj}(Q_i) = \prod_{i=1}^{3N}\chi_{v_i}(Q_i) \tag{3.17}$$

$$E_{Gj} = \sum_{i=1}^{3N}\left(v_i + \frac{1}{2}\right)h\upsilon_i$$

where $\chi_{vi}(Q_i)$ is the wavefunction of a 1D harmonic oscillator of force constant k_i and quantum number v_i, and υ_i is the frequency of this harmonic oscillator, $\upsilon_i = (1/2\pi)\sqrt{k_i}$.

Of the 3N normal coordinates, Q_i, three describe nuclear motions in which the center of mass of the molecule translates along three perpendicular directions in three-dimensional space, while at the same time the relative positions of the atoms of the molecule are fixed; these coordinates are termed the translational normal modes. Three other coordinates describe nuclear motions in which the center of mass is fixed and the relative positions of the atoms are fixed, but the inertial axes of the molecule rotate in 3D space; these coordinates are termed the rotational normal modes. The remaining 3N-6 coordinates Q_i describe nuclear motions in which the relative positions of atoms change; these coordinates are the vibrational normal modes. Since the energy of the molecule does not change when the molecule translates or rotates, the force constants and frequencies of the six translational and rotational normal modes are all zero. Only the vibrational normal modes have nonzero force constants and frequencies.

As discussed in Chapter 1, when a molecule is exposed to light, the light can be absorbed by the molecule and the molecule excited to a state of higher energy. The cause of this phenomenon is the interaction of the electric and magnetic fields of the light with the electrons and nuclei of the molecule. As discussed in Chapter 1, vibrational excitations are caused by IR light, of wavelengths (1 … 100 microns). Since IR wavelengths are much larger than organic molecules (excluding polymers) the electric and magnetic fields of an IR light wave are approximately uniform over the molecule. The electromagnetic perturbation of a molecule exposed to uniform electric and magnetic fields, **E** and **H**, is

$$H'(\mathbf{E},\mathbf{H}) = -\boldsymbol{\mu}_{el} \cdot \mathbf{E} - \boldsymbol{\mu}_{mag} \cdot \mathbf{H} \tag{3.18}$$

where $\boldsymbol{\mu}_{el}$ and $\boldsymbol{\mu}_{mag}$ are the electric and magnetic dipole moments of the molecule:

$$
\begin{aligned}
\boldsymbol{\mu}_{el} &= \boldsymbol{\mu}_{el}^e + \boldsymbol{\mu}_{el}^n \\[4pt]
\boldsymbol{\mu}_{el}^e &= -\sum_i (e)\,\mathbf{r}_i \\[4pt]
\boldsymbol{\mu}_{el}^n &= \sum_\lambda (Z_\lambda e)\mathbf{R}_\lambda \\[4pt]
\boldsymbol{\mu}_{mag} &= \boldsymbol{\mu}_{mag}^e + \boldsymbol{\mu}_{mag}^n \\[4pt]
\boldsymbol{\mu}_{mag}^e &= -\sum_i \left(\frac{e}{2mc}\right)(\mathbf{r}_i \times \mathbf{p}_i) \\[4pt]
\boldsymbol{\mu}_{mag}^n &= \sum_\lambda \left(\frac{Z_\lambda e}{2M_\lambda c}\right)(\mathbf{R}_\lambda \times \mathbf{P}_\lambda)
\end{aligned}
\tag{3.19}
$$

where $-e$ and $Z_\lambda e$, $\mathbf{r}_i$ and $\mathbf{R}_\lambda$, $\mathbf{p}_i$ and $\mathbf{P}_\lambda$ are the charge, position, and momentum of electron i and nucleus λ, respectively. **E** and **H** oscillate sinusoidally, with frequency υ.

To determine the impact of the time-dependent fields $\mathbf{E}$ and $\mathbf{H}$ on the molecule, time-dependent perturbation theory (TDPT) [2] is used. Quantitatively, the electric dipole contribution to the perturbation of the molecule by the light wave, $H'(\mathbf{E}, \mathbf{H})$, is much greater than the magnetic dipole contribution. In predicting the absorption of light by a molecule, the magnetic dipole contribution to H' is generally neglected. This approximation is termed the electric dipole approximation. Using this approximation and TDPT, the molar extinction coefficients of a sample of randomly oriented molecules for unpolarized LCP and RCP light, ε, ε_L, and ε_R, are predicted to be

$$\varepsilon(\upsilon) = \varepsilon_L(\upsilon) = \varepsilon_R(\upsilon)$$

$$= \frac{8\pi^3 N\upsilon}{(2.303)3000hc} \sum_{i,f} \alpha_i D(i \to f) f_{if}(\upsilon_{if}, \upsilon) \tag{3.20}$$

where i and f are the initial and final states in the excitation $i \to f$, of frequency υ_{if}. $D(i \to f)$ is the dipole strength of the transition $i \to f$:

$$D(i \to f) = |\langle i|\boldsymbol{\mu}_{el}|f\rangle|^2 \tag{3.21}$$

where $\langle i|\boldsymbol{\mu}_{el}|f\rangle$ is the electric dipole transition moment of the transition $i \to f$. α_i is the fractional population of state i, and $f(\upsilon_{if}, \upsilon)$ is the shape of the absorption band due to the excitation $i \to f$, which is not limited totally to the frequency $\upsilon = \upsilon_{if}$ due to the finite lifetime of the excited state f. Experimental bandshapes of vibrational IR absorption and VCD spectra are most frequently Lorentzian:

$$f_i(\upsilon, \upsilon_i) = \frac{1}{\pi\gamma_i} \frac{\gamma_i^2}{(\upsilon - \upsilon_i)^2 + \gamma_i^2} \tag{3.22}$$

In the electric dipole approximation, ε, ε_L, and ε_R are identical. The CD $\Delta\varepsilon$ is consequently predicted to be zero. It turns out that to predict nonzero CD the magnetic dipole contribution to H' must be included. Given both electric dipole and magnetic dipole contributions to H' $(\mathbf{E},\mathbf{H})$, TDPT predicts nonzero CD:

$$\Delta\varepsilon(\upsilon) = \frac{32\pi^3 N\upsilon}{(2.303)3000hc} \sum_{i,f} \alpha_i R(i \to f) f_{if}(\upsilon_{if}, \upsilon) \tag{3.23}$$

where $R(i \to f)$ is the rotational strength of the excitation $i \to f$:

$$R(i \to f) = \mathrm{Im}\left[\langle i|\boldsymbol{\mu}_{el}|f\rangle \cdot \langle f|\boldsymbol{\mu}_{mag}|i\rangle\right] \tag{3.24}$$

Thus, CD is due to the interaction of the electric dipole and magnetic dipole transition moments, $\langle i|\boldsymbol{\mu}_{el}|f\rangle$ and $\langle f|\boldsymbol{\mu}_{mag}|i\rangle$.

Within the harmonic approximation for the vibrational states of the ground electronic state, the electric dipole transition moment for excitations from the ground vibrational state, g, is only nonzero when the excited vibrational state, e, has $v = 1$ for only one normal mode, Q_i, and has $v = 0$ for all other normal modes. This excitation is referred to as the fundamental vibrational excitation of the ith normal mode. Its frequency $\upsilon_{ge} = \upsilon_i$. Thus for a molecule with N atoms and 3N-6 vibrational normal modes, 3N-6 vibrational excitations are predicted to occur, with frequencies equal to the frequencies of the vibrational normal modes. For the fundamental excitation, $0 \rightarrow 1$, of the ith normal mode, evaluation of the electric dipole transition moment, using the BO approximation harmonic approximation wavefunctions, leads to

$$\langle 0|\boldsymbol{\mu}_{el}|1\rangle_i = \left(\frac{\partial \boldsymbol{\mu}_{el}^G}{\partial Q_i}\right)_0 \left(\frac{\hbar}{4\pi\upsilon_i}\right)^{1/2} \tag{3.25}$$

where $\boldsymbol{\mu}_{el}^G$ is the electric dipole moment of the ground electronic state:

$$\boldsymbol{\mu}_{el}^G = \langle \Psi_G(\mathbf{r},\mathbf{R})|\boldsymbol{\mu}_{el}|\Psi_G(\mathbf{r},\mathbf{R})\rangle \tag{3.26}$$

and $(\partial \boldsymbol{\mu}_{el}^G/\partial Q_i)_0$ is its derivative with respect to Q_i at the equilibrium geometry $\mathbf{R}^0$. Substituting Equation 3.25 in Equation 3.21 then gives the dipole strength of the ith fundamental excitation:

$$D(0 \rightarrow 1)_i = \left(\frac{\hbar}{4\pi\upsilon_i}\right)\left|\left(\frac{\partial \boldsymbol{\mu}_{el}^G}{\partial Q_i}\right)_0\right|^2 \tag{3.27}$$

Thus the intensity of the fundamental vibrational transition of normal mode i is determined by the change in the molecular electronic ground state electric dipole moment with respect to the vibrational normal mode Q_i. Replacing Q_i in Equations 3.25 and 3.27 by the Cartesian displacement coordinates, $X_{\lambda\alpha}$, using Equation 3.11, leads to

$$\langle 0|(\boldsymbol{\mu}_{el})_\beta|1\rangle_i = \left(\frac{\hbar}{4\pi\upsilon_i}\right)^{1/2} \sum_{\lambda,\alpha} S_{\lambda\alpha,i} P_{\alpha\beta}^\lambda$$

$$D(0 \rightarrow 1)_i = \left(\frac{\hbar}{4\pi\upsilon_i}\right)\sum_\beta \sum_{\substack{\lambda,\alpha \\ \lambda',\alpha'}} [S_{\lambda\alpha,i} P_{\alpha\beta}^\lambda][S_{\lambda'\alpha',i} P_{\alpha'\beta}^{\lambda'}] \tag{3.28}$$

where

$$P_{\alpha\beta}^\lambda = \left(\partial(\boldsymbol{\mu}_{el}^G)_\beta/\partial X_{\lambda\alpha}\right)_0 \tag{3.29}$$

$P_{\alpha\beta}^\lambda$ is a second-rank tensor, termed the atomic polar tensor (APT) of nucleus λ. The electronic and nuclear contributions to $\boldsymbol{\mu}_{el}$ (Equation 3.19) give electronic and nuclear contributions to $P_{\alpha\beta}^\lambda$, $E_{\alpha\beta}^\lambda$, and $N_{\alpha\beta}^\lambda$, where

$$E_{\alpha\beta}^{\lambda} = \left[\frac{\partial}{\partial X_{\lambda\alpha}} \left\langle \Psi_G \left| (\mu_{el}^G)_\beta \right| \Psi_G \right\rangle_0 \right]$$

$$= 2\left[\left\langle \left(\frac{\partial \Psi_G}{\partial X_{\lambda\alpha}} \right)_0 \left| (\mu_{el}^e)_\beta \right| \Psi_G^0 \right\rangle_0 \right]$$

$$N_{\alpha\beta}^{\lambda} = (Z_\lambda e)\delta_{\alpha\beta} \qquad (3.30)$$

Since the rotational strength of a ground electronic state vibrational excitation is also determined by the electric dipole transition moment (Equation 3.24), the same selection rule applies to $\Delta\varepsilon$ as to ε: VCD is also predicted to occur only for fundamental vibrational transitions. However, it turns out that within the BO approximation for molecules with nondegenerate ground electronic states (which is the case for all neutral organic molecules), the magnetic dipole transition moment of a fundamental transition is zero, because the magnetic dipole moment of a nondegenerate electronic state is zero at all nuclear geometries. Thus, within the BO approximation VCD is predicted to be zero. Since, in fact, VCD is measurable, it is clear that the vibrational wavefunctions must be made more accurate than the BO wavefunctions. This can be accomplished by mixing the BO wavefunctions of the ground electronic state G with those of the excited states E using perturbation theory. Electronic transitions $G \rightarrow E$ have nonzero magnetic dipole transition moments, and mixing the BO wavefunctions of the excited electronic states with the BO wavefunctions of the ground electronic state causes the magnetic dipole transition moments of fundamental transitions in the electronic ground state to become nonzero. The resulting equation for the magnetic dipole transition moment of the fundamental transition of the ith normal mode, derived independently by Stephens [3], Galwas [4], and Buckingham et al. [5], is

$$\left\langle 0 \left| (\mu_{mag})_\beta \right| 1 \right\rangle_i = (4\pi\hbar^3 \upsilon_i)^{1/2} \sum_{\lambda,\alpha} S_{\lambda\alpha,i} M_{\alpha\beta}^{\lambda} \qquad (3.31)$$

where $M_{\alpha\beta}^{\lambda}$ is a second-rank tensor, termed the atomic axial tensor (AAT) of nucleus λ by Stephens [6], given by

$$M_{\alpha\beta}^{\lambda} = I_{\alpha\beta}^{\lambda} + J_{\alpha\beta}^{\lambda}$$

$$I_{\alpha\beta}^{\lambda} = \left\langle \left(\frac{\partial \Psi_G}{\partial X_{\lambda\alpha}} \right)_0 \left| \left(\frac{\partial \Psi_G}{\partial H_\beta} \right)_0 \right. \right\rangle$$

$$J_{\alpha\beta}^{\lambda} = \frac{i}{4\hbar c} \sum_{\gamma} (Z_\lambda e) R_{\lambda\gamma}^0 \varepsilon_{\alpha\beta\gamma} \qquad (3.32)$$

$I_{\alpha\beta}^{\lambda}$ and $J_{\alpha\beta}^{\lambda}$ are the electronic and nuclear contributions to $M_{\alpha\beta}^{\lambda}$. The derivative $(\partial\Psi_G/\partial X_{\lambda\alpha})_0$ is the same derivative as in the electronic contribution to $P_{\alpha\beta}^{\lambda}$, $E_{\alpha\beta}^{\lambda}$

(Equation 3.30). $(\partial\Psi_G/\partial H_\beta)_0$ is the derivative of the ground state electronic wavefunction with respect to an external magnetic field, H_β, whose interaction with the molecule is

$$H' = -(\mu^e_{mag})_\beta H_\beta \tag{3.33}$$

The calculation of $(\partial\Psi_G/\partial H_\beta)_0$ is also required for the calculation of paramagnetic susceptibilities:

$$\chi^p_{\alpha\beta} = 2\left\langle \Psi^0_G \; |(\mu^e_{mag})_\alpha| \left(\frac{\partial\Psi_G}{\partial H_\beta}\right)_0 \right\rangle \tag{3.34}$$

and nuclear magnetic shielding tensors:

$$\sigma^p_{\alpha\beta} = 2\left\langle \Psi^0_G \; \left| \sum_i \frac{(\mathbf{r}_i \times \mathbf{p}_i)_\alpha}{r_i^3} \right| \left(\frac{\partial\Psi_G}{\partial H_\beta}\right)_0 \right\rangle \tag{3.35}$$

Substituting Equations 3.28 and 3.31 for the electric dipole and magnetic dipole transition moments of the fundamental transition of the ith normal mode in Equation 3.24 gives the Stephens equation for the rotational strength:

$$R(0 \to 1)_i = \hbar^2 \sum_\beta \sum_{\substack{\lambda,\alpha \\ \lambda',\alpha'}} [S_{\lambda\alpha,i} P^\lambda_{\alpha\beta}][S_{\lambda'\alpha',i} M^{\lambda'}_{\alpha'\beta}] \tag{3.36}$$

Equations 3.20–3.36 define the molecular properties needed to predict IR and VCD spectra. These are υ_i, $S_{\lambda\alpha,i}$, $P^\lambda_{\alpha\beta}$, and $M^\lambda_{\alpha\beta}$. To calculate υ_i and $S_{\lambda\alpha,i}$ the equilibrium geometry, $\mathbf{R}^0$, of the electronic ground state of the molecule must first be determined. This is referred to as geometry optimization. At this geometry, the Hessian matrix $(\partial^2 W_G/\partial X_{\lambda\alpha}\partial X_{\lambda'\alpha'})_0$ must then be predicted. To calculate $P^\lambda_{\alpha\beta}$ and $M^\lambda_{\alpha\beta}$ the derivatives of the ground electronic state wavefunction Ψ_G $(\mathbf{r},\mathbf{R})$ with respect to the Cartesian displacement coordinates $X_{\lambda\alpha}$ are required. To calculate $M^\lambda_{\alpha\beta}$, the derivatives of the ground electronic state wavefunction of the molecule, perturbed by an external uniform magnetic field, with respect to the magnetic field, are also required. In Chapter 4 the calculation of these properties using *ab initio* quantum mechanical methods is discussed.

In deriving Equations 3.28 and 3.36 for the dipole and rotational strengths of vibrational transitions, several approximations have been made. First, we have assumed that the solute molecules are not perturbed by the solvent molecules that surround them; i.e., solute-solvent intermolecular interactions are neglected. As discussed in Chapter 1, experimental measurement of the IR and VCD spectra of a molecule is best carried out using a solution of the molecule in a solvent expected to interact minimally with the molecule. Second, we have used the harmonic approximation for the

PES surrounding the equilibrium geometry, Equation 3.10. Equation 3.10 neglects the higher-order terms in the Taylor expansion of $W_G(\mathbf{R})$, Equation 3.9. If they were included, Equation 3.12 would include higher-order terms with respect to the normal coordinates Q_i. The contribution of these terms to $W_G(\mathbf{R})$ is termed anharmonicity. When anharmonicity is included in the prediction of the dipole strengths of vibrational transitions, the selection rule limiting transitions to fundamental transitions changes. Two new types of transitions become allowed: (1) overtone transitions, in which a single normal mode is excited from v = 0 to v > 1; and (2) combination transitions, in which more than one normal mode is excited simultaneously. If the anharmonicity of the molecule is substantial in magnitude, its IR and VCD spectra are more complex than those predicted within the harmonic approximation. Fortunately, as will be demonstrated by the analyses of experimental IR and VCD spectra discussed in Chapters 6 and 7, in the mid-IR spectral region the perturbation of IR and VCD spectra by anharmonicity is minor. Third, we have assumed that the molecule is conformationally rigid; i.e., that there is only one minimum energy geometry on the PES that is populated. If the molecule is conformationally flexible, more than one of the minimum energy geometries on the PES are populated. In this situation the IR and VCD spectra are the sum of the contributions of all populated conformations:

$$\varepsilon = \sum_c \alpha_c \varepsilon(c) \qquad \Delta\varepsilon = \sum_c \alpha_c \Delta\varepsilon(c) \qquad (3.37)$$

where $\varepsilon(c)$ and $\Delta\varepsilon(c)$ are the IR and VCD spectra of conformation c, and α_c is the fraction of molecules occupying conformation c. Since the IR and VCD spectra of a molecule are dependent on its geometry, different conformations exhibit different spectra. As a result, the IR and VCD spectra of a conformationally flexible molecule have more bands than the spectra of one conformation; i.e., conformational flexibility leads to more complex IR and VCD spectra. In order to predict the IR and VCD spectra of conformationally flexible molecules, the fractional populations of the populated conformations must be predicted. Using Boltzmann statistics, equilibrium populations are a function of the relative free energies of the populated conformations, and of the temperature T (K):

$$N_c = N \frac{e^{-\Delta G(c)/kT}}{\sum_i e^{-\Delta G(i)/kT}} \qquad (3.38)$$

where $\Delta G(c)$ is the free energy of conformation c, relative to the free energy of the lowest free energy conformation. Since many organic molecules are conformationally flexible, predicting their IR and VCD spectra requires their conformational analysis. Conformational analysis of a molecule consists of locating the minimum energy geometries on the PES (the stable conformations of the molecule) and then calculating their relative free energies. The state-of-the-art methodology for conformational analysis of organic molecules is discussed in Chapter 5.

REFERENCES

1. M. Born and J.R. Oppenheimer, Zur Quantentheorie der Molekeln, *Ann. Physik*, 389, 457–484, 1927.
2. P.A.M. Dirac, The Basis of Statistical Quantum Mechanics, *Math. Proc. Cambr. Philos. Soc.*, 25, 62–66, 1929.
3. (a) P.J. Stephens, Theory of Vibrational Circular Dichroism, *J. Phys. Chem.*, 89, 748–752, 1985; (b) P.J. Stephens and M.A. Lowe, Vibrational Circular Dichroism, *Ann. Rev. Phys. Chem.*, 36, 213–241, 1985; (c) P.J. Stephens, Vibrational CD, Theory, *Encyclopedia of Spectroscopy and Spectrometry*, Academic Press, London, pp. 2415–2421, 1999.
4. P.A. Galwas, On the Distribution of Optical Polarization in Molecules, PhD thesis, University of Cambridge, Cambridge, 1983.
5. A.D. Buckingham, P.W. Fowler, and P.A. Galwas, Velocity-Dependent Property Surfaces and the Theory of Vibrational Circular Dichroism, *Chem. Phys.*, 112, 1–14, 1987.
6. P.J. Stephens, Gauge Dependence of Vibrational Magnetic Dipole Transition Moments and Rotational Strengths, *J. Phys. Chem.*, 91, 1712–1715, 1987.

4 *Ab Initio* Methods

To predict the infrared (IR) and vibrational circular dichroism (VCD) spectra of a molecule using the equations presented in Chapter 3, it is necessary to calculate the following ground electronic state molecular properties: (1) the equilibrium geometry, (2) the Hessian matrix, (3) the atomic polar tensors (APTs), and (4) the atomic axial tensors (AATs). The most accurate predictions of these properties are obtained using *ab initio* methods. Here we will discuss two *ab initio* methods, Hartree-Fock (HF) theory and density functional theory (DFT), for the calculation of all four properties.

Hartree-Fock theory is the oldest and most well-known *ab initio* electronic structure method. For a comprehensive presentation of HF theory, and its applications to the prediction of molecular properties, the reader is referred to the 1986 monograph [1] and references within. HF theory has the advantage that it is the simplest *ab initio* theory, and therefore is the least computationally expensive. In many cases, however, the accuracy of the HF method is not sufficient due to its incomplete description of the correlation between motions of electrons, and account must be taken of these effects in order to improve the quality of the results. More sophisticated *ab initio* methods, which include electron correlation, such as perturbation theory, configuration interaction, and coupled-cluster techniques, tend to be expensive in terms of computational resources and, at present, are not practical for large systems.

DFT has emerged in recent years as a promising alternative to these more expensive methods and has subsequently become one of the leading *ab initio* methods. The DFT approach is based upon a strategy of including electron correlation via general functionals of the electron density. Modern DFT is founded upon the Hohenberg-Kohn theorem [2], which demonstrates the existence of a unique functional that determines the ground state energy and density, although the theorem does not provide the form of this functional. Much progress, however, has been made in developing useful approximate functionals. DFT has been shown to be successful in the prediction of various molecular properties, often giving results of a quality comparable to more sophisticated methods, with a computational cost on the same order as that of HF, and hence may be applied to large molecular systems. In recognition of the importance of DFT, the 1998 Nobel Prize in Chemistry was divided equally between Walter Kohn for his developments of the density functional theory and John A. Pople for his development of computational methods in quantum chemistry. A number of recent books have discussed DFT and its application to the prediction of molecular properties [3].

The Kohn-Sham (KS) formulation [4] of DFT, which is the most well-suited for computation, is closely analogous to HF theory, and will be presented as such. Here we discuss the application of HF theory and DFT to the prediction of the equilibrium geometries, Hessian matrices, APTs, and AATs of organic molecules, as well as the implementation of these methods and their utilization.

In order to predict these properties, one needs to first compute the ground state electronic wavefunction $\Psi_G(\mathbf{r}, \mathbf{R})$ and energy $W_G(\mathbf{R})$ by solving the electronic Schrödinger equation (Equations 3.3 and 3.4) for the ground state.

$$H_{el}(\mathbf{r}, \mathbf{R})\Psi_G(\mathbf{r}, \mathbf{R}) = W_G(\mathbf{R})\Psi_G(\mathbf{r}, \mathbf{R}) \qquad (4.1)$$

As mentioned in Chapter 3, although an exact analytic solution to the Schrödinger equation for a molecule with multiple electrons and nuclei is not possible, a number of simplifying assumptions and procedures do make an approximate solution possible for a large range of molecules.

MOLECULAR ORBITAL THEORY

Molecular orbital theory decomposes $\Psi_G(\mathbf{r}, \mathbf{R})$ into a combination of one-electron functions termed spin orbitals, which are a product of spatial functions (Ψ, termed molecular orbitals), and either α or β spin components. Given the required restrictions imposed by quantum mechanics, the simplest type of wavefunction appropriate for the description of a many-electron system is a Slater determinant [5] formed from the spin orbitals. This results in the single assignment of electrons to orbitals, where n is the number of electrons and $\psi(i)$ is a function of the coordinates of electron i.

$$\Psi_G(\mathbf{r},\mathbf{R}) = \frac{1}{\sqrt{n!}} \begin{vmatrix} \psi_1(1)\alpha(1) & \psi_1(1)\beta(1)\dots\psi_{n/2}(1)\beta(1) \\ \vdots & \vdots \qquad \vdots \\ \psi_1(n)\alpha(n) & \psi_1(n)\beta(n)\dots\psi_{n/2}(n)\beta(n) \end{vmatrix} \qquad (4.2)$$

The molecular orbitals are expressed as linear combinations of a predefined set of N one-electron functions known as basis functions, χ_μ, which are usually centered on the atomic nuclei.

$$\psi_i = \sum_{\mu=1}^{N} c_{\mu i}\chi_\mu \qquad (4.3)$$

There are standard sets of basis functions for each nucleus, which depend only on the corresponding atomic number. Together, these constitute the *basis set* and will be discussed in more detail later. The unknown molecular orbital coefficients, $c_{\mu i}$, are determined so that the total electronic energy calculated from the many-electron wavefunction is minimized and, according to the variational theorem, is as close as possible to the energy corresponding to the exact solution of the Schrödinger equation. This energy and the corresponding wavefunction represent the best that can be obtained within the Hartree-Fock approximation, that is, the best given the constraints imposed by the use of a limited basis set in the orbital expansion and the use of a single assignment of electrons to orbitals. This energy minimization is carried out by solving the Roothann-Hall equations [6]. In this procedure a matrix that represents the average effect of the field of all of the electrons on each orbital, called the Fock matrix, is introduced. The Fock matrix itself depends upon the density matrix, $P_{\mu\nu}$, which is constructed from the molecular orbital coefficients as follows:

$$P_{\mu\nu} = 2\sum_{i}^{occ} c_{\mu i}^{*} c_{\nu i} \tag{4.4}$$

where the sum is over occupied molecular orbitals, the factor of 2 indicates that two electrons occupy each molecular orbital, and the asterisk denotes complex conjugation. Since the Fock matrix (via the density matrix) and the molecular orbitals depend upon the expansion coefficients, the Roothann-Hall equations must be solved iteratively. The procedure that does so is called the self-consistent field (SCF) method. At convergence the energy is a minimum and the orbitals generate a field that produces the same orbitals. The solution produces a set of occupied and virtual (unoccupied) molecular orbitals from which $\Psi_{G}(\mathbf{r}, \mathbf{R})$ is formed.

$$\Psi_{G}(\mathbf{r},\mathbf{R}) = 2\sum_{i}^{occ} \psi_{i} = 2\sum_{i}^{occ}\sum_{\mu=1}^{N} c_{\mu i}\chi_{\mu} \tag{4.5}$$

In Hartree-Fock theory, the energy has the form

$$E_{SCF}^{HF} = E_{OneEl}[P] + E_{J}[P] + E_{K}[P] + V_{nn} \equiv W_{G} \tag{4.6}$$

where $E_{OneEl}[P]$ is the one-electron energy, $E_{J}[P]$ is the classical Coulomb repulsion of the electrons, and $E_{K}[P]$ is the exchange energy resulting from the quantum nature of the electrons. In this notation, $[P]$ denotes the dependency on the density matrix. V_{nn} is the nuclear repulsion energy, which is independent of the electronic coordinates, and therefore a constant contribution to the energy for any particular nuclear configuration within the Born-Oppenheimer approximation. $E_{OneEl}[P]$, the energy of a single electron in a field of bare nuclei, is expressed in terms of an integral over the coordinates of a single electron, $\mathbf{r}_1$:

$$E_{OneEl}[P] = \sum_{\mu\nu} h_{\mu\nu} P_{\mu\nu} \qquad h_{\mu\nu} = \int \chi_{\mu}^{*}(\mathbf{r}_1)[\hat{T}_{e}(\mathbf{r}_1) + \hat{V}_{en}(\mathbf{r}_1)]\chi_{\nu}(1)d\mathbf{r}_1 \tag{4.7}$$

$\hat{T}_e(\mathbf{r}_1)$ and $\hat{V}_{en}(\mathbf{r}_1)$ are the one-electron kinetic energy and electron-nuclear potential energy operators, respectively. The Coulomb repulsion and exchange energy contributions are expressed in terms of integrals over the coordinates of two electrons, $\mathbf{r}_1$ and $\mathbf{r}_2$:

$$E_J[P] = \frac{1}{2}\sum_{\mu\nu\lambda\sigma} P_{\mu\nu}P_{\lambda\sigma}(\mu\nu \mid \lambda\sigma)$$

$$E_K[P] = -\frac{1}{2}\sum_{\mu\nu\lambda\sigma} P_{\mu\nu}P_{\lambda\sigma}(\mu\lambda \mid \sigma\nu)$$

(4.8)

where the two-electron repulsion integrals are defined as

$$(\mu\nu \mid \lambda\sigma) \equiv \iint \chi_\mu^*(\mathbf{r}_1)\chi_\nu(\mathbf{r}_1)\frac{1}{r_{12}}\chi_\lambda^*(\mathbf{r}_2)\chi_\sigma(\mathbf{r}_2)d\mathbf{r}_1 d\mathbf{r}_2$$

(4.9)

The majority of the computational effort to obtain the ground state energy and wavefunction involves evaluating these one- and two-electron integrals over basis functions.

The primary deficiency of HF theory is the inadequate treatment of the correlation between motions of electrons. In particular, single-determinant wavefunctions (Equation 4.2) do not account for correlation between electrons with opposite spin. Correlation of the motions of electrons with the same spin is partially accounted for by virtue of the determinantal form of the wavefunction.

In order to better understand the difference between the Hartree-Fock method and the KS form of DFT, it is useful to write the SCF energy as a sum of a one-electron term, the energy of the Coulomb interaction of the electrons, and the exchange-correlation (XC) energy, as follows:

$$E_{SCF} = E_{OneEl}[P]+E_J[P]+E_{XC}[P]+V_{nn} \equiv W_G$$

$$E_{XC}[P] = E_X[P]+E_C[P]$$

(4.10)

where the exchange-correlation functional, $E_{XC}[P]$, is a sum of the exchange functional, $E_X[P]$, and the correlation functional, $E_C[P]$. Hohenberg and Kohn demonstrated that this term is determined entirely by (is a functional of) the electron density. The electron density, ρ, evaluated at an arbitrary point $\mathbf{r}$ is

$$\rho(\mathbf{r}) = \sum_\mu^N\sum_\nu^N P_{\mu\nu}\chi_\mu\chi_\nu = 2\sum_\mu^N\sum_\nu^N\sum_i^{occ} c_{\mu i}^* c_{\nu i}\chi_\mu\chi_\nu$$

(4.11)

The difference between HF and KS lies in the treatment of the $E_{XC}[P]$ energy term. In HF theory, the exchange term is calculated exactly, while no correlation is included. KS theory includes both exchange and correlation at the SCF level through functionals of the density. Within the Kohn-Sham formulation, Hartree-Fock theory can be regarded as an approximation of density functional theory, with $E_X[P]$ given by the exchange integral $E_K[P]$ and $E_C[P] = 0$.

The exchange-correlation energy is given in terms of general functionals of the electron density and its derivatives. The exchange-correlation functional itself depends on a set of variables that in turn depend ultimately upon the density matrix. In general, DFT calculations proceed in the same way as Hartree-Fock calculations with the addition of the evaluation of the extra term E_{XC}. As the majority of current density functionals lead to integrals that cannot be evaluated in closed form, they are solved by numerical quadrature, which employs a grid of points in space in order to perform the numerical integration.

APPLICATION OF ANALYTICAL DERIVATIVE DFT TECHNIQUES TO THE CALCULATION OF FORCES, HESSIAN, APTs, AND AATs

As mentioned in Chapter 3 (Equation 3.9) the first derivative of the ground state energy, $W_G(\mathbf{R})$, with respect to the 3N Cartesian displacement coordinates, termed the gradients $(\partial W_G / \partial X_{\lambda\alpha})_0$, determines the forces on the nuclei and hence the equilibrium molecular geometry $\mathbf{R}^0$. The second derivative of the energy with respect to two Cartesian displacement coordinates is the Hessian matrix $(\partial^2 W_G / \partial X_{\lambda\alpha} \partial X_{\lambda'\alpha'})_0$, from which the force constants and fundamental vibrational frequencies are obtained. The electronic part of the APTs $(E_{\alpha\beta}^\lambda)$ are the derivatives of the ground state dipole moment with respect to Cartesian displacement coordinates and are often termed dipole derivatives. Since the dipole moment itself can be computed as the derivative of the ground state energy with respect to an electric field F_β, the electronic part of the APT can therefore be expressed as a second derivative of the energy with respect to Cartesian displacement coordinates and an electric field, F_β.

$$E_{\alpha\beta}^\lambda = \left[\frac{\partial}{\partial X_{\lambda\alpha}} \langle \Psi_G | (\mathbf{\mu}_{el}^G)_\beta | \Psi_G \rangle_0 \right] = 2 \left[\left\langle \left(\frac{\partial \Psi_G}{\partial X_{\lambda\alpha}} \right)_0 \middle| (\mathbf{\mu}_{el}^e)_\beta \middle| \Psi_G^0 \right\rangle_0 \right]$$

$$= \frac{\partial}{\partial X_\alpha} \left(\frac{\partial W_G}{\partial F_\beta} \right) = \frac{\partial^2 W_G}{\partial X_\alpha \partial F_\beta}$$

(4.12)

The electronic part of the AATs $(I_{\alpha\beta}^\lambda)$ as given by Stephens [7] is expressed as

$$I_{\alpha\beta}^\lambda = \left\langle \left(\frac{\partial \Psi_G}{\partial X_{\lambda\alpha}} \right)_0 \middle| \left(\frac{\partial \Psi_G}{\partial H_\beta} \right)_0 \right\rangle$$

(4.13)

In Equations 4.12 and 4.13, $\partial \Psi_G/\partial X_{\lambda\alpha}$ is the derivative of the ground state wavefunction with respect to the Cartesian displacement coordinates and $\partial \Psi_G/\partial H_\beta$ is the derivative of the ground state wavefunction, perturbed by an external uniform magnetic field, with respect to the magnetic field. ($I_{\alpha\beta}^\lambda$ can formally be written as the second derivative with respect to a magnetic field and nuclear velocity [8]). Substituting the expression for the ground state wavefunction in terms of the molecular orbitals (Equation 4.5), the wavefunction derivatives are

$$\frac{\partial \Psi_G}{\partial X_{\lambda\alpha}} = 2\sum_{i}^{occ} \frac{\partial \psi_i}{\partial X_{\lambda\alpha}} = 2\sum_{i}^{occ}\sum_{\mu=i}^{N}\left(\frac{\partial c_{\mu i}}{\partial X_{\lambda\alpha}}\chi_\mu + c_{\mu i}\frac{\partial \chi_\mu}{\partial X_{\lambda\alpha}}\right) \tag{4.14}$$

$$\frac{\partial \Psi_G}{\partial H_\beta} = 2\sum_{i}^{occ} \frac{\partial \psi_i}{\partial H_\beta} = 2\sum_{i}^{occ}\sum_{\mu=1}^{N}\left(\frac{\partial c_{\mu i}}{\partial H_\beta}\chi_\mu + c_{\mu i}\frac{\partial \chi_\mu}{\partial H_\beta}\right) \tag{4.15}$$

These expressions require the derivatives of the molecular orbital coefficients as well as the derivatives of the basis functions with respect to the perturbation (Cartesian displacement coordinates or applied magnetic field). Substituting the expressions for the ground state wavefunction derivatives, the electronic part of the AAT is

$$I_{\alpha\beta}^\lambda = \left\langle \left(\frac{\partial \Psi_G}{\partial X_{\lambda\alpha}}\right)_0 \left|\left(\frac{\partial \Psi_G}{\partial H_\beta}\right)_0\right\rangle = 2\sum_{i}\left\langle \sum_{\mu}\left(\frac{\partial c_{\mu i}}{\partial X_{\lambda\alpha}}\chi_\mu + c_{\mu i}\frac{\partial \chi_\mu}{\partial X_{\lambda\alpha}}\right)\left|\sum_{\upsilon}\left(\frac{\partial c_{\upsilon i}}{\partial H_\beta}\chi_\upsilon + c_{\upsilon i}\frac{\partial \chi_\upsilon}{\partial H_\beta}\right)\right.\right\rangle \tag{4.16}$$

The above expressions for $E_{\alpha\beta}^\lambda$ and $I_{\alpha\beta}^\lambda$ require the derivatives of the ground state energy and ground state wavefunctions with respect to Cartesian displacement coordinates and applied electric or magnetic fields. Early computations of these quantities used numerical differentiation methods. Such techniques are tedious and expensive and are hence only applicable to small molecules. Current methods evaluate these quantities using analytic derivative techniques, first introduced by Pulay [9]. Analytic derivative methods proceed by differentiating the equation for the SCF electronic energy (Equation 4.10) with respect to a perturbation, yielding equations that are easily solvable.

The analytic first derivatives of the energy can be computed by differentiating W_G (Equation 4.10) with respect to the perturbation. Due to the stationary condition of the SCF energy, it is possible to avoid the differentiation of the molecular orbital coefficients [10]. The calculation amounts to evaluating one- and two-electron integrals and their derivatives with respect to the perturbation.

Second derivatives of the energy and first derivatives of the ground state wavefunction require the derivatives of the molecular orbital coefficients with respect to one of the perturbations. These coefficient derivatives are obtained by a method called coupled perturbed Hartree-Fock (CPHF) [10–13] or for DFT, coupled perturbed Kohn-Sham (CPKS) [14]. The CP equations are obtained by differentiating the SCF equations [12,15,16] and have been used to compute a variety of perturbations [10,11,15,17,18]. In this procedure, the derivatives of the molecular orbital coefficients are expressed as linear combinations of the original set:

$$\frac{\partial c_{\mu i}}{\partial a} = \sum_j U_{ji}^a c_{\mu j} \qquad (4.17)$$

where a is the variable relevant to the specific perturbation of interest. U_{ji}^a is a rotation matrix allowing for orbital mixing between occupied and virtual orbitals and is determined by solving the following set of simultaneous equations

$$\boldsymbol{A}\boldsymbol{U}^a = \boldsymbol{B} \qquad (4.18)$$

The form of the matrix $\mathbf{A}$, which is constructed from two-electron integrals, depends upon whether the perturbation is real (electric field, Cartesian displacement coordinate) or imaginary (magnetic field). The matrix $\mathbf{B}$ depends explicitly upon the perturbation, which for applied electric and magnetic fields involves the electric and magnetic dipole moments of the molecule (Equations 3.18 and 3.19), respectively. For electric and magnetic field perturbations, there are three perturbations resulting from the x, y, and z components of the field requiring the solution of three sets of CPKS equations. Cartesian displacement coordinate perturbations require the solution of 3N CPKS equations, where N is the number of atoms in the molecule.

In practice, the Hessian matrix is formed via the resulting energy second derivative expression, which requires the molecular orbital coefficient derivatives and one- and two-electron integral first and second derivatives with respect to the Cartesian displacement coordinates. The electronic part of the APTs, $E_{\alpha\beta}^{\lambda}$, are also formed via the energy second derivative expression and require the molecular orbital coefficient derivatives and one-electron integral first derivatives with respect to the Cartesian displacement coordinates. The electronic part of the AATs, $I_{\alpha\beta}^{\lambda}$, are formed from overlaps of the molecular orbital coefficient derivatives with respect to the Cartesian displacement coordinates and the molecular orbital coefficient derivatives with respect to a magnetic field. For more details on the implementation at the DFT level of theory, the reader is referred to references [19,20].

In order to determine the fundamental frequencies and vibrational normal modes, the Hessian matrix is mass weighted and diagonalized. This procedure yields a set of 3N eigenvectors and 3N eigenvalues, λ_i, (where N is the number of atoms). As discussed in Chapter 3, three correspond to translational normal modes and three correspond to rotational normal modes. The remaining 3N-6 eigenvectors are the vibrational normal modes, and define the matrix $S_{\lambda\alpha,i}$, the transformation between

Cartesian nuclear displacements and normal coordinates. The square roots of the eigenvalues are the fundamental frequencies of the molecule.

$$\upsilon_i = \sqrt{\frac{\lambda_i}{4\pi^2 c^2}} \tag{4.19}$$

The nuclear contributions to the APTs and AATs (Equations 3.30 and 3.32, respectively) are easily computed and added to form $P^\lambda_{\alpha\beta}$ and $M^\lambda_{\alpha\beta}$. The dipole strengths and rotational strengths for each fundamental vibrational transition normal mode, i, are computed according to Equations 3.28 and 3.36, repeated here as follows:

$$D(0 \to 1)_i = \left(\frac{\hbar}{4\pi\upsilon_i}\right) \sum_\beta \sum_{\substack{\lambda,\alpha \\ \lambda',\alpha'}} [S_{\lambda\alpha,i} P^\lambda_{\alpha\beta}][S_{\lambda'\alpha',i} P^{\lambda'}_{\alpha'\beta}]$$

$$R(0 \to 1)_i = \hbar^2 \sum_\beta \sum_{\substack{\lambda,\alpha \\ \lambda',\alpha'}} [S_{\lambda\alpha,i} P^\lambda_{\alpha\beta}][S_{\lambda'\alpha',i} M^{\lambda'}_{\alpha'\beta}] \tag{4.20}$$

ORIGIN DEPENDENCE OF THE AATs

While it is not immediately apparent from the above equations, the tensors $I^\lambda_{\alpha\beta}$ and $E^\lambda_{\alpha\beta}$ are inextricably interconnected. Due to the nature of the magnetic field perturbation, $I^\lambda_{\alpha\beta}$ is origin dependent. The origin dependence is given by [7]

$$\left(I^\lambda_{\alpha\beta}\right)^O = \left(I^\lambda_{\alpha\beta}\right)^{O'} + \frac{i}{4\hbar c} \sum_{\gamma\delta} \varepsilon_{\beta\gamma\delta} Y_\gamma E^\lambda_{\alpha\delta} \tag{4.21}$$

where Y is the displacement from origin O to O'. Thus, the origin dependence of $I^\lambda_{\alpha\beta}$ depends on $E^\lambda_{\alpha\beta}$. This equation is not satisfied when ordinary basis functions, χ_μ, as discussed above, are used. This can lead to large errors in the computed rotational strengths [7]. The coefficient derivatives with respect to the magnetic field, $\partial c_{\mu i}/\partial H_\beta$, are origin dependent due to a nonphysical dependence of calculated magnetic properties on the coordinate origin or, more generally, on the gauge of the magnetic vector potential. This problem was circumvented by the introduction of basis functions that depend upon the magnetic field [21]. These basis functions include a magnetic field-dependent phase factor assigned to each basis function and are called gauge-including atomic orbitals (GIAOs).

$$\chi_\mu(\mathbf{H}) = \exp\left(-\frac{i}{2c}(\mathbf{H} \times \mathbf{R}_\mu) \cdot \mathbf{r}\right) \chi_\mu(0) \tag{4.22}$$

where $\mathbf{R}_\mu$ is the position vector of basis function χ_μ, and $\chi_\mu(0)$ denotes the usual field-independent basis functions χ_μ. The derivative of a field-dependent basis function with respect to the external magnetic field direction β is

$$\frac{\partial \chi_\mu}{\partial H_\beta} = -\frac{i}{2c}(\mathbf{R}_\mu \times \mathbf{r})_\beta \chi_\mu(0) \tag{4.23}$$

which is zero if magnetic field-independent functions are used.

GIAOs were first used for computing AATs at the HF level of theory [22] and later at the DFT level of theory [19]. The use of GIAOs leads to AATs, which satisfy Equation 4.21, thereby ensuring origin-independent rotational strengths. Other sum rules are also satisfied when GIAOs are used [19,20]. In addition, the use of GIAOs leads to AATs of comparable accuracy to the APTs. As a result, vibrational rotational strengths are predicted with the same efficiency and accuracy that vibrational dipole strengths are predicted. The use of GIAOs is standard practice in computing AATs.

CLASSIFICATION OF BASIS SETS

In order to compute the molecular properties discussed in this chapter, a *basis set* must be selected. A basis set consists of a linear combination of a predefined set of atomic one-electron functions for each nucleus. One defines a particular set of basis functions associated with each nucleus, depending only on the charge of that nucleus. These functions have the symmetry properties of atomic orbitals and may be classified as s, p, d, f, etc., types accordingly. Many *ab initio* electronic structure programs use Gaussian-type atomic functions as basis functions that have the general form of powers of x, y, z multiplied by $e^{-\alpha r^2}$:

$$g(\alpha, \mathbf{r}) = c\, x^\ell y^m z^n\, e^{-\alpha r^2} \tag{4.24}$$

where $\mathbf{r}$ specifies the distance from the center of the function, α is a constant determining the size, or radial extent, of the function, and c is a constant for normalization, so that the integral of $g(\alpha, \mathbf{r})^2$ over all space is equal to 1. For example, the normalized forms of the s and p_y Gaussian functions are

$$g_s(\alpha, \mathbf{r}) = \left(\frac{2\alpha}{\pi}\right)^{3/4} e^{-\alpha r^2}$$

$$\tag{4.25}$$

$$g_{p_y}(\alpha, \mathbf{r}) = \left(\frac{128\alpha^5}{\pi^3}\right)^{1/4} y\, e^{-\alpha r^2}$$

The specific forms of higher angular momentum Gaussian functions can be found in reference [1] as well as most quantum chemistry textbooks.

Linear combinations of primitive Gaussian functions are used to form the actual basis functions, termed contracted Gaussian functions, and have the form

$$\chi_\mu = \sum_p d_{\mu p} g_p \qquad (4.26)$$

where the $d_{\mu p}$ are fixed constants within a given basis set. The molecular orbital is then written as

$$\psi_i = \sum_\mu^N c_{\mu i} \chi_\mu = \sum_\mu c_{\mu i} \left(\sum_p d_{\mu p} g_p(\alpha, r) \right) \qquad (4.27)$$

The basis set can be interpreted as restricting each electron to a particular region of space. Larger basis sets impose fewer constraints on the electrons and more accurately approximate exact molecular orbitals. Larger basis sets also require correspondingly more computational resources.

The simplest type of basis sets, termed minimal basis sets, use fixed-size atomic-type orbitals—an example being the STO-3G basis set, which uses three Gaussian primitives per basis function (the 3G). STO stands for Slater-type orbitals, and the STO-3G basis set approximates Slater-type atomic orbitals [23] with Gaussian functions. Minimal basis sets contain the minimum number of basis functions needed for each atom. For example, hydrogen contains: $1s$; carbon: $1s$, $2s$, $2p_x$, $2p_y$, $2p_z$; and chlorine: $1s$, $2s$, $2p_x$, $2p_y$, $2p_z$, $3s$, $3p_x$, $3p_y$, $3p_z$. As a consequence, the molecular orbitals (Equation 4.27) will have only limited flexibility, which is why minimal basis sets are not used in practice. For larger basis sets, the number of adjustable coefficients in the variational procedure increases, and an improved description of the molecular orbitals is obtained.

The first way that a basis set can be made larger is to increase the number of basis functions per atom. Split valence basis sets have two or more sizes of basis functions for each valence orbital. The 6-31G basis set, introduced by Pople and coworkers [24], has two (therefore termed double-zeta) sizes of basis functions for each valence orbital. For example, hydrogen and carbon are represented as H: $1s$, $1s'$; C: $1s$, $2s$, $2s'$, $2p_x$, $2p'_x$, $2p_y$, $2p'_y$, $2p_z$, $2p'_z$, where the primed and unprimed orbitals differ in size. The 6-31G basis set consists of six primitive Gaussian functions in the core function and two functions in the valence region—one consisting of three primitive Gaussian functions and the other consisting of one primitive Gaussian function. An example of a triple-zeta split valence basis set is 6-311G, which consists of six primitive Gaussian functions in the core, three functions in the valence region—one consisting of three primitive Gaussians and two consisting of one primitive Gaussian. Another class of basis sets is Dunning's correlation consistent basis sets [25], which include double-, triple-, and quadruple-zeta versions (cc-pVDZ, cc-pVTZ, and cc-pVQZ, respectively).

Split valence basis sets allow orbitals to change size, but not to change shape. Polarized basis sets remove this limitation by adding functions to each atom with angular momentum beyond what is required for the ground state. This means an additional p-type Gaussian function on hydrogen, an additional d-type function on carbon, and an additional f-type function on transition metals. For the Pople-style basis sets (6-31G, 6-311G, etc.), polarization functions are denoted using the convention where the first angular momentum function in the parentheses implies polarization functions on second and higher row elements, and the second angular momentum function implies polarization functions on hydrogen. Using this convention, 6-311G(2d,p) implies two additional d functions on atoms other than hydrogen and one additional p function on hydrogen. Another convention is to replace the parentheses with stars. For example, 6-31G(d) and 6-31G(d,p) are equivalently 6-31G* and 6-31G**, respectively. Polarization functions are included in the definition of the Dunning correlation consistent basis sets. Other basis sets follow different conventions. For example, the TZ2P [26] basis set, used in many VCD calculations, implies triple-zeta plus 2 polarization functions. There is a choice in defining d-type polarization functions in terms of the five pure (5d) or six Cartesian functions (6d). This applies to higher angular momentum polarization functions as well.

Another way to increase the size and flexibility of the basis set is to add what are termed diffuse functions. Diffuse functions are large-size versions of s- and p-type (and higher) functions that allow orbitals to occupy a larger region of space. This becomes important for systems where electrons are relatively far from the nucleus, such as lone pairs, anions, systems with significant negative charge, excited states, and systems with low ionization potentials. For the Pople-style basis sets, the common convention is that + denotes diffuse functions on all atoms other than hydrogen and ++ denotes diffuse functions on all atoms. For the Dunning basis sets, diffuse functions are denoted by adding the prefix *aug* (aug-cc-pVDZ, aug-cc-pVTZ, etc.) [27].

The basis sets used in the following chapters are listed in Table 4.1. This table includes the number and type of contracted functions as well as the total number of basis functions for α-pinene.

FUNCTIONALS FOR DFT

In order to compute the molecular properties discussed in this chapter using DFT, a *functional* must be chosen. The E_{XC} term in Equation 4.10 includes the exchange-correlation energy arising from the antisymmetry of the quantum mechanical wavefunction as well as dynamic correlations of the motion of the individual electrons. Hohenberg and Kohn demonstrated that this term is determined entirely by (is a functional of) the electron density. E_{XC} is usually divided into separate parts, referred to as the exchange functional and correlation functional.

Various types of approximate methods have been developed. The local density approximation (LDA), which depends only on the density, was developed to reproduce the energy of a uniform electron gas; however, it has weaknesses in describing molecular systems. Generalized gradient approximation (GGA) methods make the exchange and correlation energies dependent not only on the density but also on the

TABLE 4.1

Basis Set Labels, Composition, and Reference

Basis Set	Zeta	Contracted Set[a]	No. bfns[b]	Ref.
6-31G	2	[3s2p/2s]	122	[1,24,28]
6-31G*	2	[3s2p1d/2s] (6d)	182	[1,24,28,29]
6-31G**	2	[3s2p1d/2s1p] (6d)	230	[1,24,28,29]
6-311G**	3	[4s3p1d/3s1p] (5d)	276	[30]
6-311++G**	3	[5s4p1d/4s1p] (5d)	332	[30,31]
6-311++G(2d,2p)	3	[5s4p2d/4s2p] (5d)	430	[30–32]
cc-pVDZ	2	[3s2p1d/2s1p] (5d)	220	[25]
cc-pVTZ	3	[4s3p2d1f/3s2p1d] (5d,7f)	524	[25]
cc-pVQZ	4	[5s4p3d2f1g/4s3p2d1f] (5d,7f,9g)	1030	[25]
aug-cc-pVDZ	2	[4s3p2d/3s2p] (5d)	374	[27]
aug-cc-pVTZ	3	[5s4p3d2f/4s3p2d] (5d,7f)	828	[27]
TZ2P	3	[5s4p2d/3s2p] (6d)	434	[26]

[a] Of the form [1/2], where the first term refers to the contracted set on carbon and oxygen and the second refers to the contracted set on hydrogen. The numbers in parentheses refer to the number of functions used for each type of polarization function.

[b] Number of basis functions (bfns) for α-pinene.

gradient of the density. These are also referred to as *nonlocal*, as opposed to *local*, methods. GGA methods represent a significant improvement over the local methods. Becke formulated a gradient-corrected exchange functional based on the LDA exchange functional in 1988 [33], which is now in wide use. Meta-GGA functionals depend explicitly on higher-order density gradients, or typically on the kinetic energy density. These functionals are also referred to as τ-dependent functionals, an example being τHCTH [34].

Pure DFT methods are defined by pairing an exchange functional with a correlation functional. The well-known BLYP functional pairs Becke's gradient-corrected exchange functional [33] (B) with the gradient-corrected correlation functional of Lee, Yang, and Parr [35] (LYP). The BPW91 functional pairs Becke's gradient-corrected exchange functional [33] (B) with the gradient-corrected correlation functional of Perdew and Wang [36] (PW91).

As discussed earlier, Hartree-Fock theory also includes an exchange term as part of its formulation (termed exact exchange). Becke [37,38] formulated functionals that include a mixture of Hartree-Fock and DFT exchange along with DFT correlation, conceptually defining E_{XC} as

$$E_{XC}^{hybrid} = c_{HF}E_X^{HF} + c_{DFT}E_{XC}^{DFT} \qquad (4.28)$$

where the c's are constants. Functionals that include a mixture of Hartree-Fock and DFT exchange are termed *hybrid functionals*. Hybrid functionals provide a significant improvement over GGAs for many molecular properties. There are several variations of Becke's original three-parameter hybrid functional. Two of the most popular are B3LYP [33,35,37,39] and B3PW91 [33,36,37]. The specific form of the B3LYP functional is:

$$A * E_X^{Slater} + (1 - A) * E_X^{HF} + B * E_X^{Becke88} + C * E_C^{LYP} + (1 - C) * E_C^{VWN} \quad (4.29)$$

where E_X^{Slater}, E_X^{HF}, and $E_X^{Becke88}$ are the Slater [2,4,40], Hartree-Fock, and (nonlocal) Becke88 [33] exchange contributions, respectively, and E_C^{LYP} and E_C^{VWN} are the (nonlocal) LYP [35,41] and local VWN [42] correlation contributions, respectively. The constants $A = 0.20$, $B = 0.72$, and $C = 0.81$ are those determined by Becke. The B3PW91 functional replaces the above nonlocal and local correlation contributions with those of Perdew and Wang [36] (PW91). Similarly, the B3P86 functional replaces the above nonlocal and local correlation contributions with those of Perdew [43] (P86). There are also hybrid versions of Meta-GGAs, such as B1B95 [44].

Another class of functionals are the range-separated hybrid GGAs, which include CAM-B3LYP [45] and HSE1PBE [46]. Range-separated functionals optimize the partition of Hartree-Fock and DFT exchange, considering the length scale of the electron-electron interaction. The CAM-B3LYP functional has short-range DFT exchange and long-range HF exchange, while HSE1PBE has short-range HF exchange and long-range DFT exchange.

Although the list has grown since 2007, the reader is referred to reference [47] for a list of popular functionals. The functionals used in the following chapters, along with their classification, are listed in Table 4.2.

COMPUTATIONAL PROCEDURES

The first step in computing the equilibrium geometry, IR, and VCD spectra for a molecule using *ab initio* methods is to select the level of theory. For DFT, this amounts to choosing a functional and basis set. The level of theory is denoted using the common convention functional/basis set. The most reliable choices for these will be addressed in the following chapters. The hybrid functionals, such as B3LYP, B3PW91, and B3P86, are typically the most accurate, although due to the inherent nature of DFT, no one functional is consistently the most reliable. It is often useful to try more than one functional and see which one compares best with the experimental IR and VCD spectra. The 6-31G* basis set is the smallest basis set that provides reasonable results for these properties, and is often the first choice, especially for large molecules. Larger basis sets, such as TZ2P and cc-pVTZ, significantly improve the accuracy of the computed IR and VCD spectra, but also require more computational resources.

The second step is to perform a geometry optimization. Geometry optimizations, which are also called minimizations, attempt to locate minima on the potential energy surface (PES). At minima, the first derivatives of the energy with respect to the Cartesian displacement coordinates (the gradients) are zero. Since the gradient is the negative of the force, the forces on the nuclei are zero at this point. A point on

TABLE 4.2
Functional Name, Type, and Reference

Functional	Type[a]	Reference
BLYP	GGA	[33,35]
BPW91	GGA	[33,36]
BP86	GGA	[33,43]
HCTH	GGA	[48]
OLYP	GGA	[49]
τHCTH	M-GGA	[34]
BB95	M-GGA	[44]
B3LYP	H-GGA	[33,35,37,39]
B3PW91	H-GGA	[33,36,37]
B3P86	H-GGA	[37,43]
PBE1PBE	H-GGA	[50]
O3LYP	H-GGA	[51]
MPW1PW91	H-GGA	[52]
τHCTHHYB	HM-GGA	[34]
B1B95	HM-GGA	[44]
BMK	HM-GGA	[53]
M05-2X	HM-GGA	[54]
CAMB3LYP	H-X	[45]
HSE1PBE	H-S	[46,55]

[a] GGA, generalized gradient approximation; M-GGA, meta-generalized gradient approximation; H-GGA, hybrid generalized gradient approximation; HM-GGA, hybrid meta-generalized gradient approximation; H-X, range-separated hybrid (with short-range DFT exchange and long-range HF exchange); H-S, range-separated hybrid (short-range HF exchange and long-range DFT exchange).

the PES surface where the forces are zero is called a stationary point. The gradient indicates the direction along the PES in which the energy decreases most rapidly from the current point as well as the steepness of the slope. Most geometry optimization algorithms also estimate (or compute) the value of the second derivative of the energy with respect to Cartesian displacement coordinates, updating the matrix of force constants (Hessian matrix), which specify the curvature of the surface. For conformationally flexible molecules, it is necessary to find all minimum energy geometries on the PES that are populated. The technique that does this is termed conformational analysis and will be discussed in Chapter 5. For more detail regarding geometry optimizations, the reader is referred to reference [56].

The next step is to perform a separate vibrational frequency calculation for each of the minimum energy geometries obtained from the conformational analysis. The Hessian matrix, APTs, and AATs are computed at this point. These quantities correspondingly yield the fundamental vibrational frequencies, dipole strengths, and rotational strengths for each vibrational normal mode. Various thermodynamic quantities, including the enthalpy and free energy, are also computed. A vibrational frequency calculation *must* use the same level of theory (functional and basis set) as was used to determine the minimum energy geometries (see Equation 3.10). Geometry optimizations converge to a structure on the potential energy surface where the forces on the nuclei are essentially zero. However, the final structure may correspond to a minimum on the PES, or it may represent a saddle point (which is a minimum with respect to some directions on the surface and a maximum in one or more others). The number of *imaginary frequencies* and the normal modes corresponding to these imaginary frequencies characterize the nature of the stationary point. A structure that has n imaginary frequencies is an nth-order saddle point. First-order saddle points (maxima in exactly one direction and a minimum in all other orthogonal directions) correspond to a transition state structure linking two minima. Whenever a structure yields an imaginary frequency, it means that there is some geometric distortion for which the energy of the system is lower than it is at the current structure. If one or more imaginary frequencies are found, one should continue searching by unconstraining the molecular symmetry or distort along the normal mode corresponding to the imaginary frequency. For more practical advice, the reader is referred to reference [57].

INTEGRATION GRIDS

The DFT exchange-correlation energy terms for most current functionals cannot be evaluated analytically and are therefore computed using numerical quadrature. This technique employs a grid of points in space in order to perform the numerical integration. Grids are specified as a number of radial shells around each atom, each of which contains a set number of integration points. For example, the (99,590) grid has 99 radial shells and 590 angular points per shell centered on each atom. It is advisable to use the same grid for all calculations where energy differences are of interest. The (99,590) grid is recommended for molecules containing multiple tetrahedral centers and for computing very low frequency modes of systems. This grid is also useful for geometry optimizations of larger molecules with many soft modes, such as methyl rotations, making such optimizations more reliable. As expected, a larger grid will potentially increase the computational time. However, for large molecules, the accuracy provided by a larger grid oftentimes decreases the actual number of geometry optimization steps and can also decrease the number of CPKS iterations [58]. All of the results presented in this book were obtained using the (99,590) integration grid.

IR AND VCD SPECTRA CALCULATION

The calculated harmonic frequencies, dipole strengths, and rotational strengths for the fundamental transitions of modes 6–18 for (S)-methyloxirane are given in Table 4.3. Frequencies are given in units of cm^{-1}, dipole strengths in 10^{-40} esu^2 cm^2, and rotational strengths in 10^{-44} esu^2 cm^2. These quantities were computed using the two hybrid density functionals, B3LYP and B3PW91, and the cc-pVTZ basis set.

Since experimental bandshapes of vibrational IR absorption and VCD spectra are most frequently Lorentzian, calculated spectra are plotted using the Lorentzian lineshape function (Equation 3.22)

$$f_i(\upsilon, \upsilon_i) = \frac{1}{\pi \gamma_i} \frac{\gamma_i^2}{(\upsilon - \upsilon_i)^2 + \gamma_i^2} \qquad (4.30)$$

and typically use a value of 4.0 cm^{-1} for the bandwidth, γ.

Simulations of the IR spectra of methyloxirane due to fundamentals 5–18, using Lorentzian bandshapes ($\gamma = 4.0$ cm^{-1}), are shown in Figure 4.1, together with the experimental IR spectrum in the frequency range 800–1,600 cm^{-1}. The molar extinction coefficient ε (in units of M^{-1} cm^{-1}), as a function of frequency, is obtained using (Equation 3.20 and reference [59])

TABLE 4.3

Calculated Frequencies, Dipole Strengths, and Rotational Strengths for (S)-Methyloxirane[a]

| Mode | Calculation[b] | | | | | |
| | B3LYP | | | B3PW91 | | |
	υ	D	R	υ	D	R
18	1,531	13.2	−5.2	1,531	20.0	−6.4
17	1,498	14.6	1.1	1,490	16.1	0.4
16	1,484	13.6	−2.0	1,475	15.3	−2.1
15	1,440	60.2	−13.4	1,441	58.3	−11.9
14	1,407	9.8	−1.9	1,396	11.3	−1.8
13	1,295	18.2	9.2	1,300	17.6	12.2
12	1,188	2.2	−0.2	1,186	3.8	−0.4
11	1,165	13.8	13.2	1,165	15.6	11.0
10	1,156	3.8	−4.3	1,156	1.6	−2.5
9	1,130	23.4	6.1	1,130	26.6	5.8
8	1,042	33.2	−4.0	1,042	47.5	−7.0
7	974	62.6	33.3	989	59.2	32.6
6	908	11.1	−24.9	907	11.9	−25.1

[a] Frequencies υ in cm^{-1}; dipole strengths D in 10^{-40} esu^2 cm^2; rotational strengths R in 10^{-44} esu^2 cm^2.

[b] Using the cc-pVTZ basis set.

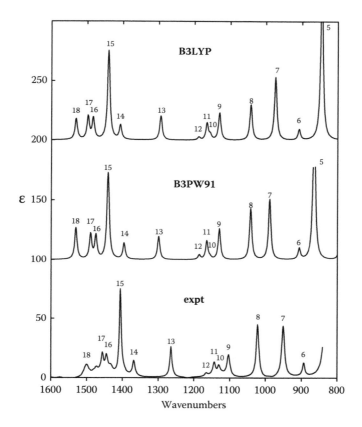

FIGURE 4.1 Comparison of the B3LYP/cc-pVTZ and B3PW91/cc-pVTZ IR spectra of methyloxirane to the experimental IR spectrum of (–)-methyloxirane. The assignment of the experimental spectrum is based on both the B3PW91/cc-pVTZ and B3LYP/cc-pVTZ spectra.

$$\varepsilon(\upsilon) = \frac{8\pi^3 N\upsilon}{(2.303)3000hc} \sum_i D_i f(\upsilon, \upsilon_i) \qquad (4.31)$$

Simulations of the VCD spectra of (S)-methyloxirane due to fundamentals 5–18, using Lorentzian bandshapes ($\gamma = 4.0$ cm^{-1}), are shown in Figure 4.2, together with the experimental VCD spectrum in the frequency range 800–1,600 cm^{-1}, where $\Delta\varepsilon$, as a function of frequency, is obtained using (Equation 3.23 and reference [59])

$$\Delta\varepsilon(\upsilon) = \frac{32\pi^3 N\upsilon}{(2.303)3000hc} \sum_i R_i f(\upsilon, \upsilon_i) \qquad (4.32)$$

A more detailed comparison between the calculated and experimental IR and VCD spectra for this molecule is presented in Chapter 6.

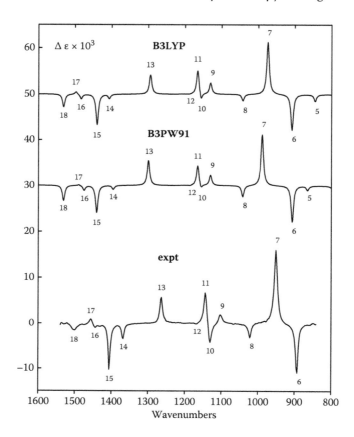

FIGURE 4.2 Comparison of the B3LYP/cc-pVTZ and B3PW91/cc-pVTZ VCD spectra of (S)-methyloxirane to the experimental VCD spectrum of (–)-methyloxirane. The assignment of the experimental spectrum is based on both the B3PW91/cc-pVTZ and B3LYP/cc-pVTZ spectra.

All *ab initio* calculations presented in this book were performed using the Gaussian series of programs [60].

REFERENCES

1. W.J. Hehre, L. Radom, P.V.R. Schleyer, and J.A. Pople, *Ab Initio Molecular Orbital Theory*, John Wiley & Sons, New York, 1986.
2. P. Hohenberg and W. Kohn, Inhomogeneous Electron Gas, *Phys. Rev.* 136 (3B), B864, 1964.
3. W. Koch and M.C. Holthausen, *A Chemist's Guide to Density Functional Theory*, 2nd ed. Wiley, 2001; G.E. Scuseria and V.N. Staroverov, Development of Approximate Exchange-Correlation Functionals. In *Theory and Applications of Computational Chemistry: The First Forty Years*, ed. C.E. Dykstra, G. Frenking, K.S. Kim, et al. Elsevier, Amsterdam, 2005, chap. 24, p. 669.
4. W. Kohn and L.J. Sham, Self-Consistent Equations Including Exchange and Correlation Effects, *Phys. Rev.* 140 (4A), A1133, 1965.
5. J.C. Slater, The Theory of Complex Spectra, *Phys. Rev.* 34, 1293, 1929; J.C. Slater, Cohesion in Monovalent Metals, *Phys. Rev.* 35, 509, 1930.

6. C.C.J. Roothaan, New Developments in Molecular Orbital Theory, *Rev. Mod. Phys.* 23 (2), 69, 1951; G.G. Hall, The Molecular Orbital Theory of Chemical Valency. VIII. A Method of Calculating Ionization Potentials, *Proc. Roy. Soc. (London)* A205, 541, 1951.

7. P.J. Stephens, Gauge Dependence of Vibrational Magnetic Dipole Transition Moments and Rotational Strengths, *J. Phys. Chem.* 91 (7), 1712, 1987.

8. L.A. Nafie, Velocity-Gauge Formalism in the Theory of Vibrational Circular Dichroism and Infrared Absorption, *J. Chem. Phys.* 96 (8), 5687, 1992; L.A. Nafie, *Vibrational Optical Activity: Principles and Applications*, Wiley, UK, 2011.

9. P. Pulay, *Ab Initio* Calculation of Force Constants and Equilibrium Geometries in Polyatomic Molecules. I. Theory, *Mol. Phys.* 17 (2), 197, 1969.

10. J.A. Pople, K. Raghavachari, H.B. Schlegel, and J.S. Binkley, Derivative Studies in Hartree-Fock and Møller-Plesset Theories, *Int. J. Quantum Chem. Quant. Chem. Symp.* S13, 225, 1979.

11. J. Gerratt and I.M. Mills, Force Constants and Dipole-Moment Derivatives of Molecules from Perturbed Hartree-Fock Calculations. I, *J. Chem. Phys.* 49 (4), 1719, 1968.

12. C.E. Dykstra and P.G. Jasien, Derivative Hartree-Fock Theory to All Orders, *Chem. Phys. Lett.* 109 (4), 388, 1984.

13. M.J. Frisch, M. Head-Gordon, and J.A. Pople, Direct Analytic SCF Second Derivatives and Electric Field Properties, *Chem. Phys.* 141 (2–3), 189, 1990.

14. B.G. Johnson and M.J. Frisch, An Implementation of Analytic Second Derivatives of the Gradient-Corrected Density Functional Energy, *J. Chem. Phys.* 100 (10), 7429, 1994.

15. R. McWeeny, Some Recent Advances in Density Matrix Theory, *Rev. Mod. Phys.* 32 (2), 335, 1960.

16. R. McWeeny, Perturbation Theory for Fock-Dirac Density Matrix, *Phys. Rev.* 126 (3), 1028, 1962; J.L. Dodds, R. McWeeny, W.T. Raynes, and J.P. Riley, SCF Theory for Multiple Perturbations, *Mol. Phys.* 33 (3), 611, 1977; J.L. Dodds, R. McWeeny, and A.J. Sadlej, Self-Consistent Perturbation Theory: Generalization for Perturbation-Dependent Non-Orthogonal Basis Set, *Mol. Phys.* 34 (6), 1779, 1977.

17. R.M. Stevens, R.M. Pitzer, and W.N. Lipscomb, Perturbed Hartree-Fock Calculations. 1. Magnetic Susceptibility and Shielding in LiH Molecule, *J. Chem. Phys.* 38 (2), 550, 1963.

18. R.D. Amos, N.C. Handy, K.J. Jalkanen, and P.J. Stephens, Efficient Calculation of Vibrational Magnetic Dipole Transition Moments and Rotational Strengths, *Chem. Phys. Lett.* 133 (1), 21, 1987.

19. J.R. Cheeseman, M.J. Frisch, F.J. Devlin, and P.J. Stephens, *Ab Initio* Calculation of Atomic Axial Tensors and Vibrational Rotational Strengths Using Density Functional Theory, *Chem. Phys. Lett.* 252, 211, 1996.

20. P.J. Stephens, C.S. Ashvar, F.J. Devlin, J.R. Cheeseman, and M.J. Frisch, *Ab Initio* Calculation of Atomic Axial Tensors and Vibrational Rotational Strengths Using Density Functional Theory, *Mol. Phys.* 89 (2), 579, 1996.

21. F. London, The Quantum Theory of Inter-Atomic Currents in Aromatic Combinations, *J. Phys. Radium* 8, 397, 1937; R. Ditchfield, Self-Consistent Perturbation Theory of Diamagnetism. 1. Gauge-Invariant LCAO Method for N.M.R. Chemical Shifts, *Mol. Phys.* 27 (4), 789, 1974.

22. K.L. Bak, P. Jørgensen, T. Helgaker, K. Ruud, and H.J.A. Jensen, Gauge-Origin Independent Multiconfigurational Self-Consistent-Field Theory for Vibrational Circular-Dichroism, *J. Chem. Phys.* 98 (11), 8873, 1993; K.L. Bak, P. Jørgensen, T. Helgaker, K. Ruud, and H.J.A. Jensen, Basis Set Convergence of Atomic Axial Tensors Obtained from Self-Consistent Field Calculations Using London Atomic Orbitals, *J. Chem. Phys.* 100 (9), 6620, 1994.

23. J.C. Slater, Atomic Shielding Constants, *Phys. Rev.* 36, 57, 1930.

24. W.J. Hehre, R. Ditchfield, and J.A. Pople, Self-Consistent Molecular Orbital Methods. 12. Further Extensions of Gaussian-Type Basis Sets for Use in Molecular-Orbital Studies of Organic-Molecules, *J. Chem. Phys.* 56 (5), 2257, 1972.

25. T.H. Dunning Jr., Gaussian Basis Sets for Use in Correlated Molecular Calculations. I. The Atoms Boron through Neon and Hydrogen, *J. Chem. Phys.* 90, 1007, 1989.

26. P.J. Stephens, K.J. Jalkanen, R.D. Amos, P. Lazzeretti, and R. Zanasi, *Ab Initio* Calculations of Atomic Polar and Axial Tensors for HF, H_2O, NH_3, and CH_4, *J. Phys. Chem.* 94 (5), 1811, 1990.

27. R.A. Kendall, T.H. Dunning Jr., and R.J. Harrison, Electron Affinities of the First-Row Atoms Revisited. Systematic Basis Sets and Wave Functions, *J. Chem. Phys.* 96, 6796, 1992.

28. M.M. Francl, W.J. Pietro, W.J. Hehre, J.S. Binkley, D.J. DeFrees, J.A. Pople, and M.S. Gordon, Self-Consistent Molecular Orbital Methods. 23. A Polarization-Type Basis Set for 2nd-Row Elements, *J. Chem. Phys.* 77 (7), 3654, 1982.

29. P.C. Hariharan and J.A. Pople, Basis Sets, *Theor. Chim. Acta* 28, 213, 1973.

30. R. Krishnan, J.S. Binkley, R. Seeger, and J.A. Pople, Basis Sets, *J. Chem. Phys.* 72, 650, 1980; A.D. McLean and G.S. Chandler, Contracted Gaussian-Basis Sets for Molecular Calculations. 1. 2nd Row Atoms, Z=11–18, *J. Chem. Phys.* 72 (10), 5639, 1980.

31. T. Clark, J. Chandrasekhar, G.W. Spitznagel, and P.V.R. Schleyer, Efficient Diffuse Function-Augmented Basis-Sets for Anion Calculations. 3. The 3-21+G Basis Set for 1st-Row Elements, Li-F, *J. Comp. Chem.* 4 (3), 294, 1983; P.M.W. Gill, B.G. Johnson, J.A. Pople, and M.J. Frisch, The Performance of the Becke-Lee-Yang-Parr (B-Lyp) Density Functional Theory with Various Basis Sets, *Chem. Phys. Lett.* 197 (4,5), 499, 1992.

32. M.J. Frisch, J.A. Pople, and J.S. Binkley,Self-Consistent Molecular Orbital Methods. 25. Supplementary Functions for Gaussian Basis Sets, *J. Chem. Phys.* 80 (7), 3265, 1984.

33. A.D. Becke, Density-Functional Exchange-Energy Approximation with Correct Asymptotic-Behavior, *Phys. Rev. A* 38 (6), 3098, 1988.

34. A.D. Boese and N.C. Handy, New Exchange-Correlation Density Functionals: The Role of the Kinetic-Energy Density, *J. Chem. Phys.* 116 (22), 9559, 2002.

35. C. Lee, W. Yang, and R.G. Parr, Development of the Colle-Salvetti Correlation-Energy Formula into a Functional of the Electron Density, *Phys. Rev. B* 37 (2), 785, 1988.

36. J.P. Perdew and Y. Wang, Accurate and Simple Analytic Representation of the Electron Gas Correlation Energy, *Phys. Rev. B* 45 (23), 13244, 1992.

37. A.D. Becke, Density-Functional Thermochemistry. III. The Role of Exact Exchange, *J. Chem. Phys.* 98 (7), 5648, 1993.

38. A.D. Becke, A New Mixing of Hartree-Fock and Local Density-Functional Theories, *J. Chem. Phys.* 98 (2), 1372, 1993.

39. P.J. Stephens, F.J. Devlin, M.J. Frisch, and C.F. Chabalowski, *Ab Initio* Calculation of Vibrational Absorption and Circular Dichroism Spectra Using Density Functional Force Fields, *J. Phys. Chem.* 98 (45), 11623, 1994.

40. J.C. Slater, *The Self-Consistent Field for Molecular and Solids* McGraw-Hill, New York, 1974.

41. B. Miehlich, A. Savin, H. Stoll, and H. Preuss, Results Obtained with the Correlation-Energy Density Functionals of Becke and Lee, Yang and Parr, *Chem. Phys. Lett.* 157 (3), 200, 1989.

42. S.H. Vosko, L. Wilk, and M. Nusair, Accurate Spin-Dependent Electron Liquid Correlation Energies for Local Spin Density Calculations: A Critical Analysis, *Can. J. Phys.* 58 (8), 1200, 1980.

43. J.P. Perdew, Density-Functional Approximation for the Correlation Energy of the Inhomogeneous Electron Gas, *Phys. Rev. B* 33 (12), 8822, 1986.

44. A.D. Becke, Density-Functional Thermochemistry. IV. A New Dynamical Correlation Functional and Implications for Exact-Exchange Mixing, *J. Chem. Phys.* 104 (3), 1040, 1996.

45. T. Yanai, D. Tew, and N. Handy, A New Hybrid Exchange-Correlation Functional Using the Coulomb-Attenuating Method (Cam-B3lyp), *Chem. Phys. Lett.* 393, 51, 2004.

46. T.M. Henderson, A.F. Izmaylov, G. Scalmani, and G.E. Scuseria, Can Short-Range Hybrids Describe Long-Range-Dependent Properties? *J. Chem. Phys.* 131, 044108, 2009.

47. S.F. Sousa, P.A. Fernandes, and M.J. Ramos,General Performance of Density Functionals, *J. Phys. Chem. A* 111 (42), 10439, 2007.

48. F.A. Hamprecht, A. Cohen, D.J. Tozer, and N.C. Handy, Development and Assessment of New Exchange-Correlation Functionals, *J. Chem. Phys.* 109 (15), 6264, 1998; A.D. Boese, N.L. Doltsinis, N.C. Handy, and M. Sprik, New Generalized Gradient Approximation Functionals, *J. Chem. Phys.* 112 (4), 1670, 2000; A.D. Boese and N.C. Handy, A New Parametrization of Exchange-Correlation Generalized Gradient Approximation Functionals, *J. Chem. Phys.* 114 (13), 5497, 2001.

49. N.C. Handy and A.J. Cohen, Left-Right Correlation Energy, *Mol. Phys.* 99 (5), 403 (2001); W.-M. Hoe, A. Cohen, and N.C. Handy, Assessment of a New Local Exchange Functional Optx, *Chem. Phys. Lett.* 341, 319, 2001.

50. J.P. Perdew, K. Burke, and M. Ernzerhof, Generalized Gradient Approximation Made Simple, *Phys. Rev. Lett.* 77 (18), 3865, 1996; C. Adamo and V. Barone, Toward Reliable Density Functional Methods without Adjustable Parameters: The Pbe0 Model, *J. Chem. Phys.* 110 (13), 6158, 1999.

51. A.J. Cohen and N.C. Handy, Dynamic Correlation, *Mol. Phys.* 99 (7), 607, 2001.

52. C. Adamo and V. Barone, Exchange Functionals with Improved Long-Range Behavior and Adiabatic Connection Methods without Adjustable Parameters: The mPW and mPW1PW Models, *J. Chem. Phys.* 108 (2), 664, 1998.

53. A.D. Boese and J.M.L. Martin, Development of Density Functionals for Thermochemical Kinetics, *J. Chem. Phys.* 121 (8), 3405, 2004.

54. Y. Zhao, N.E. Schultz, and D.G. Truhlar, Design of Density Functionals by Combining the Method of Constraint Satisfaction with Parametrization for Thermochemistry, Thermochemical Kinetics, and Noncovalent Interactions, *J. Chem. Theory Comput.* 2, 364, 2006.

55. J. Heyd, G. Scuseria, and M. Ernzerhof, Hybrid Functionals Based on a Screened Coulomb Potential, *J. Chem. Phys.* 118 (18), 8207, 2003.

56. H.B. Schlegel, Geometry Optimization, *Wiley Interdisciplinary Rev. Comput. Mol. Sci.* 1 (5), 790. http://dx.doi.org/10.1002/wcms.34; H.P. Hratchian and H.B. Schlegel, in *Theory and Applications of Computational Chemistry. The First Forty Years*, ed. C.E. Dykstra, G. Frenking, K.S. Kim, et al., Elsevier, Amsterdam, 2005, chap. 10, p. 195.

57. J.B. Foresman and Æ. Frisch, *Exploring Chemistry with Electronic Structure Methods*, 2nd ed., Gaussian, Pittsburgh, PA, 1996.

58. R.E. Stratmann, G.E. Scuseria, and M.J. Frisch, Achieving Linear Scaling in Exchange-Correlation Density Functional Quadratures, *Chem. Phys. Lett.* 257 (3–4), 213, 1996.

59. R.W. Kawiecki, F.J. Devlin, P.J. Stephens, R.D. Amos, and N.C. Handy, Vibrational Circular Dichroism of Propylene Oxide, *Chem. Phys. Lett.* 145 (5), 411, 1998.

60. M.J. Frisch, G.W. Trucks, H.B. Schlegel, G.E. Scuseria, M.A. Robb, J.R. Cheeseman, G. Scalmani, V. Barone, B. Mennucci, G.A. Petersson, H. Nakatsuji, M. Caricato, X. Li, H.P. Hratchian, A.F. Izmaylov, J. Bloino, G. Zheng, J.L. Sonnenberg, M. Hada, M. Ehara, K. Toyota, R. Fukuda, J. Hasegawa, M. Ishida, T. Nakajima, Y. Honda, O. Kitao, H. Nakai, T. Vreven, J.A. Montgomery Jr., J.E. Peralta, F. Ogliaro, M.J. Bearpark, J. Heyd, E.N. Brothers, K.N. Kudin, V.N. Staroverov, R. Kobayashi, J. Normand, K. Raghavachari, A.P. Rendell, J.C. Burant, S.S. Iyengar, J. Tomasi, M. Cossi, N. Rega, N.J. Millam, M. Klene, J.E. Knox, J.B. Cross, V. Bakken, C. Adamo, J. Jaramillo, R. Gomperts, R.E. Stratmann, O. Yazyev, A.J. Austin, R. Cammi, C. Pomelli, J.W. Ochterski, R.L. Martin, K. Morokuma, V.G. Zakrzewski, G.A. Voth, P. Salvador, J.J. Dannenberg, S. Dapprich, A.D. Daniels, Ö. Farkas, J.B. Foresman, J.V. Ortiz, J. Cioslowski, and D.J. Fox, *Gaussian 09*, Gaussian, Wallingford, CT, 2009.

5 Conformational Analysis

The objectives of the conformational analysis of a molecule are to determine the number of minima on the potential energy surface (PES) of the molecule, i.e., the number of stable conformations, and the geometries and energies of each stable conformation. The number, geometries, and energies of the stable conformations of a molecule depend on the methodology used to calculate the PES. Ideally, when infrared (IR) and vibrational circular dichroism (VCD) spectra are to be calculated using density functional theory (DFT), conformational analysis should also be carried out using DFT.

For very small molecules this is straightforward. For example, consider the chiral substituted oxiranes, methyl-oxirane, **1**, and phenyl-oxirane, **2**.

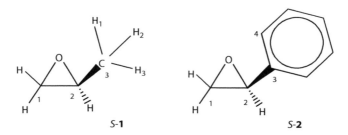

The oxirane ring is conformationally rigid, as are also the methyl and phenyl substituents. However, in **1** the methyl group can rotate about the C–C bond, connecting it to the oxirane ring, and in **2**, the phenyl group can rotate about the C–C bond, connecting it to the oxirane ring. To determine the number of stable conformations of **1** and **2**, relaxed scans of their PESs with respect to rotation of the CH_3 and C_6H_5 groups are carried out. Specifically, the dihedral angles $C_1C_2C_3H_1$ of **1** and $C_1C_2C_3C_4$ of **2** are varied from 0 to 360°. For each value of the dihedral angle, optimization of the molecule is carried out, giving the relaxed energy. The plot of the relaxed energies vs. the dihedral angle values is the relaxed PES scan.

The relaxed PES scan of *S*-1 calculated at the B3LYP/6-31G* level with $C_1C_2C_3H_1$ being varied in steps of 10° is shown in Figure 5.1. Three valleys in the PES are observed, and therefore, three stable conformations are predicted. In Table 5.1, $C_1C_2C_3H_1$ dihedral angles and the relative energies of the lowest-energy structures are listed. To determine the geometries and energies of the three actual minima of the PES, completely unconstrained (i.e., full) optimizations are then carried out, starting from the lowest-energy structures of the PES scan, with the results given in Table 5.1 and Figure 5.1. The three minima have $C_1C_2C_3H_1$ dihedral angles of 96.0, 215.6, and 335.5, and identical energies. To determine the stability of these minima and to permit the calculation of their free energies, harmonic frequency calculations were also carried out on these structures. The largest zero frequencies (the frequencies corresponding to

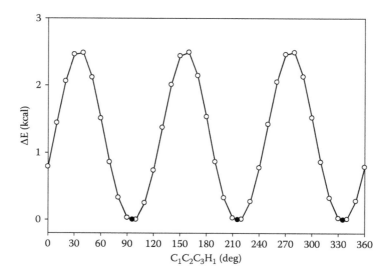

FIGURE 5.1   Variations in the energy of *S*-1 as the methyl group is rotated about the C_2–C_3 bond. The minima for the B3LYP/6-31G* optimized structures are indicated (•).

TABLE 5.1
Dihedral Angles, Relative Energies, and Relative Free Energies of *S*-1

	B3LYP/6-31G*		MMFF94	
	1D Scan	**Full Optimizations[d]**	**Search[e]**	**DFT Optimization[f]**
$C_1C_2C_3H_1$[a]	100.0, 220.0, 340.0	96.0, 215.6, 335.5	330.9	335.5
ΔE[b]	0.00, 0.01, 0.01	0.00, 0.00, 0.00		
ΔG[c]		0.00, 0.00, 0.00		

[a]   Dihedral angles in degrees.
[b]   Relative energies in kcal/mol.
[c]   Relative free energies in kcal/mol.
[d]   All molecular parameters, including the $C_1C_2C_3H_1$ dihedral angle, are allowed to vary during the optimization.
[e]   Monte Carlo and systematic conformer searches found the same structure.
[f]   DFT optimization of the MMFF94 structure found in the search.

---

the three translational and three rotational degrees of freedom) predicted were 4.47, 4.47, and 4.46 cm^{-1}, respectively, and no imaginary frequencies were found. The relative free energies of the lowest-energy structures are given in Table 5.1. Thus, rotation of the CH$_3$ group about the CC bond leads to three equivalent stable conformations, with identical molecular structures, energies, and free energies. Effectively, therefore, **1** is a conformationally rigid molecule since the properties of all three conformations are identical, and only one of the conformations needs to be considered in predicting the properties of **1**.

The relaxed PES scan of *S*-**2** at the B3LYP/6-31G* level with $C_1C_2C_3C_4$ being varied in steps of 10° is shown in Figure 5.2. Two valleys in the PES are observed, and therefore two stable conformations are predicted. In Table 5.2 the $C_1C_2C_3C_4$ dihedral angles and the relative energies of the lowest-energy structures are listed. Further unconstrained optimizations of these two structures lead to the dihedral angles and relative energies in Table 5.2 and Figure 5.2. The two minima have $C_1C_2C_3C_4$ dihedral angles of 126.4 and 307.1° and identical energies. Harmonic frequency calculations for these two structures yielded maximum zero frequencies of

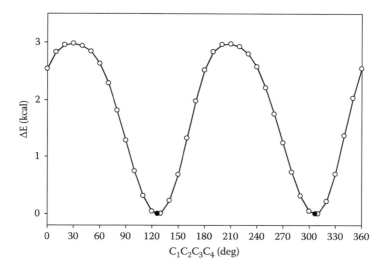

FIGURE 5.2 Variations in the energy of *S*-**2** as the phenyl group is rotated about the C_2–C_3 bond. The minima for the B3LYP/6-31G* optimized structures are indicated (•).

TABLE 5.2
Dihedral Angles, Relative Energies, and Relative Free Energies of *S*-2

	B3LYP/6-31G*		MMFF94	
	1D Scan	**Full Optimizations[d]**	**Search[e]**	**DFT Optimization[f]**
$C_1C_2C_3C_4$[a]	130.0, 310.0	126.4, 307.1	252.2	307.1
ΔE[b]	0.00, 0.00	0.00, 0.00		
ΔG[c]		0.00, 0.00		

[a] Dihedral angles in degrees.
[b] Relative energies in kcal/mol.
[c] Relative free energies in kcal/mol.
[d] All molecular parameters, including the $C_1C_2C_3C_4$ dihedral angle, are allowed to vary during the optimization.
[e] Monte Carlo and systematic conformer searches found the same structure.
[f] DFT optimization of the MMFF94 structure found in the search.

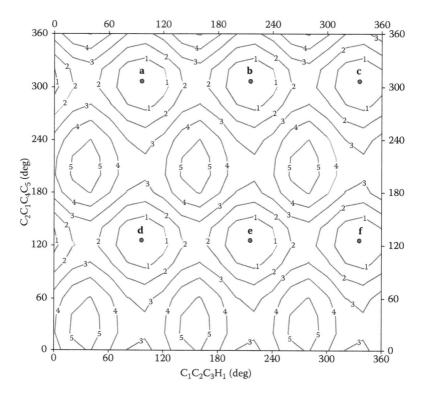

FIGURE 5.3 The B3LYP/6-31G* PES of (1S,2S)-**3**. The dihedral angles $C_1C_2C_3H_1$ and $C_2C_1C_4C_5$ were varied in 20° steps. Contours are shown at 1 kcal/mol intervals. Minima resulting from unconstrained optimizations of the lowest-energy structures in the valleys of the PES are indicated (•).

6.69 and 6.69 cm^{-1}, respectively, and gave no imaginary frequencies, confirming that both conformations are stable. The relative free energies of the lowest-energy structures are given in Table 5.2. Thus, rotation of the C_6H_5 group leads to two equivalent conformations with identical molecular structures, energies, and free energies. As with **1**, **2** is effectively a conformationally rigid molecule.

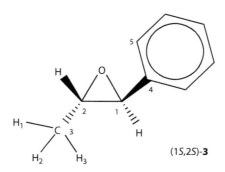

(1S,2S)-**3**

TABLE 5.3
Dihedral Angles, Relative Energies, and Relative Free Energies of (1S,2S)-3

| | B3LYP/6-31G* | | | | | | | MMFF94 | | | |
| | 2D Scan | | | Full Optimizations[d] | | | | Search[e] | | DFT Optimization[f] | |
	$C_1C_2C_3H_1{}^a$	$C_2C_1C_4C_5{}^a$	ΔE^b	$C_1C_2C_3H_1{}^a$	$C_2C_1C_4C_5{}^a$	ΔE^b	ΔG^c	$C_1C_2C_3H_1{}^a$	$C_2C_1C_4C_5{}^a$	$C_1C_2C_3H_1{}^a$	$C_2C_1C_4C_5{}^a$
a	100.0	300.0	0.01	96.4	306.2	0.00	0.00	330.7	252.1	335.7	306.2
b	220.0	300.0	0.02	215.9	306.2	0.00	0.00				
c	340.0	300.0	0.01	335.7	306.2	0.00	0.00				
d	100.0	120.0	0.00	96.4	125.5	0.00	0.00				
e	220.0	120.0	0.00	215.9	125.5	0.00	0.00				
f	340.0	120.0	0.01	335.7	125.5	0.00	0.00				

a Dihedral angles in degrees.
b Relative energies in kcal/mol.
c Relative free energies in kcal/mol.
d All molecular parameters, including the $C_1C_2C_3H_1$ and $C_2C_1C_4C_5$ dihedral angles, are allowed to vary during the optimization.
e Monte Carlo and systematic conformer searches found the same structure.
f DFT optimization of the MMFF94 structures found in the search.

In the case of the trans-methyl-phenyl-oxirane **3**, both methyl and phenyl substituents can rotate about the C–C bonds connecting them to the oxirane ring. If the substituents interact significantly, their orientations relative to the oxirane ring will differ from the values in **1** and **2**. To predict the stable conformations of **3**, a relaxed PES scan, varying the two dihedral angles $C_1C_2C_3H_1$ and $C_2C_1C_4C_5$ simultaneously, is required. The results of this two-dimensional (2D) PES scan at the B3LYP/6-31G* level are shown in Figure 5.3. Six valleys in the PES are observed. The lowest-energy valleys are **a**, **b**, **c**, **d**, **e**, and **f**. In Table 5.3, the $C_1C_2C_3H_1$ and $C_2C_1C_4C_5$ dihedral angles and the relative energies of the six lowest-energy structures in valleys **a–f** are listed. Further unconstrained optimizations of these six structures lead to the dihedral angles and relative energies in Table 5.3. Harmonic frequency calculations for these six structures all gave maximum zero frequencies of <5.0 cm^{-1} and had no imaginary frequencies, proving that all six conformations are stable. The relative free energies of the lowest-energy conformations are given in Table 5.3. Thus, simultaneous rotation of both the methyl and phenyl groups leads to six equivalent conformations with identical molecular structures, energies, and free energies. As with **1** and **2**, **3** is effectively a conformationally rigid molecule. Furthermore, since the methyl dihedral angles in **1** and **3** differ by <0.5° and the phenyl dihedral angles in **2** and **3** differ by <1.0°, these substituents must have the same orientations relative to the oxirane ring as in **1** and **2**. This suggests that there is no significant interaction between the methyl and phenyl groups in **3**.

For molecules more conformationally flexible than **1–3**, DFT PES scans must be carried out as a function of more than two dihedral angles. Unfortunately, such scans are often too demanding computationally to be practicable. Consequently, another approach to conformational analysis has to be used. At this time, the majority of conformational analyses of large organic molecules use a Molecular Mechanics Force Field (MMFF). A MMFF is an algebraic equation for the molecular PES, as a function of the bond lengths, bond angles, dihedral angles, and nonbonding interatomic distances of the molecule. Many MMFFs have been developed since the 1970s [1,2]. The parameters of a MMFF, which determine the quantitative variation of the molecular potential energy resulting from changes in molecular bond lengths, bond angles, dihedral angles, and nonbonding interatomic distances, are generally determined in one of two ways: (1) using experimental properties such as equilibrium geometries and vibrational frequencies, or (2) using properties calculated using *ab initio* methods.

How does one search for the stable conformations of a molecule predicted by a MMFF? Three types of approaches have been utilized: (1) calculation of the energies for a range of geometries (systematic searching), an approach analogous to the DFT PES scans discussed above; (2) Monte Carlo (M-C) searching; and (3) molecular dynamics (MD).

M-C searching methods explore the PES by randomly probing the geometry of the molecule. MD methods simulate the time evolution of the molecular system and are beyond the scope of this book.

Given a specific choice of MMFF and of conformational searching procedure, the geometries and relative energies of the stable conformations of a molecule are predicted. Under the assumption that the MMFF PES and DFT PES predict stable conformations with similar geometries and relative energies, the DFT PES stable conformations are then predicted by DFT optimizations of the MMFF conformations, followed by harmonic frequency calculations to confirm that the conformations are stable.

Here, we illustrate this protocol using the Merck Molecular Mechanics Force Field (MMFF94) [2] and the two conformational searching procedures of the SPARTAN program [3]: (1) systematic searching, in which the energy of the molecule is calculated for a range of molecular geometries, and (2) Monte Carlo searching. The MMFF94 force field is parameterized primarily from the results of *ab initio* calculations and has been shown to perform well for a wide range of small organic molecules [2]. For molecules 1–3 conformational analyses using MMFF94 predict only one stable conformation within a window of 20 kcal/mol, consistent with the B3LYP/6-31G* DFT PES scans discussed above. The MMFF94 dihedral angles of the conformations of 1–3 are compared to those obtained thence by DFT at the B3LYP/6-31G* level in Tables 5.1–5.3. In 1 and 3 the orientation of the methyl group, predicted by MMFF94, only differs by about 5° in the DFT optimized structures. However, the MMFF94 orientation of the phenyl substituent relative to the oxirane ring in 2 and 3 differs by more than 50° in the DFT optimized structures.

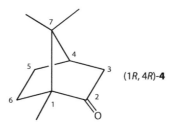

(1R, 4R)-**4**

The 1R,4R enantiomer of camphor, **4**, is expected to be conformationally rigid, given the bridge between the C_1 and C_4 atoms of the cyclohexanone ring. Conformational analysis using MMFF94 confirms this expectation: only one stable conformation is predicted within a 20 kcal/mol window. Dihedral angles of this conformation are given in Table 5.4. DFT optimization of this geometry at the B3LYP/6-31G* level leads to a stable conformation with dihedral angles, also given in Table 5.4, which are very similar to the MMFF94 dihedral angles.

The same protocol for (1R,5R)-α-pinene, **5**, also predicts conformational rigidity, with the very similar MMFF94 and DFT B3LYP/6-31G* dihedral angles given in Table 5.5.

TABLE 5.4
Key Dihedral Angles of (1R,4R)-4

Dihedrals[a]	MMFF94	B3LYP/6-31G*
$C_1C_2C_3C_4$	−2.2	−0.7
$C_2C_3C_4C_5$	−71.0	−72.0
$C_3C_4C_5C_6$	73.4	70.8
$C_4C_5C_6C_1$	1.3	2.5
$C_5C_6C_1C_2$	−69.8	−71.5
$C_6C_1C_2C_3$	73.5	71.0
$C_7C_1C_2C_3$	−33.4	−34.0
$C_7C_4C_3C_2$	37.8	35.7
$C_7C_1C_6C_5$	34.1	32.5
$C_7C_4C_5C_6$	−36.2	−36.9

[a] Dihedral angles are in degrees.

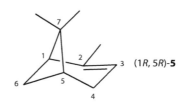

(1R, 5R)-5

TABLE 5.5
Key Dihedral Angles of (1R,5R)-5

Dihedrals[a]	MMFF94	B3LYP/6-31G*
$C_1C_2C_3C_4$	−1.5	−2.1
$C_2C_3C_4C_5$	0.4	2.5
$C_3C_4C_5C_6$	48.4	45.6
$C_4C_5C_6C_1$	−82.4	−82.7
$C_5C_6C_1C_2$	80.5	81.8
$C_6C_1C_2C_3$	−45.5	−45.9
$C_7C_1C_2C_3$	48.8	47.5
$C_7C_5C_4C_3$	−47.0	−49.1
$C_7C_1C_6C_5$	−33.4	−28.7
$C_7C_5C_6C_1$	33.5	28.9

[a] Dihedral angles are in degrees.

Since organic chemical intuition leads to the expectation that both **4** and **5** are conformationally rigid molecules, there is little reason to be concerned that the MMFF94 conformational analysis might have missed one or more stable conformations. For more flexible molecules, however, this possibility can appear more likely. In such a situation, the reliability of the MMFF94 conformational analysis

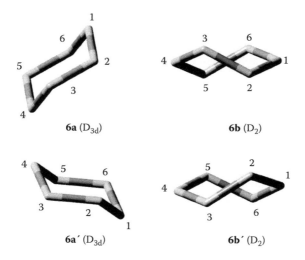

6a (D_{3d})   **6b** (D_2)

6a' (D_{3d})   **6b'** (D_2)

FIGURE 5.4 Chair (**6a**) and twist-boat (**6b**) MMFF94 conformations of **6**. Mirror image structures **6a'** and **6b'** found in the 2D scan. The hydrogen atoms are not shown for clarity.

TABLE 5.6
Ring Dihedral Angles, Relative Energies, Relative Free Energies, and Room Temperature Populations of the Conformers of 6

					B3LYP/6-31G*			
	MMFF94		Full Optimization		2D Scan			
Dihedrals[a]	6a	6b	6a	6b	6a	6a'	6b	6b'
$C_1C_2C_3C_4$	54.3	60.1	54.7	63.5	54.7	−54.7	63.5	−63.5
$C_2C_3C_4C_5$	−54.3	−29.2	−54.7	−30.6	−54.7	54.7	−30.6	30.6
$C_3C_4C_5C_6$	54.3	−29.2	54.7	−30.6	54.7	−54.7	−30.6	30.6
$C_4C_5C_6C_1$	−54.3	60.1	−54.7	63.5	−54.7	54.7	63.5	−63.5
$C_5C_6C_1C_2$	54.3	−29.2	54.7	−30.6	54.7	−54.7	−30.6	30.6
$C_6C_1C_2C_3$	−54.3	−29.2	−54.7	−30.6	−54.7	54.7	−30.6	30.6
$C_2C_3C_5C_6$	0.0	−54.3	0.0	−57.4	0.0	0.0	−57.4	57.4
ΔE^{b}	0.00	5.93	0.00	6.48	0.00	0.00	6.48	6.48
ΔG^{c}			0.00	5.82	0.00	0.00	5.82	5.81
$P(\%)^{d}$			99.99	0.01	99.99	99.99	0.01	0.01

[a] Dihedral angles in degrees.
[b] Relative energies in kcal/mol.
[c] Relative free energies in kcal/mol.
[d] Populations based on ΔG values, T = 298 K.

can be probed by carrying out 1D or 2D DFT PES scans. For example, consider the achiral molecule, cyclohexane, **6**, whose conformational flexibility is famous [4]. Conformational analysis of **6** using MMFF94 leads to two stable conformations **a** and **b**, within a 20 kcal/mol window. The structures and symmetries of these conformations are shown in Figure 5.4. The CCCC dihedral angles of the C_6 rings and the relative energies of **6a** and **6b** are given in Table 5.6. Conformations **6a** and **6b** are termed chair and twist-boat conformations, respectively [5]. DFT optimizations of the MMFF94 conformations of **6** lead to stable conformations with the dihedral angles and relative energies given in Table 5.6. The MMFF94 and B3LYP/6-31G* dihedral angles are very similar. The MMFF94 and B3LYP/6-31G* relative energies are also very similar. Calculation of harmonic vibrational frequencies for **6a** and **6b** confirmed their stability and led to the relative free energies given in Table 5.6. These in turn allowed the room temperature equilibrium populations to be calculated (Equation 3.38), with the results also given in Table 5.6. Clearly, **6a** is the dominant conformer, and thus cyclohexane exists almost exclusively (>99.9%) in the chair conformation at ambient temperature.

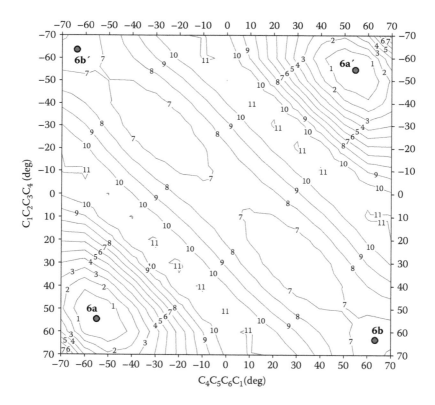

FIGURE 5.5   The B3LYP/6-31G* PES of **6**. The dihedral angles $C_4C_5C_6C_1$ and $C_1C_2C_3C_4$ were varied in 10° steps. Contours are shown at 1 kcal/mol intervals. Conformers resulting from unconstrained optimizations of the lowest-energy structures in the valleys of the PES are indicated (•).

To probe the reliability of the MMFF94 conformational analysis of **6**, it is useful to carry out a DFT PES scan of **6**, to see if additional conformations can be found. In Figure 5.5 the results of a 2D B3LYP/6-31G* PES scan of **6** as a function of the two dihedral angles, $C_1C_2C_3C_4$ and $C_4C_5C_6C_1$, are shown. Four valleys are present in the PES. B3LYP/6-31G* optimization of the lowest-energy geometry in each valley leads to four stable conformations, with the dihedral angles and relative energies given in Table 5.6. The two lowest-energy conformations **6a** and **6a′**, which are mirror images, are equivalent, having identical energies and dihedral angles of identical magnitudes, and are identical to the chair conformation **6a** obtained from the B3LYP/6-31G* optimization of the MMFF94 chair conformation (see Figure 5.4). The two higher-energy conformations **6b** and **6b′**, which are mirror images, are also equivalent, and identical to the twist-boat conformation **6b** obtained from the B3LYP/6-31G* optimization of the MMFF94 twist-boat conformation (see Figure 5.4). Thus, the 2D PES scan of **6** finds no new conformations, supporting the reliability of the MMFF94/DFT protocol for **6**. In particular, the untwisted-boat conformation **6c** is not predicted to be a stable conformation of **6**. The 2D PES scan predicts that the untwisted-boat conformation is the structure of the peak of the barrier separating the two twist-boat conformations. Confirmation of this conclusion is provided by B3LYP/6-31G* optimization of the untwisted-boat conformation, which leads to an optimized geometry of C_{2v} symmetry, 0.84 kcal/mol higher in energy than the twist-boat conformations, for which a harmonic frequency calculation leads to an imaginary frequency of -93.0 cm^{-1}, proving that it is not a stable conformation. To interconvert the chair conformers (**6a** and **6a′**) the ring must pass through a twist-boat conformation and through two half-chair transition states. The half-chair structure contains four carbon atoms located on a plane in which two bonds are fully eclipsed and is about 11 kcal/mol higher in energy than the chair conformation.

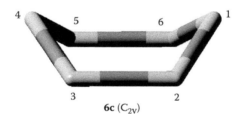

6c (C_{2v})

In order to determine the effects of substituting a carbonyl group on cyclohexane, the conformational analysis of cyclohexanone, **7**, has been carried out identically to the conformational analysis of cyclohexane, **6**. Figure 5.6 shows that the stable conformations of **7** (with the carbonyl group on C_1) are very similar to those of **6** in Figure 5.4. The dihedral angles and relative energies of the four conformations **7a**, **7a′**, **7b**, and **7b′**, given in Table 5.7, are qualitatively similar to those of conformations **6a**, **6a′**, **6b**, and **6b′**, in Table 5.6, but not quantitatively identical, demonstrating that the carbonyl group does have an impact on the conformations of **7**. This is also shown by the B3LYP/6-31G* PES scan of **7**, Figure 5.7, which differs substantially from that of **6**, Figure 5.5. As was the case for **6**, the room temperature population of

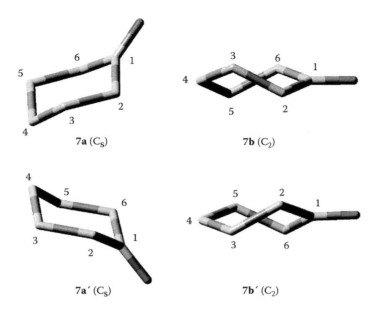

FIGURE 5.6 Chair (**7a**) and twist-boat (**7b**) MMFF94 conformations of **7**. Mirror image structures **7a′** and **7b′** of **7a** and **7b** found in the 2D scan. The oxygen atom is bonded to C_1. The hydrogen atoms are not shown for clarity.

TABLE 5.7

Ring Dihedral Angles, Relative Energies, Relative Free Energies, and Room Temperature Populations of the Conformers of 7

			B3LYP/6-31G*					
	MMFF94		Full Optimization		2D Scan			
Dihedrals[a]	7a	7b	7a	7b	7a	7a′	7b	7b′
$C_1C_2C_3C_4$	52.0	58.9	51.9	59.4	51.9	−51.9	59.4	−59.4
$C_2C_3C_4C_5$	−57.0	−29.4	−56.4	−28.9	−56.4	56.4	−28.9	28.9
$C_3C_4C_5C_6$	57.0	−29.4	56.4	−28.9	56.4	−56.4	−28.9	28.9
$C_4C_5C_6C_1$	−52.0	58.9	−51.9	59.4	−51.9	51.9	59.4	−59.4
$C_5C_6C_1C_2$	49.7	−28.5	48.9	−28.8	48.9	−48.9	−28.8	28.8
$C_6C_1C_2C_3$	−49.7	−28.5	−48.9	−28.8	−48.9	48.9	−28.8	28.8
$C_2C_3C_5C_6$	0.0	−54.4	0.0	−53.6	0.0	0.0	−53.5	53.5
ΔE^b	0.00	2.97	0.00	3.72	0.00	0.00	3.72	3.72
ΔG^c			0.00	3.54	0.00	0.00	3.18	3.18
$P(\%)^d$			99.75	0.25	99.54	99.54	0.46	0.46

[a] Dihedral angles in degrees.
[b] Relative energies in kcal/mol.
[c] Relative free energies in kcal/mol.
[d] Populations based on ΔG values, T = 298 K.

the chair conformation of **7** is >99%, and the twist-boat conformation is not significantly populated at room temperature.

In order to determine the effects of substituting a methyl group on cyclohexanone, the conformational analysis of *R*-**3**-methylcyclohexanone, *R*-**8**, has been carried out identically to the conformational analysis of cyclohexanone, **7**. The MMFF94 conformations, **8a–8e**, are shown in Figure 5.8; the dihedral angles and relative energies are given in Table 5.8. The B3LYP/6-31G* optimizations of the MMFF94 conformations gave the dihedral angles and relative energies of **8a–8e** in Table 5.8. The B3LYP/6-31G* relative free energies and room temperature populations, given in Table 5.8, were also obtained. Conformation **8a** is the very lowest-energy conformation, and is predicted to have a population of ~95% at room temperature [6].

In Figure 5.9, the 2D PES scan of *R*-**8** is shown. Conformations **8a**, **8b**, **8c**, and **8d** are also identified. Since conformation **8e** does not have a well-defined valley on the PES, it is omitted from Figure 5.9.

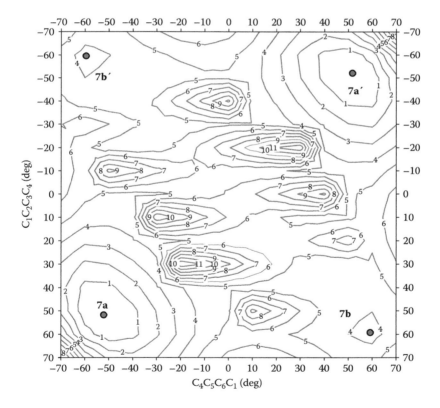

FIGURE 5.7 The B3LYP/6-31G* PES of **7**. The dihedral angles $C_4C_5C_6C_1$ and $C_1C_2C_3C_4$ were varied in 10° steps. Contours are shown at 1 kcal/mol intervals. Conformers resulting from unconstrained optimizations of the lowest-energy structures in the valleys of the PES are indicated (●).

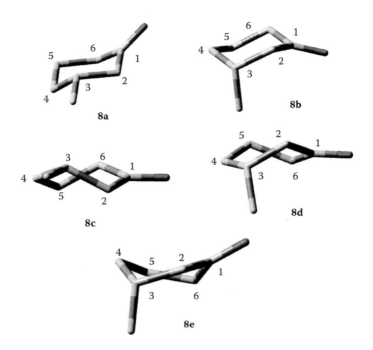

FIGURE 5.8 eq-chair (**8a**), ax-chair (**8b**), eq-twist-boat (**8c**), ax-twist-boat (**8d**) and ax-twist-boat (**8e**) MMFF94 conformations of *R*-**8**. The oxygen atom is bonded to C_1. The hydrogen atoms are not shown for clarity.

The chiral alkane, anti-trans-anti-trans-anti-trans-perhydrotriphenylene, **9**, contains four cyclohexane rings. Conformational analysis of **9** using the MMFF94 force field leads to four conformations within a 20 kcal/mol window, **9a**, **9b**, **9c**, and **9d**. In

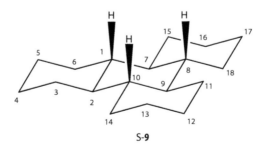

S-**9**

conformation **9a** all four cyclohexane rings have chair conformations. In **9b**, there are three chair rings and one twist-boat ring. In **9c**, there are two chair rings and two twist-boat rings. In **9d**, the central ring has a chair conformation and the three surrounding rings have twist-boat conformations. The relative energies of the MMFF94 conformations are given in Table 5.9. B3LYP/6-31G* optimizations of the MMFF94 conformations lead to the relative energies in Table 5.9. The MMFF94 and

TABLE 5.8

Ring Dihedral Angles, Relative Energies, Relative Free Energies, and Room Temperature Populations of the Conformers of R-8

| | MMFF94 | | | | | B3LYP/6-31G* | | | | | | | | | |
| | | | | | | Full Optimization | | | | | 2D Scan[e] | | | |
Dihedrals[a]	8a	8b	8c	8d	8e	8a	8b	8c	8d	8e	8a	8b	8c	8d
$C_1C_2C_3C_4$	52.4	-50.0	59.8	-53.2	-33.6	51.8	-52.5	59.9	-43.0	-35.1	51.8	-52.5	59.9	-43.0
$C_2C_3C_4C_5$	-57.1	54.0	-31.8	14.9	59.8	-56.1	55.6	-30.7	-8.5	63.8	-56.1	55.6	-30.6	-8.5
$C_3C_4C_5C_6$	57.3	-55.2	-27.1	40.9	-31.0	56.7	-55.3	-26.7	56.1	-36.9	56.7	-55.3	-26.7	56.1
$C_4C_5C_6C_1$	-52.3	51.6	58.8	-58.6	-21.2	-51.6	50.9	58.4	-51.6	-17.1	-51.6	50.9	58.4	-51.6
$C_5C_6C_1C_2$	50.1	-50.7	-30.0	19.3	48.1	48.8	-50.0	-29.4	0.2	45.4	48.8	-49.9	-29.4	0.2
$C_6C_1C_2C_3$	-50.5	50.3	-27.7	35.7	-19.1	-49.5	51.4	-28.4	48.2	-17.5	-49.6	51.3	-28.5	48.2
$C_2C_3C_5C_6$	-0.2	-0.7	-54.8	51.4	26.1	-0.2	1.0	-53.5	43.1	24.8	-0.2	1.0	-53.5	43.1
ΔE^b	0.00	0.48	3.16	4.11	6.04	0.00	1.52	3.83	4.88	6.37	0.00	1.52	3.83	4.88
ΔG^c						0.00	1.77	3.49	5.25	5.78	0.00	1.77	3.50	5.25
$P(\%)^d$						94.93	4.78	0.26	0.01	0.01	94.94	4.78	0.26	0.01

[a] Dihedral angles in degrees.
[b] Relative energies in kcal/mol.
[c] Relative free energies in kcal/mol.
[d] Populations based on ΔG values, T = 298 K.
[e] Since conformation **8e** does not have a well-defined valley, it cannot be located on the PES.

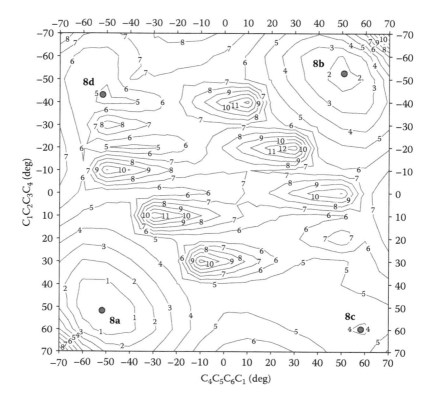

FIGURE 5.9 The B3LYP/6-31G* PES of *R*-**8**. The dihedral angles $C_4C_5C_6C_1$ and $C_1C_2C_3C_4$ were varied in 10° steps. Contours are shown at 1 kcal/mol intervals. Conformers resulting from unconstrained optimizations of the lowest-energy structures in the valleys of the PES are indicated (•).

B3LYP/6-31G* relative energies are very similar. The B3LYP/6-31G* structures of conformations **9a–9d** are shown in Figure 5.10. Harmonic frequency calculations proved that the four conformations are stable, and gave their relative free energies and room temperature populations, given in Table 5.9. As anticipated from the conformational analysis of cyclohexane, the conformation of **9** with all chair rings, **9a**, is predicted to have a room temperature population of >99.9%. **9** is therefore predicted to be a conformationally rigid molecule.

The conformational analysis of the chiral sulfoxide, 1-thiochroman-4-one-sulfoxide, **10**, is presented in Figures 5.11 and 5.12, and Table 5.10. Conformational analysis of **10** using MMFF94 shows that for this bicyclic molecule, one ring (the phenyl ring) is planar, whereas the other is significantly puckered and has two possible conformations **10a** and **10b**, shown in Figure 5.11. In order to verify the MMFF94 conformational analysis of **10**, and to determine whether additional conformations exist, a 2D PES scan was carried out varying simultaneously the dihedral angles $C_{10}C_6S_1C_2$ and $C_9C_5C_4C_3$. The resulting contour plot is shown in Figure 5.12. Only two wells were found in the PES. Geometry optimization at the

TABLE 5.9
Ring Dihedral Angles, Relative Energies, Relative Free Energies, and Populations of the Conformers of S-9

Dihedrals[a]	MMFF94				B3LYP/6-31G*			
	9a	9b	9c	9d	9a	9b	9c	9d
$C_1C_2C_3C_4$	54.4	29.7	29.4	29.4	54.6	29.1	30.1	29.3
$C_2C_3C_4C_5$	−55.6	29.9	30.0	29.9	−55.2	32.2	31.1	31.8
$C_3C_4C_5C_6$	55.3	−61.4	−61.3	−61.2	54.3	−64.6	−63.6	−64.0
$C_4C_5C_6C_1$	−55.6	29.9	29.7	29.9	−55.2	32.2	31.7	31.8
$C_5C_6C_1C_2$	54.4	29.7	29.6	29.4	54.6	29.1	29.3	29.3
$C_6C_1C_2C_3$	−53.1	−60.3	−60.0	−59.9	−53.2	−60.9	−61.4	−61.0
$C_2C_3C_5C_6$	0.3	−27.8	−27.5	−27.6	0.1	−28.7	−28.7	−28.4
$C_8C_9C_{10}C_2$	57.0	56.9	50.8	51.1	54.6	54.9	48.8	49.4
$C_9C_{10}C_2C_1$	−57.0	−52.4	−49.0	−51.2	−54.6	−51.1	−47.3	−49.4
$C_{10}C_2C_1C_7$	57.0	50.1	50.8	51.1	54.6	48.9	48.8	49.4
$C_2C_1C_7C_8$	−57.0	−52.4	−54.6	−51.2	−54.6	−51.1	−52.6	−49.4
$C_1C_7C_8C_9$	57.0	56.9	56.5	51.1	54.6	54.9	54.9	49.4
$C_7C_8C_9C_{10}$	−57.0	−59.1	−54.6	−51.2	−54.6	−56.4	−52.6	−49.4
$C_9C_{10}C_1C_7$	0.0	−2.4	1.9	−0.1	0.0	−2.0	1.9	0.0
$C_{11}C_{12}C_{13}C_{14}$	55.3	55.0	−61.3	−61.2	54.3	54.3	−63.6	−64.0
$C_{12}C_{13}C_{14}C_{10}$	−55.6	−55.8	30.0	29.9	−55.2	−55.1	31.1	31.8
$C_{13}C_{14}C_{10}C_9$	54.4	55.4	29.4	29.4	54.6	54.9	30.1	29.3
$C_{14}C_{10}C_9C_{11}$	−53.1	−53.8	−60.0	−59.9	−53.2	−53.4	−61.4	−61.0

(Continued)

TABLE 5.9 (CONTINUED)
Ring Dihedral Angles, Relative Energies, Relative Free Energies, and Populations of the Conformers of S-9

Dihedrals[a]	MMFF94				B3LYP/6-31G*			
	9a	9b	9c	9d	9a	9b	9c	9d
$C_{10}C_9C_{11}C_{12}$	54.4	54.4	29.6	29.4	54.6	54.4	29.3	29.3
$C_9C_{11}C_{12}C_{13}$	-55.6	-55.1	29.7	29.9	-55.2	-55.1	31.7	31.8
$C_{12}C_{13}C_{10}C_9$	-0.6	-0.1	55.7	55.5	-0.3	-0.2	57.6	57.5
$C_{17}C_{18}C_8C_7$	54.4	54.4	55.5	29.4	54.6	54.4	54.8	29.3
$C_{18}C_8C_7C_{15}$	-53.1	-53.8	-54.8	-59.9	-53.2	-53.4	-53.8	-61.0
$C_8C_7C_{15}C_{16}$	54.4	55.4	55.5	29.4	54.6	54.9	54.8	29.3
$C_7C_{15}C_{16}C_{17}$	-55.6	-55.8	-55.4	29.9	-55.2	-55.1	-54.9	31.8
$C_{15}C_{16}C_{17}C_{18}$	55.3	55.0	54.7	-61.2	54.3	54.3	54.0	-64.0
$C_{16}C_{17}C_{18}C_8$	-55.6	-55.1	-55.4	29.9	-55.2	-55.1	-54.9	31.8
$C_{18}C_8C_{15}C_{16}$	0.3	0.5	-0.2	-27.3	0.1	0.3	-0.1	-28.2
ΔE^b	0.00(C_3)	4.72(C_2)	9.94(C_2)	15.45(C_3)	0.00(D_3)	5.10(C_2)	11.03(C_2)	16.98(D_3)
ΔG^c					0.00	4.69	9.89	16.59
$P(\%)^d$					99.96	0.04	0.00	0.00

a   Dihedral angles in degrees.
b   Relative energies in kcal/mol. Symmetries in parentheses.
c   Relative free energies in kcal/mol.
d   Populations based on ΔG values, T = 298 K.

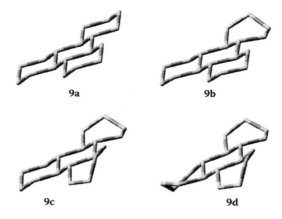

FIGURE 5.10 B3LYP/6-31G* conformations of *S*-**9**. The hydrogen atoms are not shown for clarity.

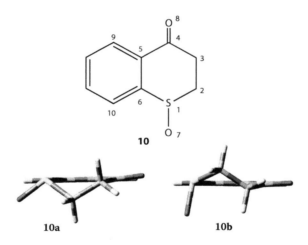

FIGURE 5.11 (SEE COLOR INSERT.) MMFF94 conformations of *S*-**10**. The oxygen atoms are red and the sulfur atom is yellow.

B3LYP/6-31G* level led to two stable conformers, whose dihedral angles are very similar to the MMFF94 dihedral angles, with **10a** being lower in energy by 0.64 kcal/mol. Atoms C_3, C_4 and S_1 are essentially coplanar with the phenyl ring but C_2 and O_7 deviate substantially from this plane. In **10a**, C_2 and O_7 are on the same side of the plane; in **10b** they are on opposite sides. The B3LYP/6-31G* relative free energies and room temperature equilibrium populations of these structures are given in Table 5.10. The lowest-energy conformation, **10a**, is predicted to have a population of ~78% at room temperature [7].

The chiral heterocyclic amine, Tröger's base, **11**, contains two stereogenic N atoms. Conformational analysis of **11** using MMFF94 leads to a single, C_2-symmetric,

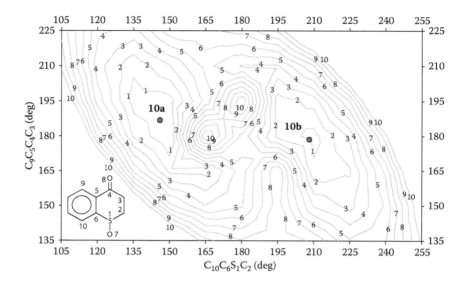

FIGURE 5.12 The B3LYP/6-31G* PES of *S*-**10**. The dihedral angles $C_{10}C_6S_1C_2$ and $C_9C_5C_4C_3$ were varied in 15° steps. Contours are shown at 1 kcal/mol intervals. Conformers resulting from unconstrained optimizations of the lowest-energy structures in the valleys of the PES are indicated (•).

conformer within a 20 kcal/mol window. Thus, **11** is a conformationally rigid molecule. The MMFF94 structure of **11** is shown in Figure 5.13. Key dihedral angles of this conformation are given in Table 5.11. DFT optimization of this structure at the B3LYP/6-31G* level leads to a stable conformation with dihedral angles, also given in Table 5.11, which are very similar to the MMFF94 dihedral angles.

Methyllactate, **12**, is a highly conformationally flexible chiral organic molecule. The hydroxyl group can rotate about the C_4–O_5 bond, the methoxy group can rotate about the C_3–O_2 bond, and the acetate group can rotate about the C_4–C_3 bond. Conformational analysis of **12** using the MMFF94 force field finds seven stable conformations within a 20 kcal/mol window. The MMFF94 conformations, **12a–12g**, are shown in Figure 5.14; the dihedral angles, bond lengths, and relative energies are given in Table 5.12. Conformations **12a–12d** have a *cis* acetate structure, whereas **12e–12g** have the less favorable *trans* acetate structure (see figure 1 of [8]).

TABLE 5.10

Key Dihedral Angles, Relative Energies, Relative Free Energies, and Room Temperature Populations of the Conformers of S-10

			B3LYP/6-31G*			
	MMFF94		Full Optimization		2D Scan	
Dihedrals[a]	10a	10b	10a	10b	10a	10b
$S_1C_2C_3C_4$	−65.1	67.9	−64.7	67.6	−64.8	67.6
$C_2C_3C_4C_5$	28.8	−42.3	27.1	−33.4	27.1	−33.4
$C_3C_4C_5C_6$	4.9	6.4	4.6	−0.3	4.6	−0.3
$C_4C_5C_6S_1$	1.7	1.4	6.3	−4.0	6.3	−4.0
$C_5C_6S_1C_2$	−32.0	20.4	−36.9	31.4	−36.9	31.4
$C_6S_1C_2C_3$	61.6	−53.2	62.5	−60.0	62.5	−60.0
$C_{10}C_6S_1O_7$	38.7	89.4	33.1	99.6	33.1	99.5
$C_9C_5C_4O_8$	7.5	4.8	6.7	−2.0	6.7	−2.0
$C_{10}C_6S_1C_2$	149.4	−159.6	146.1	−151.8	146.1	−151.8
$C_9C_5C_4C_3$	−174.9	−173.4	−173.2	−181.5	−173.3	178.5
ΔE^b	0.00	1.26	0.00	0.64	0.00	0.64
ΔG^c			0.00	0.76	0.00	0.76
$P(\%)^d$			78.3	21.7	78.3	21.7

[a] Dihedral angles in degrees.
[b] Relative energies in kcal/mol.
[c] Relative free energies in kcal/mol.
[d] Populations based on ΔG values, T = 298 K.

In conformers **12a** and **12b** there is an internal H-bond to the carbonyl oxygen atom (H_6O_7). However, in **12c** and **12d** the H-bond is to the methoxy oxygen (H_6O_2). The five-membered ring formed as a result of this H-bonding is either envelope up (**12c**) or envelope down (**12d**). Conformer **12e** also has an internal H-bond to its carbonyl oxygen (H_6O_7). Conformers **12f** and **12g** exhibit no internal H-bonding. In **12g** the OH group has rotated about 120° from its orientation in **12f**. B3LYP/6-31G* optimizations of the MMFF94 conformations lead to the dihedral angles, bond lengths, and relative energies in Table 5.12. The MMFF94 and B3LYP/6-31G* relative energies are very similar. However, at the DFT level, conformations **12a** and **12b** are equivalent. Harmonic frequency calculations proved that the six DFT conformations are stable, and gave their relative free energies and room temperature populations, also given in Table 5.12. Conformation **12a** is the most favorable conformation, and is predicted to have a population of >95% at room temperature [9].

For the molecules considered here, the conformations obtained using the MMFF94 force field have very similar geometries to those predicted by DFT optimizations at the B3LYP/6-31G* level as evidenced by the excellent agreement of their dihedral angles, shown in Table 5.13 and Figure 5.15. Furthermore, the relative energies of the various conformations of these molecules, obtained using the MMFF94 force field, are also in reasonably good agreement with those predicted by DFT, as can be seen in Table 5.14 and Figure 5.16. In particular, for each molecule the lowest-energy MMFF94 conformation is the same as that predicted by DFT and the order of the

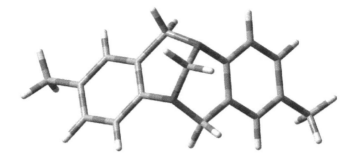

FIGURE 5.13   MMFF94 structure of (R,R)-**11**.

TABLE 5.11
Key Dihedral Angles of (R,R)-11

Dihedrals[a]	MMFF94	B3LYP/6-31G*
$N_1C_2C_3C_4$	–0.6	3.1
$N_5C_6C_7C_8$	–0.6	3.1
$C_2C_3C_4 N_5$	8.7	10.9
$C_6C_7C_8 N_1$	8.7	10.9
$C_3C_4N_5C_6$	85.9	78.6
$C_7C_8N_1C_2$	85.9	78.6
$C_4N_5C_6C_7$	–101.7	–103.6
$C_8N_1C_2C_3$	–101.7	–103.6
$N_1C_9N_5C_6$	–60.0	–55.3
$C_2N_1C_9N_5$	–60.0	–55.3
$C_9N_5C_6C_7$	26.8	18.1
$C_3C_2N_1C_9$	26.8	18.1
$C_7C_8N_1C_9$	–43.2	–44.7
$C_9N_5C_4C_3$	–43.2	–44.7
$C_8N_1C_9N_5$	69.8	69.9
$N_1C_9N_5C_4$	69.8	69.9

[a]   Dihedral angles in degrees.

relative energies of the MMFF94 conformers is also identical to that obtained from the DFT optimizations. This, however, will not always be the case. Oftentimes, the most stable MMFF94 conformation will not correspond to the global minimum on the DFT PES and the order of the relative energies of the DFT conformers will be different from those obtained using the MMFF94 force field. However, our hope and expectation is that the MMFF94 conformational search will at least identify the most important conformations of the particular molecule being studied. We do, of course, have both the experimental IR and VCD spectra to help guide the search. If the calculated IR and VCD spectra, which are the population-weighted sum of the

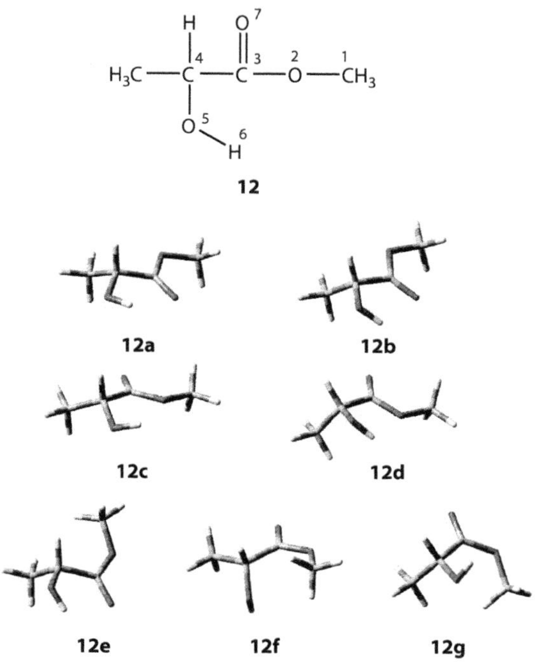

FIGURE 5.14   MMFF94 conformations of *R*-**12**.

spectra of the individual conformers, are in poor agreement with the experimental IR and VCD spectra, then either the initial structure of the molecule is wrong or we have an incorrect set of conformations. On the other hand, if the agreement between the predicted spectra and the experimental spectra is good, then we can be more confident that the conformational analysis is correct. As we will see in some of the applications presented in Chapter 7, conformational splittings are often observed in the experimental IR and VCD spectra, allowing the assignment of specific bands to individual conformers which further supports the reliability of the conformational analysis.

In summary, in order to deduce molecular structure from molecular spectra we must first propose a structure and then evaluate its reliability by comparing the spectra predicted for that structure to the experimental spectra. Thus, we need to determine the stable structure(s) of the molecule. In the case of rigid molecules, where only a single conformer exists, geometry optimization is straightforward. However, many chiral organic molecules are conformationally flexible and multiple conformations are present in equilibrium at the temperature at which the experimental VCD spectra are measured. In such cases, we must carry out a conformational analysis (CA) to determine the geometries, relative free energies, and equilibrium populations of the possible conformations of the molecule. For molecules containing a small number of dihedral angles with respect to which internal rotation can occur, the most reliable method to find their conformations is to carry out PES scans using DFT. At the present time, scans with respect to one or two degrees of freedom are generally practicable, however, DFT PES scanning

TABLE 5.12
Key Dihedral Angles, Bond Lengths, Relative Energies, Relative Free Energies, and Room Temperature Populations of the Conformers of R-12

Dihedrals[a]	12a		12b		12c		12d		12e	
	MMFF94	B3LYP[f]	MMFF94	B3LYP[f]	MMFF94	B3LYP[f]	MMFF94	B3LYP[f]	MMFF94	B3LYP[f]
$C_1O_2C_3C_4$	178.9	180.8	179.7	180.8	−179.7	−177.6	179.6	177.8	22.3	12.4
$O_2C_3C_4O_5$	174.5	186.9	−171.8	−173.1	−29.5	−25.2	21.3	29.1	−162.7	−158.3
$C_3C_4O_5H_6$	25.7	−7.9	−19.4	−7.8	43.9	39.2	−41.2	−44.5	−30.3	−25.1
$C_1O_2C_3O_7$	1.9	0.2	−0.1	0.2	−2.1	−0.6	2.0	0.8	−160.8	−171.0
$O_5C_4C_3O_7$	−8.4	7.5	7.9	7.4	152.9	157.8	−161.1	−153.9	20.3	25.1
H_6O_7[b]	2.226	2.051	2.199	2.051	3.722	3.679	3.725	3.704	2.181	2.060
H_6O_2[b]					2.160	2.136	2.157	2.188		
ΔE[c]	0.00	0.00	0.16	0.00	1.72	2.28	2.42	2.39	9.60	8.54
ΔG[d]		0.00		0.00		2.12		2.24		9.18
P(%)[e]		95.17		95.17		2.66		2.17		0.00

Dihedrals[a]	12f		12g	
	MMFF94	B3LYP[f]	MMFF94	B3LYP[f]
$C_1O_2C_3C_4$	-8.3	-1.7	20.6	2.7
$O_2C_3C_4O_5$	-52.5	-70.4	24.6	36.6
$C_3C_4O_5H_6$	-65.2	-65.5	66.1	70.3
$C_1O_2C_3O_7$	172.3	181.5	-161.4	-176.0
$O_5C_4C_3O_7$	126.8	106.4	-153.4	-144.7
H_6O_7[b]	3.457	3.132	3.739	3.642
H_6O_2[b]				
ΔE[c]	10.51	10.38	12.41	12.85
ΔG[d]		10.82		12.92
$P(\%)$[e]		0.00		0.00

[a] Dihedral angles in degrees.
[b] Bond lengths in angstroms.
[c] Relative energies in kcal/mol.
[d] Relative free energies in kcal/mol.
[e] Populations based on ΔG values, T = 298 K.
[f] Calculated using the 6-31G* basis set.

TABLE 5.13
MMFF94 and B3LYP/6-31G* Dihedral
Angles for Molecules 1–12

| Molecule | Dihedrals[a] | |
	MMFF94	B3LYP[b]
1	330.9	335.5
2	252.2	307.1
3	330.7	335.7
	252.1	306.2
4	−2.2	−0.7
	−71.0	−72.0
	73.4	70.8
	1.3	2.5
	−69.8	−71.5
	73.5	71.0
	−33.4	−34.0
	37.8	35.7
	34.1	32.5
	−36.2	−36.9
5	−1.5	−2.1
	0.4	2.5
	48.4	45.6
	−82.4	−82.7
	80.5	81.8
	−45.5	−45.9
	48.8	47.5
	−47.0	−49.1
	−33.4	−28.7
	33.5	28.9
6a	54.3	54.7
	−54.3	−54.7
	0.0	0.0
6b	60.1	63.5
	−29.2	−30.6
	−54.3	−57.4
7a	52.0	51.9
	−57.0	−56.4
	57.0	56.4
	−52.0	−51.9
	49.7	48.9
	−49.7	−48.9
	0.0	0.0
7b	58.9	59.4
	−29.4	−28.9

(Continued)

TABLE 5.13 (CONTINUED)
MMFF94 and B3LYP/6-31G* Dihedral
Angles for Molecules 1–12

Molecule	Dihedrals[a]	
	MMFF94	B3LYP[b]
	−29.4	−28.9
	58.9	59.4
	−28.5	−28.8
	−28.5	−28.8
	−54.4	−53.6
8a	52.4	51.8
	−57.1	−56.1
	57.3	56.7
	−52.3	−51.6
	50.1	48.8
	−50.5	−49.5
	−0.2	−0.2
8b	−50.0	−52.5
	54.0	55.6
	−55.2	−55.3
	51.6	50.9
	−50.7	−50.0
	50.3	51.4
	−0.7	1.0
8c	59.8	59.9
	−31.8	−30.7
	−27.1	−26.7
	58.8	58.4
	−30.0	−29.4
	−27.7	−28.4
	−54.8	−53.5
8d	−53.2	−43.0
	14.9	−8.5
	40.9	56.1
	−58.6	−51.6
	19.3	0.2
	35.7	48.2
	51.4	43.1
8e	−33.6	−35.1
	59.8	63.8
	−31.0	−36.9
	−21.2	−17.1
	48.1	45.4
	−19.1	−17.5

(Continued)

TABLE 5.13 (CONTINUED)
MMFF94 and B3LYP/6-31G* Dihedral
Angles for Molecules 1–12

Molecule	Dihedrals[a]	
	MMFF94	B3LYP[b]
	26.1	24.8
9a	54.4	54.6
	−55.6	−55.2
	55.3	54.3
	−55.6	−55.2
	54.4	54.6
	−53.1	−53.2
	0.3	0.1
	57.0	54.6
	−57.0	−54.6
	57.0	54.6
	−57.0	−54.6
	57.0	54.6
	−57.0	−54.6
	0.0	0.0
	55.3	54.3
	−55.6	−55.2
	54.4	54.6
	−53.1	−53.2
	54.4	54.6
	−55.6	−55.2
	−0.6	−0.3
	54.4	54.6
	−53.1	−53.2
	54.4	54.6
	−55.6	−55.2
	55.3	54.3
	−55.6	−55.2
	0.3	0.1
9b	29.7	29.1
	29.9	32.2
	−61.4	−64.6
	29.9	32.2
	29.7	29.1
	−60.3	−60.9
	−27.8	−28.7
	56.9	54.9
	−52.4	−51.1
	50.1	48.9

(Continued)

TABLE 5.13 (CONTINUED)
MMFF94 and B3LYP/6-31G* Dihedral
Angles for Molecules 1–12

	Dihedrals[a]	
Molecule	MMFF94	B3LYP[b]
	−52.4	−51.1
	56.9	54.9
	−59.1	−56.4
	−2.4	−2.0
	55.0	54.3
	−55.8	−55.1
	55.4	54.9
	−53.8	−53.4
	54.4	54.4
	−55.1	−55.1
	−0.1	−0.2
	54.4	54.4
	−53.8	−53.4
	55.4	54.9
	−55.8	−55.1
	55.0	54.3
	−55.1	−55.1
	0.5	0.3
9c	29.4	30.1
	30.0	31.1
	−61.3	−63.6
	29.7	31.7
	29.6	29.3
	−60.0	−61.4
	−27.5	−28.7
	50.8	48.8
	−49.0	−47.3
	50.8	48.8
	−54.6	−52.6
	56.5	54.9
	−54.6	−52.6
	1.9	1.9
	−61.3	−63.6
	30.0	31.1
	29.4	30.1
	−60.0	−61.4
	29.6	29.3
	29.7	31.7
	55.7	57.6

(Continued)

TABLE 5.13 (CONTINUED)
MMFF94 and B3LYP/6-31G* Dihedral
Angles for Molecules 1–12

Molecule	Dihedrals[a]	
	MMFF94	B3LYP[b]
	55.5	54.8
	−54.8	−53.8
	55.5	54.8
	−55.4	−54.9
	54.7	54.0
	−55.4	−54.9
	−0.2	−0.1
9d	29.4	29.3
	29.9	31.8
	−61.2	−64.0
	29.9	31.8
	29.4	29.3
	−59.9	−61.0
	−27.6	−28.4
	51.1	49.4
	−51.2	−49.4
	51.1	49.4
	−51.2	−49.4
	51.1	49.4
	−51.2	−49.4
	−0.1	0.0
	−61.2	−64.0
	29.9	31.8
	29.4	29.3
	−59.9	−61.0
	29.4	29.3
	29.9	31.8
	55.5	57.5
	29.4	29.3
	−59.9	−61.0
	29.4	29.3
	29.9	31.8
	−61.2	−64.0
	29.9	31.8
	−27.3	−28.2
10a	−65.1	−64.7
	28.8	27.1
	4.9	4.6
	1.7	6.3

(Continued)

TABLE 5.13 (CONTINUED)
MMFF94 and B3LYP/6-31G* Dihedral
Angles for Molecules 1–12

Molecule	Dihedrals[a]	
	MMFF94	B3LYP[b]
	−32.0	−36.9
	61.6	62.5
	38.7	33.1
	7.5	6.7
	149.4	146.1
	−174.9	−173.2
10b	67.9	67.6
	−42.3	−33.4
	6.4	−0.3
	1.4	−4.0
	20.4	31.4
	−53.2	−60.0
	89.4	99.6
	4.8	−2.0
	−159.6	−151.8
	−173.4	−181.5
11	−0.6	3.1
	8.7	10.9
	85.9	78.6
	−101.7	−103.6
	−60.0	−55.3
	26.8	18.1
	−43.2	−44.7
	69.8	69.9
12a	178.9	180.8
	174.5	186.9
	25.7	−7.9
	1.9	0.2
	−8.4	7.5
12b	179.7	180.8
	−171.8	−173.1
	−19.4	−7.8
	−0.1	0.2
	7.9	7.4
12c	−179.7	−177.6
	−29.5	−25.2
	43.9	39.2
	−2.1	−0.6
	152.9	157.8
12d	179.6	177.8
	21.3	29.1
	−41.2	−44.5
	2.0	0.8

(Continued)

TABLE 5.13 (CONTINUED)
MMFF94 and B3LYP/6-31G* Dihedral
Angles for Molecules 1–12

| | Dihedrals[a] | |
Molecule	MMFF94	B3LYP[b]
	−161.1	−153.9
12e	22.3	12.4
	−162.7	−158.3
	−30.3	−25.1
	−160.8	−171.0
	20.3	25.1
12f	−8.3	−1.7
	−52.5	−70.4
	−65.2	−65.5
	172.3	181.5
	126.8	106.4
12g	20.6	2.7
	24.6	36.6
	66.1	70.3
	−161.4	−176.0
	−153.4	−144.7

[a] Dihedral angles in degrees.
[b] Calculated using the 6-31G* basis set.

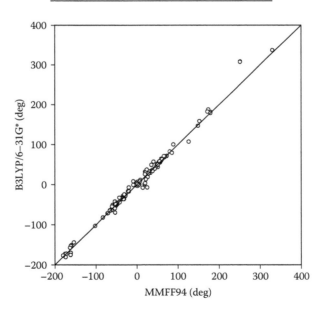

FIGURE 5.15 Comparison of MMFF94 and B3LYP/6-31G* dihedral angles for molecules **1–12**.

TABLE 5.14
MMFF94 and B3LYP/6-31G* Relative
Energies for Molecules 6–10,12

Molecule	ΔE^a	
	MMFF94	B3LYP[b]
6b	5.93	6.48
7b	2.97	3.72
8b	0.48	1.52
8c	3.16	3.83
8d	4.11	4.88
8e	6.04	6.37
9b	4.72	5.10
9c	9.94	11.03
9d	15.45	16.98
10b	1.26	0.64
12b	0.16	0.00
12c	1.72	2.28
12d	2.42	2.39
12e	9.60	8.54
12f	10.51	10.38
12g	12.41	12.85

[a] Relative energies in kcal/mol.
[b] Calculated using the 6-31G* basis set.

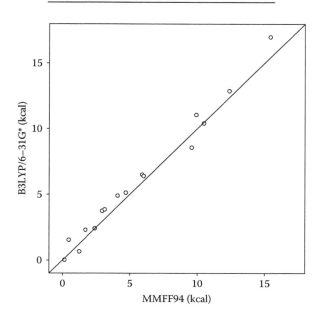

FIGURE 5.16 Comparison of MMFF94 and B3LYP/6-31G* relative energies for molecules **6–12**, excluding molecule **11**.

becomes more time-consuming as the flexibility of the molecule increases. For large flexible organic molecules the most efficient CA procedure is to perform a Monte Carlo search using a MMFF to determine the most stable conformations of the molecule with energies within 20 kcal/mol of the global minimum conformation. These structures are then optimized using DFT and their harmonic frequencies calculated to prove that they are, indeed, stable conformations and to allow their relative free energies, and room temperature equilibrium populations to be calculated. Specific examples of this protocol are given in the applications presented in Chapter 7.

REFERENCES

1. (a) Wertz, D.H., Allinger, N.L., Conformational Analysis-CI: The Gauche-Hydrogen Interaction as the Basis of Conformational-Analysis, *Tetrahedron*, 30, 1579, 1974. (b) Allinger, N.L., Conformational-Analysis. MM2-Hydrocarbon Force-Field Utilizing V1 and V2 Torsional Terms, *J. Am. Chem. Soc.*, 99, 8127–8134, 1977. (c) Allinger, N.L., Yuh, Y.H., Lii, J.-H., Molecular Mechanics—The MM3 Force-Field for Hydrocarbons 1, *J. Am. Chem. Soc.*, 111, 8551–8566, 1989. (d) Nevins, N., Allinger, N.L., Molecular Mechanics (MM4) Calculations on Alkenes, *J. Comput. Chem.*, 17, 669–694, 1996. (e) Cornell, W. D., Cieplak, P., Bayly, C. I., Gould, I. R., Merz, K.M. J., Ferguson, D.M., Spellmeyer, D.C., Fox, T., Caldwell, J. W., Kollman, P.A., A 2nd Generation Force-Field for the Simulation of Proteins, Nucleic-Acids, and Organic-Molecules, *J. Am. Chem. Soc.*, 117, 5179–5197, 1995. (f) Brooks, B.R., Bruccoleri, R.E., Olafson, B.D., States, D.J., Swaminathan, S., Karplus, M., CHARMM—A Program of Macromolecular Energy, Minimization, and Dynamics Calculations, *J. Comput. Chem.*, 4, 187–217, 1983. (g) Carlson, H.A., Nguyen, T.B., Orozco, M., Jorgensen, W.L., Accuracy of Free-Energies of Hydration for Organic Molecules from 6-31G* Derived Partial Charges, *J. Comput. Chem.*, 14, 1240–1249, 1993.
2. Halgren, T.A., Merck Molecular Force Field. I. Basis, Form, Scope, Parameterization, and Performance of MMFF94, *J. Comput. Chem.*, 17, 490–519, 1996.
3. Spartan 02, Wavefunction, www.wavefun.com.
4. Barton, D.H.R., The Principles of Conformational Analysis, *Science*, 169, 539–544, 1970.
5. Johnson, W.S., Bauer, V.J., Margrave, J.L., Frisch, M.A., Dreger, L.H., and Hubbard, W.N., The Energy Difference between the Chair and Boat Forms of Cyclohexane. The Twist Conformation of Cyclohexane, *J. Am. Chem. Soc.*, 83, 606–614, 1961.
6. Devlin, F.J., Stephens, P.J., Conformational Analysis Using *ab Initio* Vibrational Spectroscopy: 3-Methylcyclohexanone, *J. Am. Chem. Soc.*, 121, 7413–7414, 1999.
7. (a) Devlin, F.J., Stephens, P.J., Scafato, P., Superchi, S., Rosini, C., Determination of Absolute Configuration Using Vibrational Circular Dichroism Spectroscopy: The Chiral Sulfoxide 1-Thiochromanone S-oxide, *Chirality*, 14, 400–406, 2002. (b) Devlin, F.J., Stephens, P.J., Scafato, P., Superchi, S., Rosini, C., Conformational Analysis Using Infrared and Vibrational Circular Dichrosim Spectroscopies: The Chiral Cyclic Sulfoxides 1-Thiochroman-4-one S-oxide, 1-Thiaindan S-oxide and 1-Thiochroman S-oxide, *J. Phys. Chem. A*, 106, 10510–10524, 2002.
8. Devlin, F.J., Stephens, P.J., Österle, C., Wiberg, K.B., Cheeseman, J.R., Frisch, M.J., Configurational and Conformational Analysis of Chiral Molecules using IR and VCD Spectroscopies: Spiropentylcarboxylic Acid Methyl Ester and Spiropentyl Acetate, *J. Org. Chem.*, 67, 8090–8096, 2002.
9. Stephens, P.J., Devlin, F.J., Determination of the Structure of Chiral Molecules Using *ab initio* Vibrational Circular Dichroism Spectroscopy, *Chirality*, 12, 172–179, 2000.

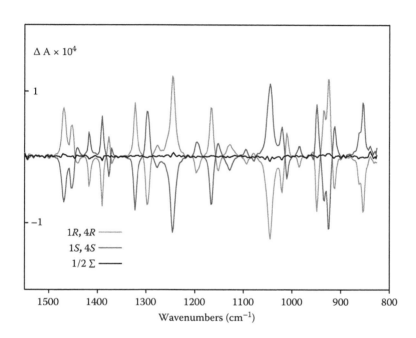

FIGURE 1.2 The mid-IR VCD spectra of 0.38 M CCl_4 solutions of 1R,4R and 1S,4S camphor, using a cell of pathlength 236 μ. Σ is the sum of the spectra. The measurement of the spectra is discussed in Chapter 2.

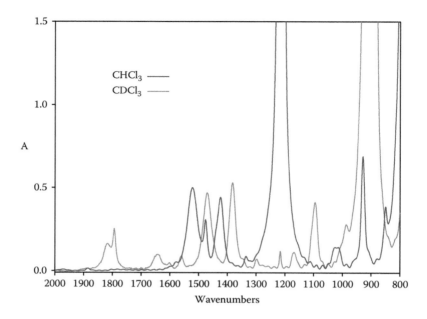

FIGURE 2.2   The IR absorption spectra of CHCl₃ and CDCl₃; pathlength 236 μ.

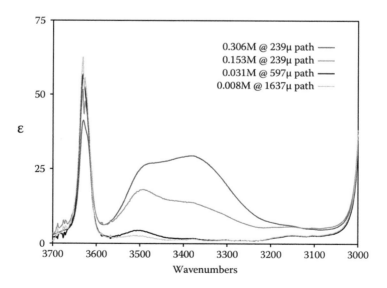

FIGURE 2.4   The concentration dependence of the IR absorption in the O–H stretching region of *endo*-borneol in CCl₄ solution.

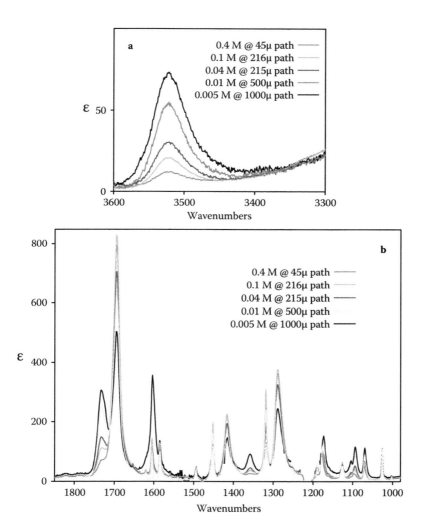

FIGURE 2.5 The concentration dependence of the IR absorption spectrum of benzoic acid in CHCl$_3$ solution, in the O–H stretching region (a) and in the mid-IR (b).

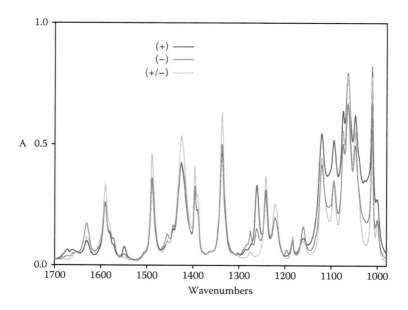

FIGURE 2.6 The IR absorption spectra of 0.04 M CDCl₃ solutions of (+), (−), and (±) (8-(4-bromophenyl)-8-ethoxy-5-methyl-8H-[1,4]thiazino[3,4-c][1,2,4]-oxadiazol-3-one; pathlength 597 μ.

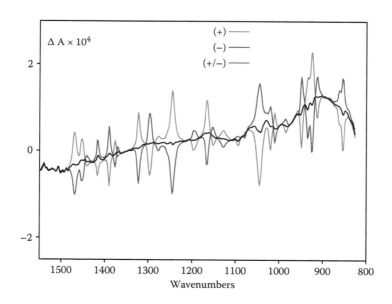

FIGURE 2.10 The VCD spectra of 0.38 M CCl₄ solutions of (+), (−), and (±) camphor; pathlength 236 μ.

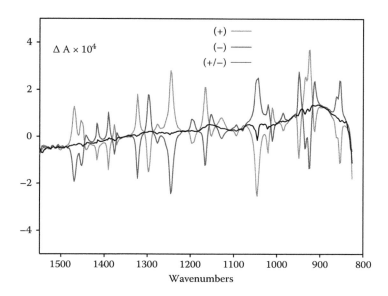

FIGURE 2.11   The VCD spectra of 0.38 M CCl$_4$ solutions of (+), (−), and (±) camphor; pathlength 546 μ.

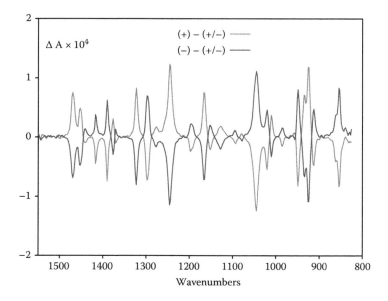

FIGURE 2.13   The (±)-baseline-subtracted VCD spectra of 0.38 M CCl$_4$ solutions of (+) and (−) camphor; pathlength 236 μ.

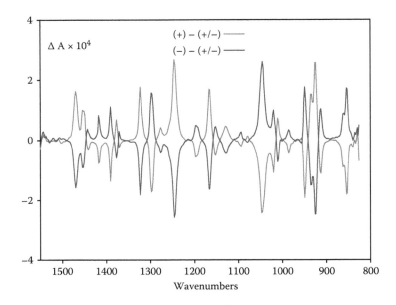

FIGURE 2.14 The (±)-baseline-subtracted VCD spectra of 0.38 M CCl₄ solutions of (+) and (−) camphor; pathlength 546 μ.

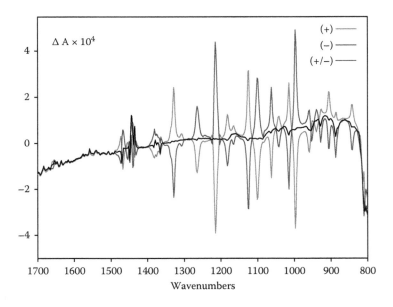

FIGURE 2.26 The VCD spectra of 1.87 M CCl₄ solutions of (+), (−), and (±) α-pinene; pathlength 236 μ.

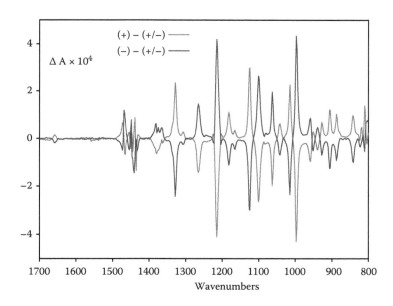

FIGURE 2.27 The (±)-baseline-subtracted VCD spectra of 1.87 M CCl$_4$ solutions of (+) and (−) α-pinene; pathlength 236 μ.

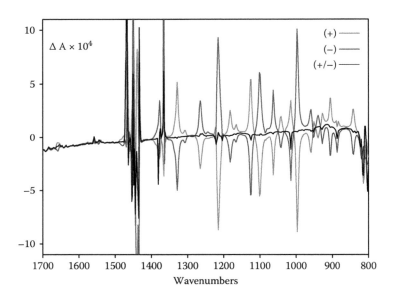

FIGURE 2.30 The VCD spectra of 1.87 M CCl_4 solutions of (+), (−), and (±) α-pinene; pathlength 546 μ.

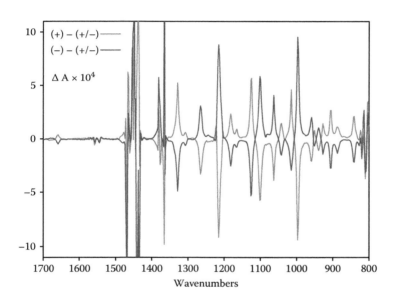

FIGURE 2.31 The (±)-baseline-subtracted VCD spectra of 1.87 M CCl₄ solutions of (+) and (–) α-pinene; pathlength 546 μ.

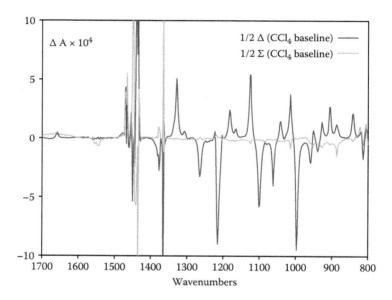

FIGURE 2.33 The half-difference and half-sum VCD spectra, obtained from the CCl₄-baseline-subtracted VCD spectra of (+) and (−) α-pinene; pathlength 546 μ.

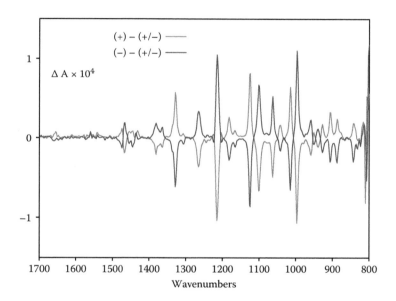

FIGURE 2.35 The (±)-baseline-subtracted VCD spectra of 0.93 M CCl$_4$ solutions of (+) and (−) α-pinene; pathlength 113 μ.

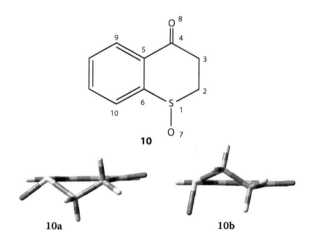

FIGURE 5.11 MMFF94 conformations of *S*-**10**. The oxygen atoms are red and the sulfur atom is yellow.

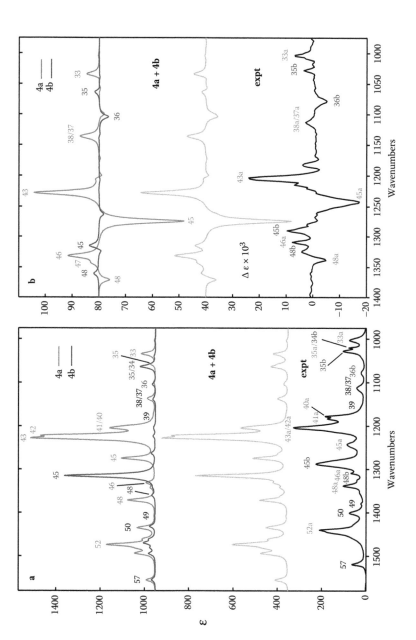

FIGURE 7.29 (a) Comparison of the conformationally averaged B3PW91/TZ2P IR spectrum of **4** (green) to the experimental IR spectrum of (+)-**4** (black). The population-weighted B3PW91/TZ2P IR spectra for each conformer (**4a** (red) and **4b** (blue)) are also shown. (b) Comparison of the conformationally averaged B3PW91/TZ2P VCD spectrum of (+)-**4** (green) to the experimental VCD spectrum of (2*S*,3*R*)-**4** (black). The population-weighted B3PW91/TZ2P VCD spectra for each conformer (**4a** (red) and **4b** (blue)) are also shown. Bands assigned to the fundamentals of single conformations **4a** and **4b** are numbered in red and blue, respectively; bands assigned to both conformations are numbered in black.

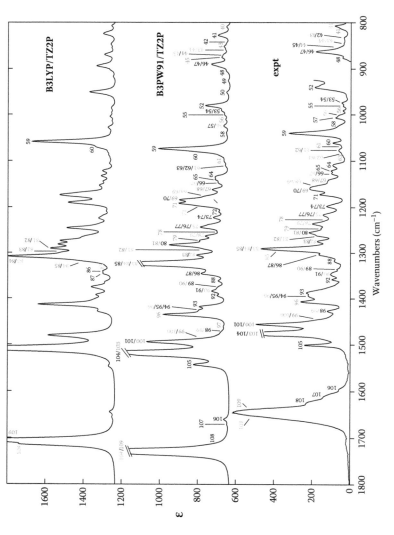

FIGURE 7.49 Comparison of the experimental and conformationally averaged B3LYP/TZ2P and B3PW91/TZ2P IR spectra of **1**. The assignment of the experimental spectrum is based on the B3PW91/TZ2P spectrum. The numbers define the fundamentals contributing to resolved bands. Red and green numbers indicate bands of **1a/1a′** and **1b/1b′**, respectively. Black numbers indicate superpositions of bands of **1a**, **1a′**, **1b**, and **1b′**. The bandshapes of the calculated spectra are Lorentzian ($\gamma = 4.0$ cm^{-1}).

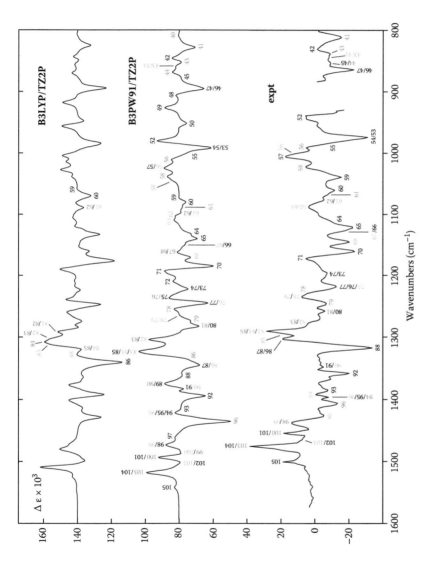

FIGURE 7.50 Comparison of the experimental VCD spectrum of (+)-**1** and the conformationally averaged B3LYP/TZ2P and B3PW91/TZ2P VCD spectra of (2R,7S,20S,21S)-**1**. Bandshapes of the calculated spectra are Lorentzian ($\gamma = 4.0\ cm^{-1}$). The assignment of the experimental spectrum is based on the B3PW91/TZ2P-calculated spectrum. Fundamentals are numbered as in Figure 7.49.

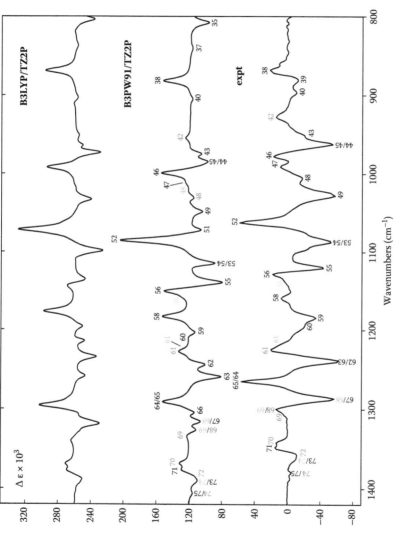

FIGURE 7.56 Comparison of the experimental and conformationally averaged B3LYP/TZ2P and B3PW91/TZ2P VCD spectra of (1R,5S,8S,9S,10S)-**1** for the range 800–1,420 cm⁻¹. Red, green, and cyan numbers indicate fundamentals of **1a**, **1b**, and **1c/1d**, respectively. Black numbers indicate superpositions of fundamentals of **1a** and **1b**. Bandshapes of the calculated spectra are Lorentzian (γ = 4.0 cm⁻¹). The assignment of the experimental spectrum is based on the B3PW91/TZ2P-calculated spectrum.

6 Analyses of the IR and VCD Spectra of Conformationally Rigid Molecules

In this chapter we discuss the utilization of *ab initio* calculations of IR and VCD spectra for the analysis of experimental IR and VCD spectra. The *ab initio* options are Hartree-Fock (HF) theory and density functional theory (DFT). In both cases, the choice of the basis set is also an option. In the case of DFT, the choice of the density functional is also an option. The evaluation of the reliability of *ab initio* calculations of IR and VCD spectra is most easily carried out by comparison of the predicted spectra to the experimental spectra of conformationally rigid molecules, since the spectra of conformationally flexible molecules are more complex. In this chapter we compare *ab initio* calculated IR and VCD spectra to the experimental IR and VCD spectra of the conformationally rigid chiral molecules, camphor, α-pinene, and methyloxirane. These three molecules have been used in testing the accuracies of VCD calculations, using the Stephens theory of VCD, many times [1–19]. The results define the relative reliabilities of the HF and DFT methods, and of multiple basis sets and density functionals, and thence the optimum choices for the prediction of IR and VCD spectra of maximum accuracy.

IR and VCD spectra are calculated using Equations 3.20–3.36 and Lorentzian bandshapes (Equation 3.22). The equilibrium geometries and harmonic frequencies, dipole strengths, and rotational strengths are calculated using the methodologies presented in Chapter 4. The experimental IR and VCD spectra are assigned on the basis of the calculated spectra exhibiting the best agreement with the experimental spectra. To define the quantitative accuracies of the calculated frequencies, dipole strengths and rotational strengths of the fundamental transitions, the experimental frequencies, dipole strengths, and rotational strengths are obtained by Lorentzian fitting of the experimental IR and VCD spectra.

The PeakFit program [20] is used to fit the experimental IR and VCD spectra. This program utilizes a nonlinear least-squares curve-fitting procedure to minimize the error between a Lorentzian fitting function and the experimental data. The fitting software employs a gradient method when the errors are large and gradually shifts over to a Taylor series method as the errors approach zero. This combined minimization procedure is implemented using the Marquardt-Levenberg algorithm [21].

First we fit the IR spectrum. The experimental IR spectrum (in ε units) is imported into PeakFit. The molar extinction values are then divided by 0.0108858 υ, where υ is the experimental frequency, so that a Lorentzian area function can be used for fitting the IR spectrum. This scale factor also ensures that the extracted dipole strength values are in 10^{-40} esu^2 cm^2 units. The experimental data are then fit using the AutoFit Residuals option while allowing the bandwidths to vary. Before the spectrum is actually fit, a Savitsky-Golay smoothing function is applied; otherwise the program will try to fit extraneous noise peaks resulting in an excessive number of component bands. PeakFit usually makes a good initial guess at finding the component bands, but as the fitting progresses, one often has to add or delete bands in order to get the best fit to the experimental data. The calculated IR spectrum should be used as a guide during the fitting process. One should have at least the same number of component bands as there are predicted frequencies. The fitting process is continued until an acceptable degree of convergence is achieved (typically $r^2 \geq 0.99$). When the IR spectrum has been fit, we then obtain the experimental frequency, dipole strength, and bandwidth for each band in the spectrum.

The VCD spectrum (in $\Delta\varepsilon \times 10^{-4}$ units) is fit in a similar fashion. Here the scale factor is 0.0435432 υ and we must allow the program to fit negative as well as positive peaks. The VCD spectrum is fit until a suitable level of convergence is achieved. This yields the VCD frequencies, rotational strengths (in 10^{-44} esu^2 cm^2), and bandwidths for each band in the spectrum. There should be good agreement between the IR frequencies and the VCD frequencies resulting from the fitting procedure. Since the VCD spectrum usually has more noise than the IR spectrum, the quality of fit is often less than that obtained for the IR spectrum. Nevertheless, one should try to achieve an r^2 value of at least 0.98 or better.

CAMPHOR

Measurements of the experimental IR and VCD spectra of camphor were presented in Chapter 2. Here, we analyze the spectra obtained using 0.38 and 0.04 M solutions of (+), (−), and (±) camphor in CCl$_4$, Figure 2.22, over the frequency range 2,000–800 cm^{-1}. As discussed in Chapter 5, camphor is predicted to be conformationally rigid. Initially, we predict the IR and VCD spectra using DFT, the basis set cc-pVTZ, and the functionals B3LYP and B3PW91. DFT is expected to be more accurate than HF theory. The basis set cc-pVTZ is expected to be a good approximation to the complete basis set, and to give minimal basis set error. The hybrid functionals B3LYP and B3PW91 are expected to be good functionals; comparison of the B3LYP and B3PW91 IR and VCD spectra to the experimental spectra enables the functional that is more accurate in predicting the spectra of camphor to be determined.

The B3LYP/cc-pVTZ and B3PW91/cc-pVTZ equilibrium geometries of (1R,4R)-camphor were obtained by reoptimization of the B3LYP/6-31G* equilibrium geometries obtained in Chapter 5. The harmonic frequencies, dipole strengths, and rotational strengths were then calculated, with the results for the fundamental transitions of modes 59–22 given in Table 6.1.

TABLE 6.1

Calculated and Experimental Frequencies, Dipole Strengths, and Rotational Strengths for Camphor[a]

| | Calculation[b] | | | | | | Experiment[c] | | | | | |
| | B3LYP | | | B3PW91 | | | IR | | | VCD | | |
Mode	ν	D	R	ν	D	R	ν	D	γ	ν	R	γ
59	1,815	522.8	16.0	1,831	524.8	16.0	1,745	650.4	4.7	1,746	25.6	3.7
58	1,528	15.1	−2.1	1,517	16.6	−2.6	1,481	18.2	5.3	1,480	−2.9	6.9
57	1,519	14.5	10.1	1,508	17.6	12.1	1,470	22.9	5.1	1,469	14.1	4.0
56	1,514	1.1	0.9	1,504	1.5	1.1	1,462	2.0	4.4	1,466	0.9	2.1
55	1,505	9.8	0.4	1,494	9.2	0.7	1,455	15.0	5.7			
54	1,500	9.3	2.7	1,489	13.1	4.3	1,453	15.1	3.0	1,453	5.5	2.9
53	1,495	15.8	5.5	1,484	15.2	5.6	1,448	30.2	4.5	1,450	6.2	3.8
52	1,490	15.9	−1.2	1,479	21.6	−2.4	1,445	22.1	4.6	1,444	−3.6	3.4
51	1,487	9.5	−0.5	1,474	9.6	−0.4	1,439	7.0	5.0	1,440	−1.9	3.2
50	1,464	31.1	−3.6	1,450	34.8	−4.3	1,417	39.9	2.5	1,417	−6.6	2.7
49	1,432	28.5	−6.7	1,418	35.6	−8.1	1,390	40.4	2.3	1,390	−11.3	2.7
48	1,417	19.5	3.6	1,406	21.2	4.4	1,377	18.5	2.5	1,377	6.5	3.0
47	1,410	18.2	−1.3	1,397	24.5	−2.2	1,371	25.5	2.4	1,371	−3.1	2.8
46	1,344	13.9	7.5	1,349	21.7	10.8	1,323	28.7	2.9	1,323	16.8	3.2
45	1,334	2.6	−6.4	1,329	4.7	−10.3	1,299	10.3	3.1	1,298	−14.3	3.2
44	1,326	7.2	−0.8	1,324	5.1	−3.3	1,294	3.5	2.9	1,294	−4.2	2.1
43	1,297	24.8	−2.8	1,300	28.4	−0.1	1,277	29.7	4.1	1,278	4.8	4.9
42	1,272	9.0	19.5	1,273	10.7	27.4	1,245	7.9	3.7	1,245	38.6	5.0
41	1,264	0.3	3.9	1,264	0.4	3.7						
40	1,244	8.1	1.4	1,245	5.1	−0.1	1,220	4.1	3.7			
39	1,225	12.7	−6.2	1,221	12.0	−5.6	1,197	15.5	3.7	1,197	−6.1	3.8
38	1,214	3.8	3.0	1,215	2.8	−1.8	1,191	3.3	3.7	1,191	−2.5	3.0
37	1,187	13.0	17.7	1,187	11.5	17.0	1,166	13.6	3.7	1,166	20.9	3.6

(Continued)

TABLE 6.1 (CONTINUED)
Calculated and Experimental Frequencies, Dipole Strengths, and Rotational Strengths for Camphor[a]

| | Calculation[b] | | | | | | Experiment[c] | | | | | |
| | B3LYP | | | B3PW91 | | | IR | | | VCD | | |
Mode	v	D	R	v	D	R	v	D	γ	v	R	γ
36	1,167	1.5	-5.8	1,171	2.6	-6.4	1,153	4.4	5.4	1,152	-8.6	4.3
35	1,145	4.5	8.7	1,148	4.0	9.1	1,128	7.0	5.6	1,128	9.2	6.2
34	1,118	14.7	-2.4	1,113	17.7	-3.1	1,093	22.5	3.4	1,095	-3.0	3.8
33	1,099	6.0	-4.4	1,095	9.7	0.1	1,079	10.2	2.5	1,079	3.1	4.2
32	1,054	65.4	-29.3	1,058	123.4	-38.1	1,045	126.0	5.4	1,046	-55.0	5.8
31	1,037	102.4	-1.2	1,038	44.0	-13.2	1,021	41.5	3.0	1,020	-13.3	3.4
30	1,030	8.4	-11.7	1,028	3.4	7.9	1,011	3.7	2.7	1,010	10.8	3.0
29	999	0.2	-0.1	1,003	1.6	-3.7	987	0.9	3.8	985	-3.4	2.5
28	961	9.9	9.6	966	15.0	-8.3	951	15.6	3.9	949	-23.4	3.0
27	955	7.7	-24.0	955	0.3	0.3						
26	943	0.8	6.7	950	11.7	4.3	935	15.3	3.4	935	19.7	3.3
25	941	15.5	25.0	938	7.5	24.3	925	10.2	3.5	925	33.9	3.1
24	921	5.3	-9.9	926	4.4	-13.3	913	4.3	3.6	913	-16.7	3.1
23	869	2.5	-8.7	878	3.2	-6.4	861	3.2	4.0	862	-10.2	3.6
22	857	8.2	-14.7	865	7.4	-13.6	854	8.9	2.1	854	-24.6	3.1

[a] Frequencies v and bandwidths γ in cm^{-1}; dipole strengths D in 10^{-40} esu^2 cm^2; rotational strengths R in 10^{-44} esu^2 cm^2. Experimental rotational strengths are for (+)-camphor; calculated rotational strengths are for (1R,4R)-camphor.

[b] Using the cc-pVTZ basis set.

[c] From Lorentzian fitting of the experimental IR and VCD spectra.

Simulations of the IR spectra of camphor due to fundamentals 59–21, using Lorentzian bandshapes (γ = 4.0 cm^{-1}), are shown in Figure 6.1, together with the experimental IR spectrum in the frequency range 1,900–800 cm^{-1}. Allowing for the higher values of the calculated frequencies, relative to the experimental frequencies, due to the neglect of anharmonicity in the calculations (as discussed in Chapters 3 and 4), comparison of the relative frequencies and intensities of the bands of the calculated spectra to the relative frequencies and intensities of the bands of the experimental spectrum leads to the assignment of the experimental spectrum. The B3LYP/cc-pVTZ and B3PW91/cc-pVTZ IR spectra are qualitatively very similar over most

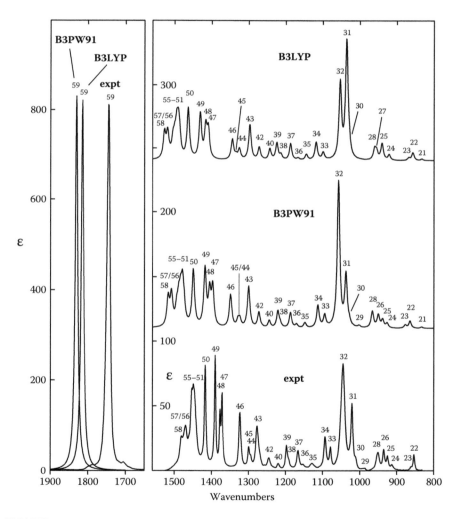

FIGURE 6.1 Comparison of the B3LYP/cc-pVTZ and B3PW91/cc-pVTZ IR spectra of camphor to the experimental IR spectrum of (+)-camphor. The assignment of the experimental spectrum is based on the B3PW91/cc-pVTZ spectrum.

of the frequency range. However, a dramatic difference is exhibited by modes 31 and 32; using B3LYP the IR absorption of mode 31 is greater than that of mode 32, while using B3PW91, the opposite is the case. A significant difference also occurs for modes 24–28. Comparison of the calculated IR spectra to the experimental IR spectrum shows that the relative frequencies and intensities of modes 24–28, 31, and 32 are in closer agreement for the B3PW91/cc-pVTZ spectrum, and in worse agreement for the B3LYP/cc-pVTZ spectrum. The assignment of the experimental IR spectrum, shown in Figure 6.1, is therefore based on the B3PW91/cc-pVTZ spectrum.

In order to evaluate the quantitative accuracies of the calculated IR spectra, the experimental frequencies and dipole strengths are obtained by Lorentzian fitting of the experimental IR spectrum. The Lorentzian fit is shown in Figure 6.2. A small number of bands in the experimental IR spectrum are not assigned to fundamental transitions. These bands can be due to overtone or combination transitions, due to anharmonicity, or to impurities in the sample of (+)-camphor. The frequencies and dipole strengths obtained for the fundamental transitions are given in Table 6.1. The experimental frequencies and dipole strengths are compared to the B3LYP/cc-pVTZ and B3PW91/cc-pVTZ frequencies and dipole strengths in Figures 6.3 and 6.4. As expected, the calculated frequencies are a few percent

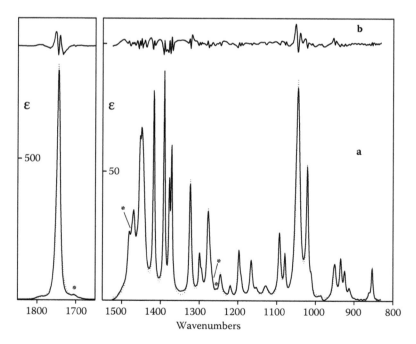

FIGURE 6.2 (a) Lorentzian fits of the experimental IR spectra of (+)-camphor over the ranges 1,850–1,650 cm^{-1} and 1,525–831 cm^{-1}. The solid lines are the experimental spectra; the dotted lines are the Lorentzian fits. (b) The difference spectra: the experimental spectra minus the Lorentzian fits. The asterisks indicate bands that are not assigned to fundamental transitions.

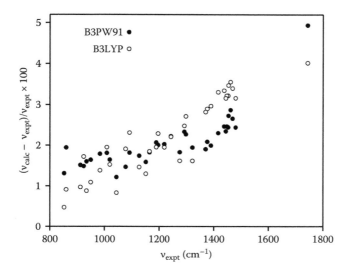

FIGURE 6.3 Comparison of the B3LYP/cc-pVTZ and B3PW91/cc-pVTZ frequencies to the experimental IR frequencies of modes 22–59.

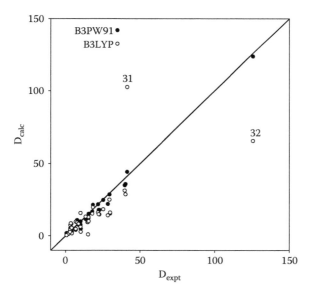

FIGURE 6.4 Comparison of the B3LYP/cc-pVTZ and B3PW91/cc-pVTZ dipole strengths to the experimental dipole strengths of modes 22–58. Dipole strengths D are in 10^{-40} esu^2 cm^2.

higher than the experimental frequencies, the percentage difference increasing with the frequency, from ~1% to 4–5% over the range 800–1,800 cm^{-1}. The

B3PW91/cc-pVTZ dipole strengths are clearly in better agreement with the experimental dipole strengths than are the B3LYP/cc-pVTZ dipole strengths, especially for modes 31 and 32. The mean absolute deviations between the calculated and experimental dipole strengths (average of the absolute deviations of calculated dipole strengths from the experimental dipole strengths) of modes 22–59 are 6.2 and 11.5×10^{-40} esu^2 cm^2 for B3PW91/cc-pVTZ and B3LYP/cc-pVTZ, respectively, confirming this conclusion.

Simulations of the VCD spectra of $(1R,4R)$-camphor due to fundamentals 21–59, using Lorentzian bandshapes ($\gamma = 4.0$ cm^{-1}), are shown in Figure 6.5, together with the experimental VCD spectrum of (+)-camphor in the frequency range 800–1,900 cm^{-1}. Although the B3LYP/cc-pVTZ and B3PW91/cc-pVTZ VCD spectra exhibit considerable qualitative similarity, significant differences in both sign and magnitude do occur, specifically for modes 27–31, 33, 38, 40, and 43. Comparison of the calculated VCD spectra to the experimental VCD spectrum shows that, for these modes, the B3PW91/cc-pVTZ VCD spectrum is in closer agreement with the experimental spectrum than is the B3LYP/cc-pVTZ VCD spectrum. The assignment of the experimental VCD spectrum, shown in Figure 6.5, is therefore based on the B3PW91/cc-pVTZ spectrum.

In order to evaluate the quantitative accuracies of the calculated VCD spectra, the experimental frequencies and rotational strengths are obtained by Lorentzian fitting of the experimental VCD spectrum. The Lorentzian fit is shown in Figure 6.6. The frequencies and rotational strengths obtained for the fundamental transitions are given in Table 6.1. For each fundamental transition, the frequency obtained from the VCD spectrum is in excellent agreement with the frequency obtained from the IR spectrum, proving that the assignments of the experimental IR and VCD spectra are consistent. (The slight differences in the frequencies and bandwidths of the fundamental transitions obtained from the IR and VCD spectra can be attributed to the use of different instruments and resolutions in measuring the IR and VCD spectra.) The experimental rotational strengths are compared to the B3LYP/cc-pVTZ and B3PW91/cc-pVTZ rotational strengths in Figure 6.7. The B3PW91/cc-pVTZ rotational strengths are clearly in better agreement with the experimental rotational strengths than are the B3LYP/cc-pVTZ rotational strengths, especially for modes 28 and 30. The mean absolute deviations between calculated and experimental rotational strengths of modes 22–59 are 4.1 and 7.0×10^{-44} esu^2 cm^2 for B3PW91/cc-pVTZ and B3LYP/cc-pVTZ, respectively, confirming this conclusion. Thus, comparison of the calculated and experimental VCD spectra and rotational strengths leads to the same conclusion as to which functional is more accurate as did the comparison of the calculated and experimental IR spectra and dipole strengths.

For camphor, the large basis set cc-pVTZ has 554 basis functions. To confirm that the basis set error in the IR and VCD spectra calculated using cc-pVTZ is minimal, the IR and VCD spectra of camphor have been calculated using the much larger basis set, cc-pVQZ, which has 1,085 basis set functions, and the B3PW91 functional. The B3PW91/cc-pVQZ and B3PW91/cc-pVTZ dipole strengths of modes 22–58 and rotational strengths of modes 22–59 are compared in Figures 6.8 and 6.9. The B3PW91/cc-pVQZ and B3PW91/cc-pVTZ IR and VCD spectra over the range

800–1,550 cm^{-1} are compared in Figure 6.10. The B3PW91/cc-pVQZ dipole strengths and rotational strengths are quantitatively identical to the B3PW91/cc-pVTZ dipole strengths and rotational strengths, confirming that the basis set error of cc-pVTZ is no greater than that of cc-pVQZ. This conclusion is further confirmed by the identicality of the B3PW91/cc-pVQZ and B3PW91/cc-pVTZ IR and VCD spectra, which further demonstrates that, in addition to the dipole and rotational strengths, the B3PW91/cc-pVQZ and B3PW91/cc-pVTZ vibrational frequencies are quantitatively identical. As a result, the assignments of the experimental IR and VCD spectra,

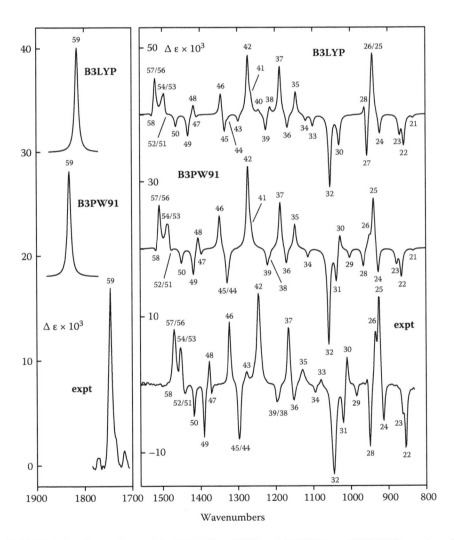

FIGURE 6.5 Comparison of the B3LYP/cc-pVTZ and B3PW91/cc-pVTZ VCD spectra of (1R,4R)-camphor to the experimental VCD spectrum of (+)-camphor. The assignment of the experimental spectrum is based on the B3PW91/cc-pVTZ spectrum.

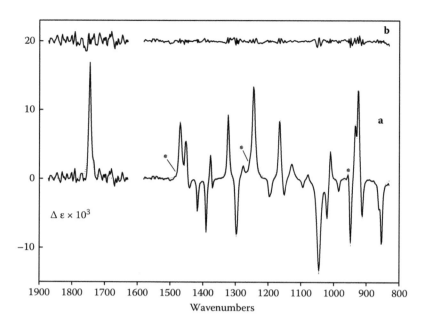

FIGURE 6.6 (a) Lorentzian fits of the experimental VCD spectra of (+)-camphor over the ranges 1,873–1,630 cm^{-1} and 1,530–831 cm^{-1}. The solid lines are the experimental spectra; the dotted lines are the Lorentzian fits. (b) The difference spectra: the experimental spectra minus the Lorentzian fits. The asterisks indicate bands that are not assigned to fundamental transitions.

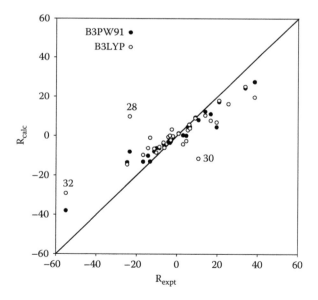

FIGURE 6.7 Comparison of the B3LYP/cc-pVTZ and B3PW91/cc-pVTZ rotational strengths to the experimental rotational strengths of modes 22–59. Rotational strengths R are in 10^{-44} esu^2 cm^2.

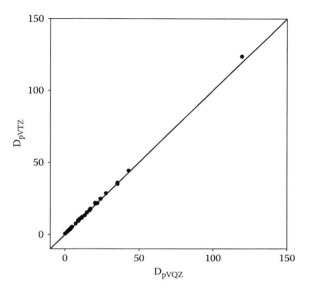

FIGURE 6.8   Comparison of the B3PW91/cc-pVQZ dipole strengths to the B3PW91/cc-pVTZ dipole strengths of modes 22–58 of camphor. Dipole strengths D are in 10^{-40} esu^2 cm^2.

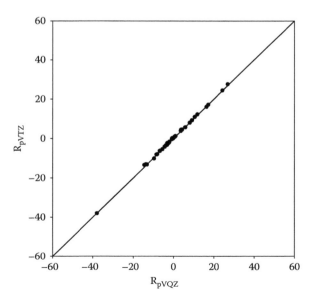

FIGURE 6.9   Comparison of the B3PW91/cc-pVQZ rotational strengths to the B3PW91/cc-pVTZ rotational strengths of modes 22–59 of (1R,4R)-camphor. Rotational strengths R are in 10^{-44} esu^2 cm^2.

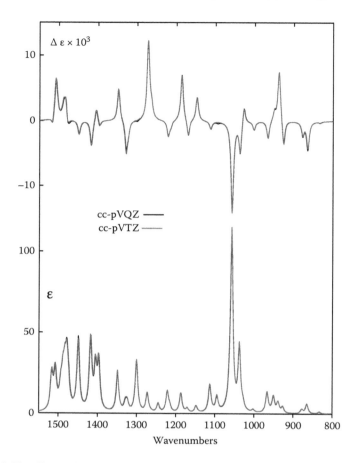

FIGURE 6.10   Comparison of the B3PW91/cc-pVQZ and B3PW91/cc-pVTZ IR and VCD spectra of (1*R*,4*R*)-camphor.

based on the B3PW91/cc-pVQZ IR and VCD spectra, are identical to the assignments based on the B3PW91/cc-pVTZ IR and VCD spectra. The mean absolute deviations of the calculated and experimental rotational strengths of modes 22–59 are 4.18 and 4.14×10^{-44} esu^2 cm^2 for cc-pVQZ and cc-pVTZ, respectively, further confirming the equal accuracies of the two basis sets.

To evaluate the accuracies of other density functionals, relative to B3PW91 and B3LYP, the calculations have been repeated using the 17 additional functionals listed in Table 6.2, recently reviewed [22], and the cc-pVTZ basis set. The values of the mean absolute deviations of the calculated and experimental rotational strengths for these functionals are given in Table 6.2, together with the values for B3PW91 and B3LYP, and plotted in Figure 6.11. The mean absolute deviation of the calculated rotational strengths from the experimental rotational strengths for modes 22–59 varies from 4.03 to 10.44×10^{-44} esu^2 cm^2. Three functionals, B3P86,

TABLE 6.2

The Functional Dependence of the Mean Absolute Deviations of cc-pVTZ-Calculated and Experimental Rotational Strengths of Modes 22–59 of Camphor

Functional	Type[a]	MAD[b]	No. Wrong Signs
BLYP	GGA	10.44	11
OLYP	GGA	7.52	7
τHCTH	M-GGA	7.06	7
B3LYP	H-GGA	6.99	5
BPW91	GGA	6.67	7
BP86	GGA	6.54	6
BB95	M-GGA	6.07	5
τHCTHHYB	HM-GGA	5.61	4
HCTH	GGA	5.59	4
O3LYP	H-GGA	5.42	4
B1B95	HM-GGA	4.73	2
CAMB3LYP	H-X	4.54	2
BMK	HM-GGA	4.41	3
PBE1PBE	H-GGA	4.28	1
M05-2X	HM-GGA	4.24	1
B3PW91	H-GGA	4.14	1
HSE1PBE	H-S	4.11	1
MPW1PW91	H-GGA	4.06	0
B3P86	H-GGA	4.03	0

[a] GGA, generalized gradient approximation; M-GGA, meta-generalized gradient approximation; H-GGA, hybrid generalized gradient approximation; HM-GGA, hybrid meta-generalized gradient approximation; H-X, range-separated hybrid (with short-range DFT exchange and long-range HF exchange); H-S, range-separated hybrid (short-range HF exchange and long-range DFT exchange).

[b] Mean absolute deviation relative to experiment in 10^{-44} esu^2 cm^2.

MPW1PW91, and HSE1PBE, give only slightly more accurate rotational strengths than B3PW91. Most of the functionals, except for BLYP, OLYP, and τHCTH, are more accurate than B3LYP. The number of modes for which the calculated rotational strengths are opposite in sign to the experimental rotational strengths increases in parallel with the mean absolute deviations, varying from 0 to 11. For B3PW91 and B3LYP, the numbers of incorrect signs are 1 and 5. Two functionals, B3P86 and MPW1PW91, have 0 incorrect signs. The functional BLYP has 11 incorrect signs, and the functionals OLYP, τHCTH, and BPW91 have 7 incorrect signs, all more than B3LYP.

The accuracies of the pure functionals BLYP and BPW91 are substantially lower than the accuracies of the related hybrid functionals B3LYP and B3PW91.

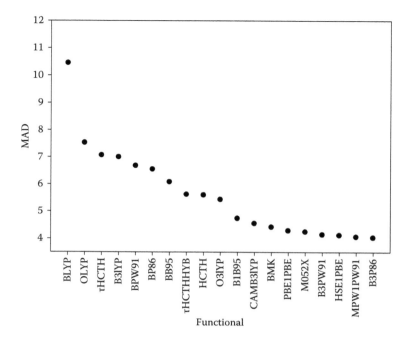

FIGURE 6.11   Mean absolute deviation functional dependence of cc-pVTZ-calculated rotational strengths of camphor.

Also, the pure functionals OLYP, BP86, and BB95 are substantially less accurate than the related hybrid functionals O3LYP, B3P86, and B1B95. These data support the conclusion that the hybridization of pure functionals increases their accuracy. Consistent with this conclusion, the 10 most accurate functionals are all hybrid.

To evaluate the accuracies of other basis sets, relative to cc-pVTZ and cc-pVQZ, calculations have been repeated using the 10 additional basis sets listed in Table 6.3, and the B3PW91 functional. The values of the mean absolute deviations of the calculated rotational strengths from the cc-pVQZ rotational strengths of modes 22–59 are given in Table 6.3. In Figure 6.12 the dependence of the mean absolute deviation on the number of basis functions is plotted. As the number of basis functions diminishes, the mean absolute deviation increases substantially.

The basis sets TZ2P and 6-31G* are frequently used in calculating VCD spectra. The B3PW91/TZ2P and B3PW91/6-31G* VCD spectra of camphor are compared to the B3PW91/cc-pVQZ VCD spectrum in Figure 6.13. The TZ2P VCD spectrum is quantitatively identical to the cc-pVQZ VCD spectrum, proving that the basis set error of TZ2P is no greater than the basis set error of cc-pVTZ. The 6-31G* VCD spectrum is quantitatively more different from the cc-pVQZ VCD spectrum, proving that the basis set error of 6-31G* is greater than the basis set errors of TZ2P and cc-pVTZ. Qualitatively, the 6-31G* VCD spectrum is similar to the cc-pVQZ and TZ2P VCD spectra.

TABLE 6.3
The Basis Set Dependence of the Mean Absolute Deviations of B3PW91-Calculated and B3PW91/cc-pVQZ Rotational Strengths of Modes 22–59 of Camphor

Basis Set	Contracted Set[a]	No. Fns.[b]	MAD[c]
cc-pVQZ	[5s4p3d2f1g/4s3p2d1f] (5d,7f,9g)	1,085	0.00
aug-cc-pVTZ	[5s4p3d2f/4s3p2d] (5d,7f)	874	0.19
cc-pVTZ	[4s3p2d1f/3s2p1d] (5d,7f)	554	0.24
6-311++G(2d,2p)	[5s4p2d/4s2p] (5d)	457	0.54
TZ2P	[5s4p2d/3s2p] (6d)	463	0.63
6-311++G**	[5s4p1d/4s1p] (5d)	354	1.11
6-31G**	[3s2p1d/2s1p] (6d)	245	1.22
6-311G**	[4s3p1d/3s1p] (5d)	294	1.27
6-31G*	[3s2p1d/2s] (6d)	197	1.32
aug-cc-pVDZ	[4s3p2d/3s2p] (5d)	397	1.87
cc-pVDZ	[3s2p1d/2s1p] (5d)	234	2.71
6-31G	[3s2p/2s]	131	4.41

[a] Of the form [1/2] where the first term refers to the contracted set on carbon and oxygen and the second refers to the contracted set on hydrogen. The numbers in parentheses refer to the number of functions used for each type of polarization function.

[b] Number of basis functions.

[c] Mean absolute deviation relative to cc-pVQZ for modes 22–59.

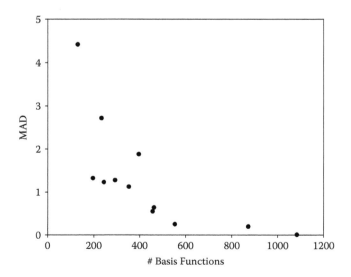

FIGURE 6.12   B3PW91 basis set dependence of camphor rotational strengths.

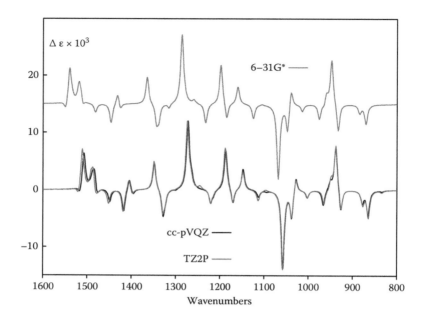

FIGURE 6.13 Comparison of the B3PW91/TZ2P and B3PW91/6-31G* VCD spectra of camphor to the B3PW91/cc-pVQZ VCD spectrum.

To confirm that DFT is more accurate than HF theory, the IR and VCD spectra have been calculated using HF theory and the cc-pVTZ basis set. The spectra obtained are compared to the experimental spectra and to the B3PW91/cc-pVTZ spectra in Figures 6.14 and 6.15. The experimental rotational strengths are compared to the HF/cc-pVTZ and B3PW91/cc-pVTZ rotational strengths in Figure 6.16. The B3PW91/cc-pVTZ rotational strengths are clearly in better agreement with the experimental rotational strengths than are the HF/cc-pVTZ rotational strengths, especially for modes 25 and 28. The mean absolute deviations between calculated and experimental rotational strengths of modes 22–59 are 4.1 and 8.1 × 10^{-44} esu^2 cm^2 for B3PW91/cc-pVTZ and HF/cc-pVTZ, respectively, confirming this conclusion.

α-PINENE

Measurements of the experimental IR and VCD spectra of α-pinene were presented in Chapter 2. Here, we analyze the spectra obtained using 1.87 and 0.93 M solutions of (+), (−), and (±) α-pinene in CCl$_4$, Figure 2.38, over the frequency range 1,700–800 cm^{-1}. As discussed in Chapter 5, α-pinene is predicted to be conformationally rigid. As with camphor, initially we predict the IR and VCD spectra using DFT, the basis set cc-pVTZ, and the functionals B3LYP and B3PW91.

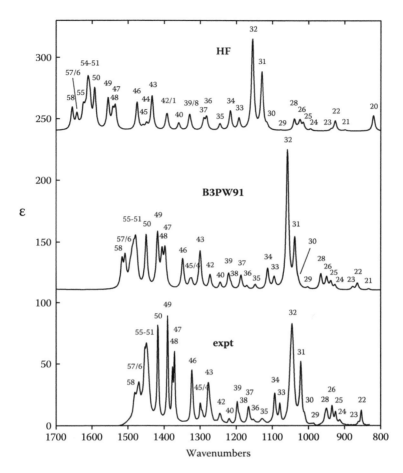

FIGURE 6.14 Comparison of the HF/cc-pVTZ and B3PW91/cc-pVTZ IR spectra of camphor to the experimental IR spectrum of (+)-camphor.

The B3LYP/cc-pVTZ and B3PW91/cc-pVTZ equilibrium geometries of $(1R,5R)$-α-pinene were obtained by reoptimization of the B3LYP/6-31G* equilibrium geometries obtained in Chapter 5. The harmonic frequencies, dipole strengths, and rotational strengths were then calculated, with the results for the fundamental transitions of modes 56–19 given in Table 6.4. Simulations of the IR spectra of α-pinene due to fundamentals 56–18, using Lorentzian bandshapes ($\gamma = 4.0$ cm^{-1}), are shown in Figure 6.17, together with the experimental IR spectrum in the frequency range 1,800–800 cm^{-1}. The B3LYP/cc-pVTZ and B3PW91/cc-pVTZ IR spectra are qualitatively very similar over the entire frequency range. The assignments of the experimental IR spectrum, based on the B3PW91/cc-pVTZ and B3LYP/cc-pVTZ spectra, are therefore identical, as shown in Figure 6.17.

In order to evaluate the quantitative accuracies of the calculated IR spectra, the experimental frequencies and dipole strengths are obtained by Lorentzian fitting of

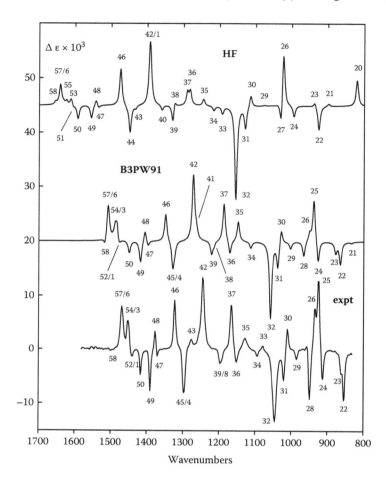

FIGURE 6.15 Comparison of the HF/cc-pVTZ and B3PW91/cc-pVTZ VCD spectra of (1*R*,4*R*)-camphor to the experimental VCD spectrum of (+)-camphor.

the experimental IR spectrum. The Lorentzian fit is shown in Figure 6.18. The frequencies and dipole strengths obtained for the fundamental transitions are given in Table 6.4. The experimental frequencies and dipole strengths are compared to the B3LYP/cc-pVTZ and B3PW91/cc-pVTZ frequencies and dipole strengths in Figures 6.19 and 6.20. As with camphor, the calculated frequencies are a few percent higher than the experimental frequencies, the percentage difference increasing with the frequency, from ~1 to 4–5% over the range 800–1,800 cm⁻¹. The B3PW91/cc-pVTZ dipole strengths are clearly in better agreement with the experimental dipole strengths than are the B3LYP/cc-pVTZ dipole strengths, especially for modes 52, 49, and 45. The mean absolute deviations between the calculated and experimental dipole strengths of modes 20–56 are 3.3 and

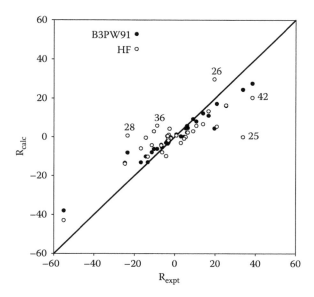

FIGURE 6.16 Comparison of the HF/cc-pVTZ and B3PW91/cc-pVTZ rotational strengths to the experimental rotational strengths of modes 22–59. Rotational strengths R are in 10^{-44} esu^2 cm^2.

4.6×10^{-40} esu^2 cm^2 for B3PW91/cc-pVTZ and B3LYP/cc-pVTZ, respectively, confirming this conclusion.

Simulations of the VCD spectra of (1R,5R)-α-pinene due to fundamentals 19–56, using Lorentzian bandshapes (γ = 4.0 cm^{-1}), are shown in Figure 6.21, together with the experimental VCD spectrum of (+)-α-pinene in the frequency range 800–1,800 cm^{-1}. Although the B3LYP/cc-pVTZ and B3PW91/cc-pVTZ VCD spectra exhibit a certain amount of qualitative similarity, a substantial number of differences in sign do occur, specifically for modes 25, 26, 29, 30, 35, 36, 48, and 49. Comparison of the calculated VCD spectra to the experimental VCD spectrum shows that, for these modes, the B3PW91/cc-pVTZ VCD spectrum is in better agreement with the experimental spectrum than is the B3LYP/cc-pVTZ VCD spectrum. The assignment of the experimental VCD spectrum, shown in Figure 6.21, is therefore based on the B3PW91/cc-pVTZ spectrum.

In order to evaluate the quantitative accuracies of the calculated VCD spectra, the experimental frequencies and rotational strengths are obtained by Lorentzian fitting of the experimental VCD spectrum. The Lorentzian fit is shown in Figure 6.22. The frequencies and rotational strengths obtained for the fundamental transitions are given in Table 6.4. For each fundamental transition, the frequency obtained from the VCD spectrum is in excellent agreement with the frequency obtained from the IR spectrum, proving that the assignments of the experimental IR and VCD spectra are consistent. The experimental rotational strengths are compared to the

TABLE 6.4

Calculated and Experimental Frequencies, Dipole Strengths, and Rotational Strengths for α-Pinene[a]

	Calculation[b]						Experiment[c]					
	B3LYP			B3PW91			IR			VCD		
Mode	ν	D	R	ν	D	R	ν	D	γ	ν	R	γ
56	1,717	3.6	1.0	1,728	4.1	1.1	1,658	3.7	3.5	1,659	0.9	4.2
55	1,519	11.3	5.6	1,509	12.6	4.5	1,471	16.9	4.4	1,474	2.5	3.5
54	1,515	19.7	-3.3	1,504	21.4	-2.9	1,468	26.1	4.8	1,467	-3.0	2.6
53	1,501	1.6	0.7	1,490	3.7	1.2	1,453	13.3	4.4	1,454	1.0	2.7
52	1,495	15.8	0.2	1,483	22.0	1.0	1,447	32.7	3.9			
51	1,490	10.7	0.7	1,480	7.8	-0.2	1,443	14.0	4.1			
50	1,485	8.8	0.9	1,474	7.2	1.1	1,441	6.3	6.7	1,445	1.4	2.9
49	1,481	8.2	0.6	1,469	21.4	-0.6	1,436	26.5	3.6	1,436	-0.3	1.7
48	1,479	19.4	-0.9	1,468	12.3	0.5	1,433	12.9	5.4	1,431	0.7	2.8
47	1,423	17.4	-2.1	1,409	21.7	-2.7	1,381	20.6	2.1	1,381	-3.7	4.0
46	1,414	6.7	-1.0	1,404	12.4	-0.6	1,375	12.2	2.0	1,374	-0.9	3.1
45	1,403	24.3	-1.0	1,391	32.9	-2.0	1,365	39.2	2.8	1,364	-1.9	2.9
44	1,370	3.6	-2.0	1,368	3.6	-1.8	1,336	6.8	3.6	1,335	1.9	3.4
43	1,357	7.1	6.2	1,358	9.5	9.2	1,328	10.8	3.6	1,328	9.0	3.0
42	1,338	3.8	2.7	1,336	2.0	1.9	1,305	4.6	7.6	1,307	1.0	3.1
41	1,294	7.7	-4.4	1,298	10.2	-4.0	1,265	17.5	5.0	1,265	-8.1	4.0
40	1,280	0.7	-1.1	1,275	1.0	-0.5	1,248	0.6	1.8	1,244	-0.4	4.0
39	1,251	14.4	-0.8	1,250	12.7	-1.3	1,221	13.0	3.6	1,224	3.1	2.8
38	1,236	8.2	-6.4	1,243	4.3	-11.3	1,215	6.2	3.1	1,215	-21.3	3.5

	ν	γ/D	R	ν[b]	D	R	ν	D	γ	ν	R	γ
37	1,230	7.8	-3.0	1,232	10.3	-3.6	1,204	14.4	2.7	1,202	1.7	2.4
36	1,210	7.0	-1.0	1,208	5.7	1.8	1,182	7.1	3.9	1,182	6.5	3.9
35	1,190	4.0	-0.8	1,191	6.0	1.3	1,165	10.4	3.4	1,166	1.7	3.0
34	1,148	14.4	14.9	1,149	12.1	14.3	1,125	16.3	3.0	1,125	16.0	3.4
33	1,127	8.1	-12.9	1,126	8.9	-13.9	1,101	11.3	4.1	1,100	-17.1	3.9
32	1,109	10.9	3.3	1,108	7.7	2.0	1,084	13.5	3.6	1,080	-0.5	4.2
31	1,084	5.5	-6.5	1,080	7.7	-11.3	1,063	7.4	2.6	1,063	-9.0	3.0
30	1,066	2.2	-1.2	1,065	4.8	4.7	1,043	6.4	4.2	1,042	4.1	3.5
29	1,056	5.1	1.6	1,058	1.7	-0.9	1,033	2.8	3.6	1,033	-0.8	3.6
28	1,036	16.0	13.1	1,031	22.2	16.9	1,015	17.8	2.8	1,014	12.1	3.0
27	1,021	1.6	-13.3	1,016	1.7	-14.3	997	0.8	1.5	997	-22.4	3.0
26	974	7.7	0.7	975	11.5	-8.2	959	6.7	3.1	959	-6.1	3.5
25	971	8.3	-1.7	969	7.5	3.1	953	13.7	3.0	951	3.0	2.9
24	950	1.7	-1.6	949	2.8	-0.2	940	1.6	3.2	939	-3.2	3.4
23	938	5.8	2.7	945	7.4	7.3	928	8.6	3.0	927	4.8	3.5
22	917	3.6	5.9	924	1.1	3.0	906	2.7	3.3	906	6.9	3.0
21	895	13.9	-1.5	903	16.4	-0.3	887	23.4	3.0	887	5.5	3.0
20	848	4.1	3.9	862	3.7	4.5	842	2.7	2.8	842	6.7	3.4
19	824	2.6	2.9	832	0.3	2.5				819	1.9	3.0

a Frequencies ν and bandwidths γ in cm⁻¹; dipole strengths D in 10^{-40} esu² cm²; rotational strengths R in 10^{-44} esu² cm². Experimental rotational strengths are for (+)-α-pinene; calculated rotational strengths are for (1R,5R)-α-pinene.

b Using the cc-pVTZ basis set.

c From Lorentzian fitting of the experimental IR and VCD spectra.

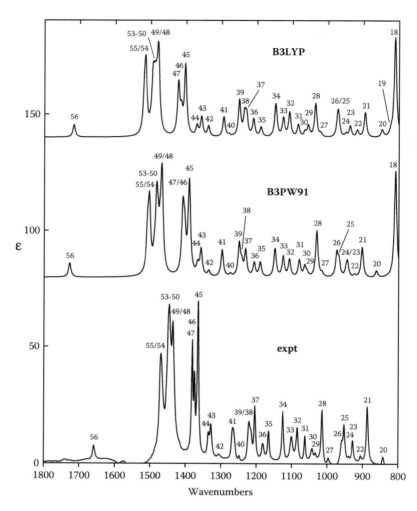

FIGURE 6.17 Comparison of the B3LYP/cc-pVTZ and B3PW91/cc-pVTZ IR spectra of α-pinene to the experimental IR spectrum of (+)-α-pinene. The assignment of the experimental spectrum is based on both the B3PW91/cc-pVTZ and B3LYP/cc-pVTZ spectra.

B3LYP/cc-pVTZ and B3PW91/cc-pVTZ rotational strengths in Figure 6.23. The B3PW91/cc-pVTZ rotational strengths are clearly in better agreement with the experimental rotational strengths than are the B3LYP/cc-pVTZ rotational strengths. The mean absolute deviations between calculated and experimental rotational strengths of modes 19–56 are 2.3 and 3.1 × 10^{-44} esu^2 cm^2 for B3PW91/cc-pVTZ and B3LYP/cc-pVTZ, respectively, confirming this conclusion. Thus, comparison of the calculated and experimental rotational strengths leads to the same conclusion as to which functional is more accurate as did the comparison of the calculated and experimental dipole strengths.

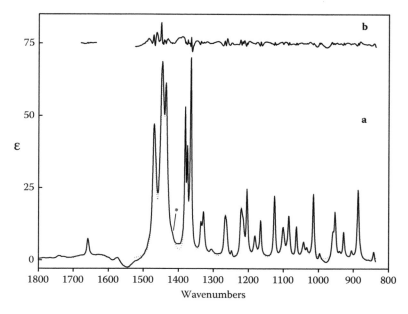

FIGURE 6.18   (a) Lorentzian fits of the experimental IR spectra of (+)-α-pinene over the ranges 1,680–1,635 cm^{-1} and 1,525–837 cm^{-1}. The solid lines are the experimental spectra; the dotted lines are the Lorentzian fits. (b) The difference spectra: the experimental spectra minus the Lorentzian fits. The asterisk indicates a band that is not assigned to a fundamental transition.

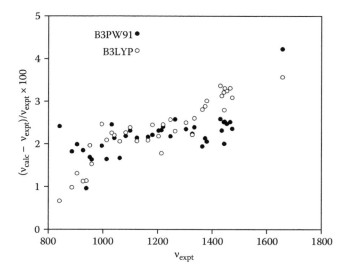

FIGURE 6.19   Comparison of the B3LYP/cc-pVTZ and B3PW91/cc-pVTZ frequencies to the experimental IR frequencies of modes 20–56.

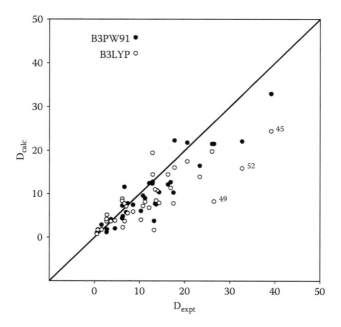

FIGURE 6.20 Comparison of the B3LYP/cc-pVTZ and B3PW91/cc-pVTZ dipole strengths to the experimental dipole strengths of modes 20–56. Dipole strengths D are in $10^{-40}\,esu^2\,cm^2$.

For α-pinene, the large basis set cc-pVTZ has 524 basis functions. To confirm that the basis set error in the IR and VCD spectra calculated using cc-pVTZ is minimal, the IR and VCD spectra of α-pinene have been calculated using the cc-pVQZ basis set, which has 1,030 basis functions, and the B3PW91 functional. The B3PW91/cc-pVQZ and B3PW91/cc-pVTZ dipole strengths of modes 20–56 and rotational strengths of modes 19–56 are compared in Figures 6.24 and 6.25. The B3PW91/cc-pVQZ and B3PW91/cc-pVTZ IR and VCD spectra over the range 750–1,750 cm^{-1} are compared in Figure 6.26. As with camphor, for α-pinene the B3PW91/cc-pVQZ dipole and rotational strengths are quantitatively identical to the B3PW91/cc-pVTZ dipole and rotational strengths, confirming that the basis set error of cc-pVTZ is no greater than that of cc-pVQZ. This conclusion is further confirmed by the identicality of the B3PW91/cc-pVQZ and B3PW91/cc-pVTZ IR and VCD spectra. The mean absolute deviations of the calculated and experimental rotational strengths of modes 19–56 are 2.25 and 2.27 $\times$ 10^{-44} esu^2 cm^2 for cc-pVQZ and cc-pVTZ, respectively, further confirming the equal accuracies of the two basis sets.

To evaluate the accuracies of other density functionals, relative to B3PW91 and B3LYP, the calculations have been repeated using the 17 additional functionals listed in Table 6.5, and the cc-pVTZ basis set. The values of the mean absolute deviations of the calculated and experimental rotational strengths for these functionals are given in Table 6.5, together with the values for B3PW91 and B3LYP,

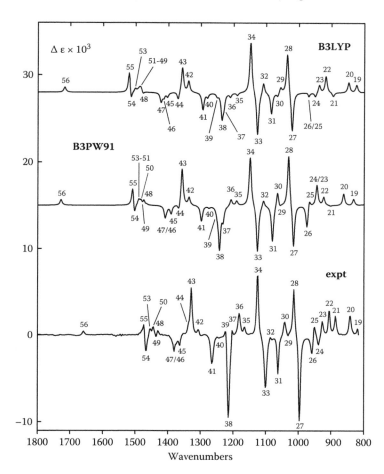

FIGURE 6.21   Comparison of the B3LYP/cc-pVTZ and B3PW91/cc-pVTZ VCD spectra of (1R,5R)-α-pinene to the experimental VCD spectrum of (+)-α-pinene. The assignment of the experimental spectrum is based on the B3PW91/cc-pVTZ spectrum.

and plotted in Figure 6.27. The mean absolute deviation of the calculated rotational strengths from the experimental rotational strengths for modes 19–56 varies from 2.13 to 3.90 × 10^{-44} esu^2 cm^2. Two functionals, B3P86 and BP86, give slightly more accurate rotational strengths than B3PW91. The number of modes for which the calculated rotational strengths are opposite in sign to the experimental rotational strengths increases in parallel with the mean absolute deviations.

The accuracies of the pure functionals BLYP, OLYP, BPW91, and BP86 are lower than the accuracies of the related hybrid functionals B3LYP, O3LYP, B3PW91, and B3P86, supporting the conclusion that the hybridization of pure functionals increases their accuracy.

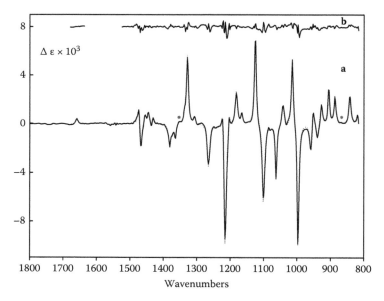

FIGURE 6.22 (a) Lorentzian fits of the experimental VCD spectra of (+)-α-pinene over the ranges 1,680–1,635 and 1,525–815 cm⁻¹. The solid lines are the experimental spectra; the dotted lines are the Lorentzian fits. (b) The difference spectra: the experimental spectra minus the Lorentzian fits. The asterisks indicate bands that are not assigned to fundamental transitions.

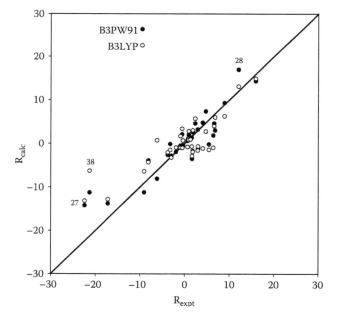

FIGURE 6.23 Comparison of the B3LYP/cc-pVTZ and B3PW91/cc-pVTZ rotational strengths to the experimental rotational strengths of modes 19–56. Rotational strengths R are in 10^{-44} esu² cm².

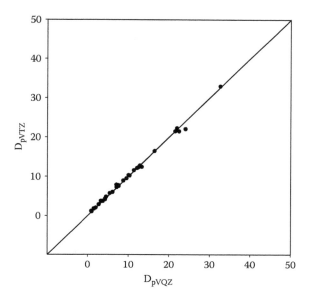

FIGURE 6.24 Comparison of the B3PW91/cc-pVQZ dipole strengths to the B3PW91/cc-pVTZ dipole strengths of modes 20–56 of α-pinene. Dipole strengths D are in 10^{-40} esu^2 cm^2.

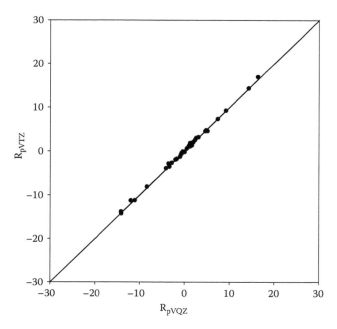

FIGURE 6.25 Comparison of the B3PW91/cc-pVQZ rotational strengths to the B3PW91/cc-pVTZ rotational strengths of modes 19–56 of (1R,5R)-α-pinene. Rotational strengths R are in 10^{-44} esu^2 cm^2.

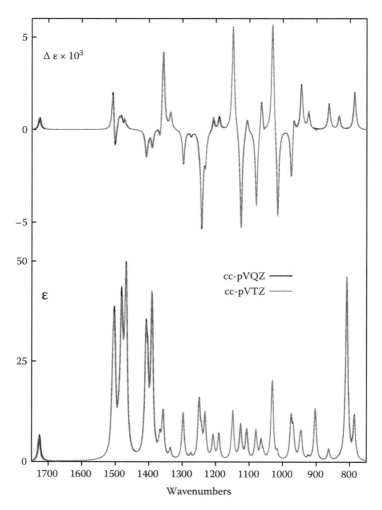

FIGURE 6.26 Comparison of the B3PW91/cc-pVQZ and B3PW91/cc-pVTZ IR and VCD spectra of (1*R*,5*R*)-α-pinene.

To evaluate the accuracies of other basis sets, relative to cc-pVTZ and cc-pVQZ, calculations have been repeated using the 10 additional basis sets listed in Table 6.6, and the B3PW91 functional. The values of the mean absolute deviations of the calculated rotational strengths from the cc-pVQZ rotational strengths of modes 19–56 are given in Table 6.6. In Figure 6.28 the dependence of the mean absolute deviation on the number of basis functions is plotted. As the number of basis functions diminishes, the mean absolute deviation increases substantially.

The basis sets TZ2P and 6-31G* are frequently used in calculating VCD spectra. The B3PW91/TZ2P and B3PW91/6-31G* VCD spectra of α-pinene are compared to

TABLE 6.5
The Functional Dependence of the Mean Absolute Deviations of cc-pVTZ-Calculated and Experimental Rotational Strengths of Modes 19–56 of α-Pinene

Functional	Type[a]	MAD[b]	No. Wrong Signs
BLYP	GGA	3.90	11
OLYP	GGA	3.53	12
BMK	HM-GGA	3.25	9
HCTH	GGA	3.14	10
B3LYP	H-GGA	3.12	13
B1B95	HM-GGA	3.05	7
O3LYP	H-GGA	3.02	10
τHCTH	M-GGA	2.83	9
CAMB3LYP	H-X	2.75	7
BB95	M-GGA	2.49	6
τHCTHHYB	HM-GGA	2.46	8
M05-2X	HM-GGA	2.42	5
PBE1PBE	H-GGA	2.40	5
BPW91	GGA	2.39	7
MPW1PW91	H-GGA	2.37	5
HSE1PBE	H-S	2.32	6
B3PW91	H-GGA	2.27	5
BP86	GGA	2.20	5
B3P86	H-GGA	2.13	3

[a] GGA, generalized gradient approximation; M-GGA, meta-generalized gradient approximation; H-GGA, hybrid generalized gradient approximation; HM-GGA, hybrid meta-generalized gradient approximation; H-X, range-separated hybrid (with short-range DFT exchange and long-range HF exchange); H-S, range-separated hybrid (short-range HF exchange and long-range DFT exchange).
[b] Mean absolute deviation relative to experiment in 10^{-44} esu^2 cm^2.

the B3PW91/cc-pVQZ VCD spectrum in Figure 6.29. The TZ2P VCD spectrum is quantitatively very nearly identical to the cc-pVQZ VCD spectrum, proving that, as for camphor, the basis set error of TZ2P is not significantly greater than the basis set error of cc-pVTZ. The 6-31G* VCD spectrum is both quantitatively and qualitatively very different from the cc-pVQZ and TZ2P VCD spectra. The qualitative difference is greatest in the frequency range 1,150–1,250 cm^{-1}. The 6-31G* basis set error is therefore much greater than the basis set errors of TZ2P and cc-pVTZ.

To confirm that DFT is more accurate than HF theory, the IR and VCD spectra have been calculated using HF theory and the cc-pVTZ basis set. The spectra obtained are compared to the experimental spectra in Figures 6.30 and 6.31. The

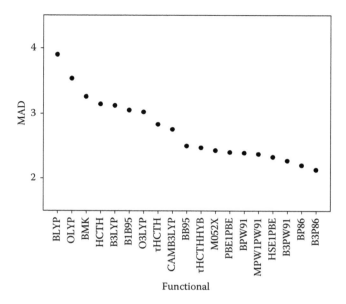

FIGURE 6.27 Mean absolute deviation (MAD) functional dependence of cc-pVTZ-calculated rotational strengths of α-pinene.

TABLE 6.6
The Basis Set Dependence of the Mean Absolute Deviations of B3PW91-Calculated and B3PW91/cc-pVQZ Rotational Strengths of Modes 19–56 of α-Pinene

Basis Set	Contracted Set[a]	No. Fns.[b]	MAD[c]
cc-pVQZ	[5s4p3d2f1g/4s3p2d1f] (5d,7f,9g)	1,030	0.00
aug-cc-pVTZ	[5s4p3d2f/4s3p2d] (5d,7f)	828	0.16
cc-pVTZ	[4s3p2d1f/3s2p1d] (5d,7f)	524	0.21
6-311++G(2d,2p)	[5s4p2d/4s2p] (5d)	430	0.32
6-311++G**	[5s4p1d/4s1p] (5d)	332	0.58
TZ2P	[5s4p2d/3s2p] (6d)	434	0.58
6-311G**	[4s3p1d/3s1p] (5d)	276	0.65
6-31G**	[3s2p1d/2s1p] (6d)	230	1.30
6-31G*	[3s2p1d/2s] (6d)	182	1.35
aug-cc-pVDZ	[4s3p2d/3s2p] (5d)	374	1.59
cc-pVDZ	[3s2p1d/2s1p] (5d)	220	2.12
6-31G	[3s2p/2s]	122	2.65

[a] Of the form [1/2] where the first term refers to the contracted set on carbon and the second refers to the contracted set on hydrogen. The numbers in parentheses refer to the number of functions used for each type of polarization function.

[b] Number of basis functions.

[c] Mean absolute deviation relative to cc-pVQZ.

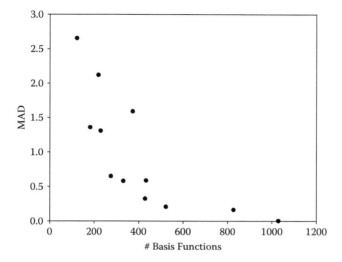

FIGURE 6.28   B3PW91 basis set dependence of α-pinene rotational strengths.

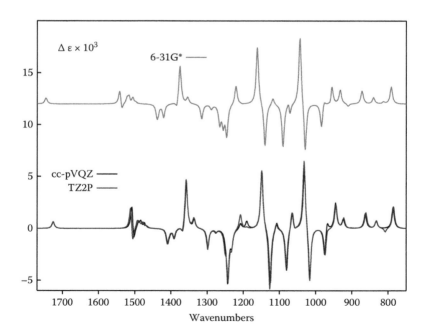

FIGURE 6.29   Comparison of the B3PW91/TZ2P and B3PW91/6-31G* VCD spectra of α-pinene to the B3PW91/cc-pVQZ VCD spectrum.

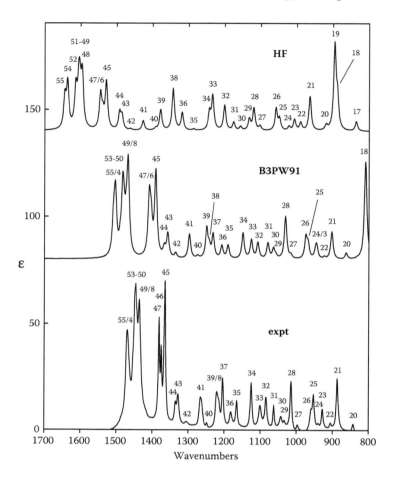

FIGURE 6.30 Comparison of the HF/cc-pVTZ and B3PW91/cc-pVTZ IR spectra of α-pinene to the experimental IR spectrum of (+)-α-pinene.

experimental rotational strengths are compared to the HF/cc-pVTZ and B3PW91/cc-pVTZ rotational strengths in Figure 6.32. The B3PW91/cc-pVTZ rotational strengths are clearly in better agreement with the experimental rotational strengths than are the HF/cc-pVTZ rotational strengths, especially for modes 33, 34, and 38. The mean absolute deviations between calculated and experimental rotational strengths of modes 19–56 are 2.3 and 4.3 × 10⁻⁴⁴ esu² cm² for B3PW91/cc-pVTZ and HF/cc-pVTZ, respectively, confirming this conclusion.

METHYLOXIRANE

The experimental IR and VCD spectra of methyloxirane, whose enantiomers are $R(+)$ and $S(-)$, have been measured many times [23]. Here, the spectra are measured

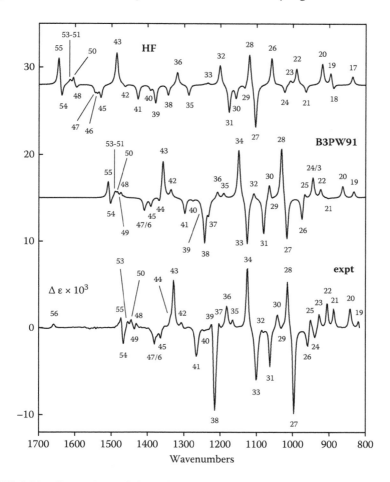

FIGURE 6.31   Comparison of the HF/cc-pVTZ and B3PW91/cc-pVTZ VCD spectra of (1*R*,5*R*)-α-pinene to the experimental VCD spectrum of (+)-α-pinene.

again, using *R*(+), *S*(–), and (±) samples from Aldrich. According to Aldrich, these samples are 99% pure, and the ee of the *R*(+) sample is 99.7%. The spectra were obtained using 0.36 M solutions in CCl_4 and a 546 μ pathlength cell. The specific rotation, $[\alpha]_D$ of the *R*(+) solution, was 19.7 (c = 2.11, CCl_4). The magnitude of the $[\alpha]_D$ of the *S*(–) solution was identical, demonstrating that the ee of the *S*(–) sample is also >99%. The IR spectra of the (+), (–), and (±) samples were identical, confirming that their purities are identical. The IR and VCD spectra of *S*(–)–methyloxirane within the range 1,600–840 cm^{-1} are shown in Figure 6.33. The baselines of the VCD spectra of the *R*(+) and *S*(–) solutions were the spectrum of the (±) solution. The VCD spectrum of *S*(–)–methyloxirane is the half-difference spectrum: $\frac{1}{2}[\Delta\varepsilon(-) - \Delta\varepsilon(+)]$.

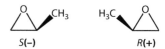

As discussed in Chapter 5, methyloxirane is predicted to be conformationally rigid. As with camphor and α-pinene, initially we predict the IR and VCD spectra using DFT, the basis set cc-pVTZ, and the functionals B3LYP and B3PW91. The B3LYP/cc-pVTZ and B3PW91/cc-pVTZ equilibrium geometries of *S*-methyloxirane were obtained by reoptimization of the B3LYP/6-31G* equilibrium geometry obtained in Chapter 5. The harmonic frequencies, dipole strengths, and rotational strengths were then calculated, with the results for the fundamental transitions of modes 6–18 given in Table 6.7. Simulations of the IR spectra of methyloxirane due to fundamentals 5–18, using Lorentzian bandshapes (γ = 4.0 cm⁻¹), are shown in Figure 6.34, together with the experimental IR spectrum in the frequency range 1,600–800 cm⁻¹. The B3LYP/cc-pVTZ and B3PW91/cc-pVTZ IR spectra are qualitatively very similar over the entire frequency range. The assignments of the experimental IR spectrum, based on the B3PW91/cc-pVTZ and B3LYP/cc-pVTZ spectra, are therefore identical, as shown in Figure 6.34.

In order to evaluate the quantitative accuracies of the calculated IR spectra, the experimental frequencies and dipole strengths are obtained by Lorentzian fitting

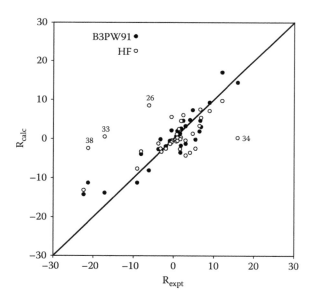

FIGURE 6.32 Comparison of the HF/cc-pVTZ and B3PW91/cc-pVTZ rotational strengths to the experimental rotational strengths of modes 19–56. Rotational strengths *R* are in 10⁻⁴⁴ esu² cm².

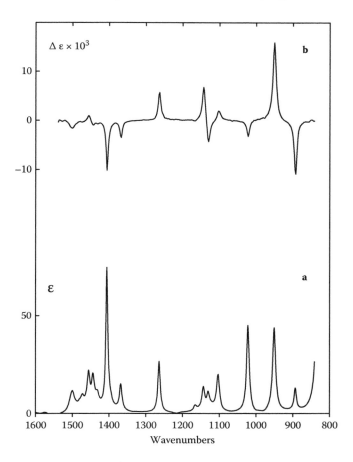

FIGURE 6.33   (a) The IR absorption spectrum of (–)-methyloxirane in CCl_4; 0.36 M, 1,600–841 cm^1; pathlength 546 μ. (b) The half-difference VCD spectrum of (–)-methyloxirane in CCl_4; 0.36 M, 1,539–841 cm^{-1}; pathlength 546 μ.

of the experimental IR spectrum. The Lorentzian fit is shown in Figure 6.35. The frequencies and dipole strengths obtained for the fundamental transitions are given in Table 6.7. The experimental frequencies and dipole strengths are compared to the B3LYP/cc-pVTZ and B3PW91/cc-pVTZ frequencies and dipole strengths in Figures 6.36 and 6.37. As with camphor and α-pinene, the calculated frequencies are a few percent higher than the experimental frequencies, from 1 to 4% over the range 800–1,600 cm^{-1}. The B3PW91/cc-pVTZ dipole strengths are in better agreement with the experimental dipole strengths than are the B3LYP/cc-pVTZ dipole strengths, especially for mode 8. The mean absolute deviations between the calculated and

TABLE 6.7

Calculated and Experimental Frequencies, Dipole Strengths, and Rotational Strengths for Methyloxirane[a]

Mode	Calculation[b]						Experiment[c]					
	B3LYP			B3PW91			IR			VCD		
	ν	D	R	ν	D	R	ν	D	γ	ν	R	γ
18	1,531	13.2	−5.2	1,531	20.0	−6.4	1,500	13.5	7.1	1,501	−6.5	8.1
17	1,498	14.6	1.1	1,490	16.1	0.4	1,456	15.8	4.7	1,456	2.3	3.7
16	1,484	13.6	−2.0	1,475	15.3	−2.1	1,444	11.6	4.3	1,444	−2.0	4.3
15	1,440	60.2	−13.4	1,441	58.3	−11.9	1,406	50.8	3.5	1,406	−20.2	4.1
14	1,407	9.8	−1.9	1,396	11.3	−1.8	1,368	16.1	6.1	1,368	−7.5	4.0
13	1,295	18.2	9.2	1,300	17.6	12.2	1,264	23.7	4.0	1,264	13.4	4.0
12	1,188	2.2	−0.2	1,186	3.8	−0.4	1,165	6.4	9.0	1,168	−0.7	4.0
11	1,165	13.8	13.2	1,165	15.6	11.0	1,144	11.7	4.4	1,143	21.1	4.5
10	1,156	3.8	−4.3	1,156	1.6	−2.5	1,130	16.9	8.0	1,130	−15.4	4.5
9	1,130	23.4	6.1	1,130	26.6	5.8	1,104	25.9	5.4	1,102	7.1	5.3
8	1,042	33.2	−4.0	1,042	47.5	−7.0	1,022	54.7	4.4	1,022	−10.9	4.6
7	974	62.6	33.3	989	59.2	32.6	951	64.3	5.0	951	66.2	5.5
6	908	11.1	−24.9	907	11.9	−25.1	894	16.3	4.7	893	−43.8	5.0

[a] Frequencies ν and bandwidths γ in cm^{-1}; dipole strengths D in 10^{-40} esu^2 cm^2; rotational strengths R in 10^{-44} esu^2 cm^2. Experimental rotational strengths are for (−)-methyloxirane; calculated rotational strengths are for (S)-methyloxirane.

[b] Using the cc-pVTZ basis set.

[c] From Lorentzian fitting of the IR and the VCD spectra.

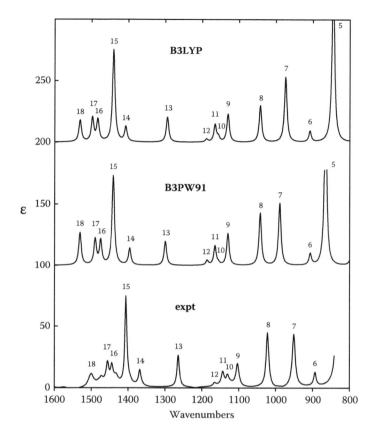

FIGURE 6.34   Comparison of the B3LYP/cc-pVTZ and B3PW91/cc-pVTZ IR spectra of methyloxirane to the experimental IR spectrum of (–)-methyloxirane. The assignment of the experimental spectrum is based on both the B3PW91/cc-pVTZ and B3LYP/cc-pVTZ spectra.

experimental dipole strengths are 5.2 and 5.8×10^{-40} esu^2 cm^2 for B3PW91/cc-pVTZ and B3LYP/cc-pVTZ, respectively, confirming this conclusion.

Simulations of the VCD spectra of S-methyloxirane due to fundamentals 5–18, using Lorentzian bandshapes ($\gamma = 4.0$ cm^{-1}), are shown in Figure 6.38, together with the experimental VCD spectrum of (–)-methyloxirane in the frequency range 1,539–841 cm^{-1}. The B3LYP/cc-pVTZ and B3PW91/cc-pVTZ VCD spectra are quali-tatively very similar over the entire frequency range. The assignments of the experi-mental VCD spectrum, based on the B3PW91/cc-pVTZ and B3LYP/cc-pVTZ spectra, are therefore identical, as shown in Figure 6.38.

In order to evaluate the quantitative accuracies of the calculated VCD spectra, the experimental frequencies and rotational strengths are obtained by Lorentzian fitting of the experimental VCD spectrum. The Lorentzian fit is shown in Figure 6.39. The

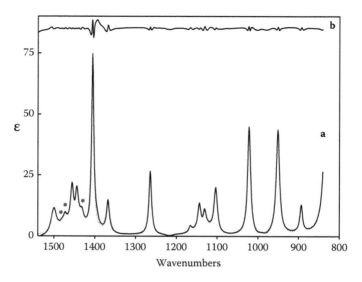

FIGURE 6.35 (a) Lorentzian fit of the experimental IR spectrum of (–)-methyloxirane over the range 1,533–839 cm⁻¹. The solid line is the experimental spectrum; the dotted line is the Lorentzian fit. (b) The difference spectrum: the experimental spectrum minus the Lorentzian fit. The asterisks indicate bands that are not assigned to fundamental transitions.

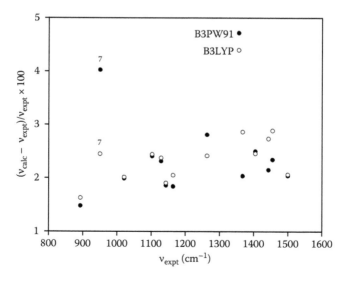

FIGURE 6.36 Comparison of the B3LYP/cc-pVTZ and B3PW91/cc-pVTZ frequencies to the experimental IR frequencies of modes 6–18.

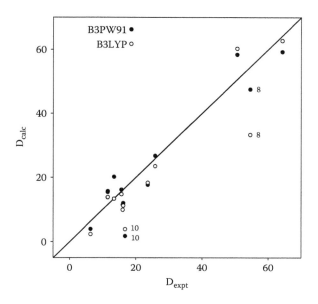

FIGURE 6.37 Comparison of the B3LYP/cc-pVTZ and B3PW91/cc-pVTZ dipole strengths to the experimental dipole strengths of modes 6–18. Dipole strengths D are in 10^{-40} esu^2 cm^2.

frequencies and rotational strengths obtained for the fundamental transitions are given in Table 6.7. For each fundamental transition, the frequency obtained from the VCD spectrum is in excellent agreement with the frequency obtained from the IR spectrum, proving that the assignments of the experimental IR and VCD spectra are consistent. The experimental rotational strengths are compared to the B3LYP/cc-pVTZ and B3PW91/cc-pVTZ rotational strengths in Figure 6.40. The B3PW91/cc-pVTZ and B3LYP/cc-pVTZ rotational strengths are clearly in comparable agreement to the experimental rotational strengths. The mean absolute deviations between calculated and experimental rotational strengths of modes 6–18 are 7.5 and 7.6 × 10^{-44} esu^2 cm^2 for B3PW91/cc-pVTZ and B3LYP/cc-pVTZ, respectively, confirming this conclusion.

For methyloxirane, the large basis set cc-pVTZ has 230 basis functions. To confirm that the basis set error in the IR and VCD spectra calculated using cc-pVTZ is minimal, the IR and VCD spectra of methyloxirane have been calculated using the cc-pVQZ basis set, which has 400 basis functions, and the B3PW91 functional. The B3PW91/cc-pVQZ and B3PW91/cc-pVTZ dipole strengths of modes 6–18, and rotational strengths of modes 6–18, are compared in Figures 6.41 and 6.42. The B3PW91/cc-pVQZ and B3PW91/cc-pVTZ IR and VCD spectra over the range 1,600–800cm^{-1} are compared in Figure 6.43. As with camphor and α-pinene the B3PW91/cc-pVQZ dipole and rotational strengths of methyloxirane are quantitatively

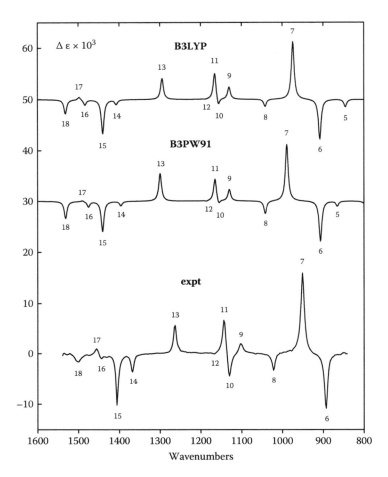

FIGURE 6.38 Comparison of the B3LYP/cc-pVTZ and B3PW91/cc-pVTZ VCD spectra of (S)-methyloxirane to the experimental VCD spectrum of (–)-methyloxirane. The assignment of the experimental spectrum is based on both the B3PW91/cc-pVTZ and B3LYP/cc-pVTZ spectra.

identical to the B3PW91/cc-pVTZ dipole and rotational strengths, confirming that the basis set error of cc-pVTZ is no greater than that of cc-pVQZ. This conclusion is further confirmed by the identicality of the B3PW91/cc-pVQZ and B3PW91/cc-pVTZ IR and VCD spectra. The mean absolute deviations of the calculated and experimental rotational strengths of modes 6–18 are 7.16 and 7.54 × 10^{-44} esu^2 cm^2 for cc-pVQZ and cc-pVTZ, respectively, confirming the comparable accuracies of the two basis sets.

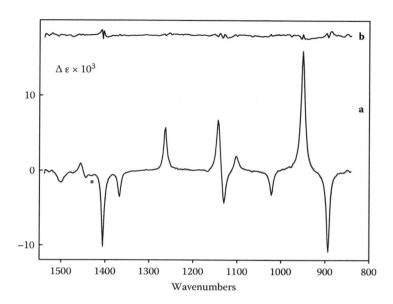

FIGURE 6.39 (a) Lorentzian fit of the experimental VCD spectrum of (–)-methyloxirane over the range 1,533–839 cm^{-1}. The solid line is the experimental spectrum; the dotted line is the Lorentzian fit. (b) The difference spectrum: the experimental spectrum minus the Lorentzian fit. The asterisk indicates a band that is not assigned to a fundamental transition.

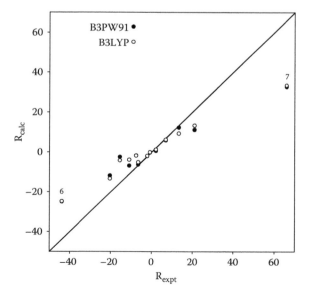

FIGURE 6.40 Comparison of the B3LYP/cc-pVTZ and B3PW91/cc-pVTZ rotational strengths to the experimental rotational strengths of modes 6–18. Rotational strengths R are in 10^{-44} esu^2 cm^2.

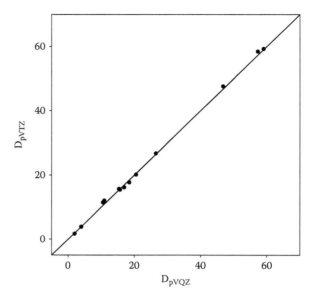

FIGURE 6.41 Comparison of the B3PW91/cc-pVQZ dipole strengths to the B3PW91/cc-pVTZ dipole strengths of modes 6–18 of methyloxirane. Dipole strengths D are in 10^{-40} esu² cm².

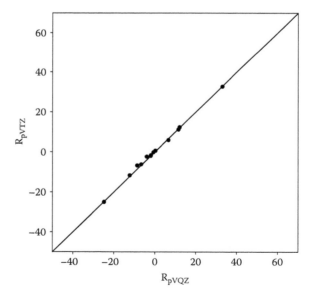

FIGURE 6.42 Comparison of the B3PW91/cc-pVQZ rotational strengths to the B3PW91/cc-pVTZ rotational strengths of modes 6–18 of (S)-methyloxirane. Rotational strengths R are in 10^{-44} esu² cm².

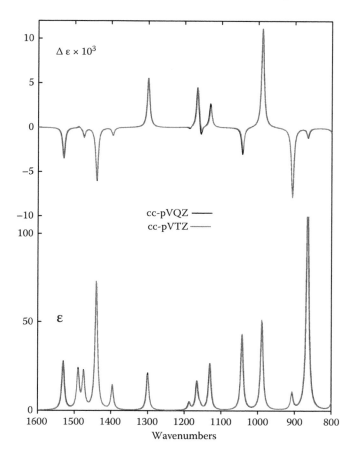

FIGURE 6.43   Comparison of the B3PW91/cc-pVQZ and B3PW91/cc-pVTZ IR and VCD spectra of (*S*)-methyloxirane.

To evaluate the accuracies of other density functionals, relative to B3PW91 and B3LYP, the calculations have been repeated using the 17 additional functionals listed in Table 6.8, and the cc-pVTZ basis set. The values of the mean absolute deviations of the calculated and experimental rotational strengths for these functionals are given in Table 6.8, together with the values for B3PW91 and B3LYP, and plotted in Figure 6.44. The mean absolute deviation of the calculated rotational strengths from the experimental rotational strengths for modes 6–18 varies from 7.4 to 8.4×10^{-44} $esu^2\ cm^2$. Four functionals, B3P86, PBE1PBE, HSE1PBE, and M05-2X, give slightly more accurate rotational strengths than B3PW91 and B3LYP. The number of modes for which the calculated rotational strengths are opposite in sign to the experimental rotational strengths is 0 or 1 for all functionals.

TABLE 6.8

The Functional Dependence of the Mean Absolute Deviations of cc-pVTZ-Calculated and Experimental Rotational Strengths of Modes 6–18 of Methyloxirane

Functional	Type[a]	MAD[b]	No. Wrong Signs
BMK	HM-GGA	8.41	0
BB95	M-GGA	8.29	0
OLYP	GGA	8.27	1
O3LYP	H-GGA	8.01	1
BP86	GGA	7.89	0
BPW91	GGA	7.86	0
CAMB3LYP	H-X	7.78	1
τHCTH	M-GGA	7.76	1
BLYP	GGA	7.76	0
HCTH	GGA	7.73	1
τHCTHHYB	HM-GGA	7.67	0
B1B95	HM-GGA	7.57	0
B3LYP	H-GGA	7.56	0
MPW1PW91	H-GGA	7.55	0
B3PW91	H-GGA	7.54	0
B3P86	H-GGA	7.49	0
PBE1PBE	H-GGA	7.40	1
HSE1PBE	H-S	7.39	0
M05-2X	HM-GGA	7.37	1

[a] GGA, generalized gradient approximation; M-GGA, meta-generalized gradient approximation; H-GGA, hybrid generalized gradient approximation; HM-GGA, hybrid meta-generalized gradient approximation; H-X, range-separated hybrid (with short-range DFT exchange and long-range HF exchange); H-S, range-separated hybrid (short-range HF exchange and long-range DFT exchange).

[b] Mean absolute deviation relative to experiment in 10^{-44} esu^2 cm^2.

The accuracies of the pure functionals OLYP, BP86, BPW91, and BLYP are lower than the accuracies of the related hybrid functionals O3LYP, B3P86, B3PW91, and B3LYP, supporting the conclusion that the hybridization of pure functionals increases their accuracy.

To evaluate the accuracies of other basis sets, relative to cc-pVTZ and cc-pVQZ, calculations have been repeated using the 10 additional basis sets listed in Table 6.9, and the B3PW91 functional. The values of the mean absolute deviations of the calculated rotational strengths from the cc-pVQZ rotational strengths of modes 6–18 are given in Table 6.9. In Figure 6.45 the dependence of the mean absolute deviation

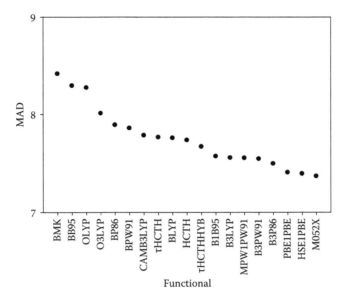

FIGURE 6.44   MAD functional dependence of cc-pVTZ-calculated rotational strengths of methyloxirane.

on the number of basis functions is plotted. As the number of basis functions diminishes, the mean absolute deviation increases substantially.

The basis sets TZ2P and 6-31G* are frequently used in calculating VCD spectra. The B3PW91/TZ2P and B3PW91/6-31G* VCD spectra of methyloxirane are compared to the B3PW91/cc-pVQZ VCD spectrum in Figure 6.46. The TZ2P VCD spectrum is quantitatively very nearly identical to the cc-pVQZ VCD spectrum, proving that, as for camphor and α-pinene, the basis set error of TZ2P is not significantly greater than the basis set error of cc-pVTZ. The 6-31G* VCD spectrum is both quantitatively and qualitatively very different from the cc-pVQZ and TZ2P VCD spectra. The qualitative difference is greatest in the frequency range 1,000–1,200 cm^{-1}. The 6-31G* basis set error is therefore much greater than the basis set errors of TZ2P and cc-pVTZ.

To confirm that DFT is more accurate than HF theory, the IR and VCD spectra have been calculated using HF theory and the cc-pVTZ basis set. The spectra obtained are compared to the experimental spectra in Figures 6.47 and 6.48. The experimental rotational strengths are compared to the HF/cc-pVTZ and B3PW91/cc-pVTZ rotational strengths in Figure 6.49. The B3PW91/cc-pVTZ rotational strengths are clearly in better agreement with the experimental rotational strengths than are the HF/cc-pVTZ rotational strengths, especially for modes 7 and 10. The mean absolute deviations between calculated and experimental rotational strengths of modes 6–18 are 7.5 and 10.7 × 10^{-44} esu^2 cm^2 for B3PW91/cc-pVTZ and HF/cc-pVTZ, respectively, confirming this conclusion.

TABLE 6.9
The Basis Set Dependence of the Mean Absolute Deviations of B3PW91-Calculated and B3PW91/cc-pVQZ Rotational Strengths of Modes 6–18 of Methyloxirane

Basis Set	Contracted Set[a]	No. Fns.[b]	MAD[c]
cc-pVQZ	[5s4p3d2f1g/4s3p2d1f] (5d,7f,9g)	400	0.00
aug-cc-pVTZ	[5s4p3d2f/4s3p2d] (5d,7f)	322	0.37
cc-pVTZ	[4s3p2d1f/3s2p1d] (5d,7f)	230	0.51
6-311++G(2d,2p)	[5s4p2d/4s2p] (5d)	168	0.66
TZ2P	[5s4p2d/3s2p] (6d)	170	0.87
6-311++G**	[5s4p1d/4s1p] (5d)	130	0.90
6-311G**	[4s3p1d/3s1p] (5d)	108	1.37
aug-cc-pVDZ	[4s3p2d/3s2p] (5d)	146	1.54
6-31G**	[3s2p1d/2s1p] (6d)	90	2.45
cc-pVDZ	[3s2p1d/2s1p] (5d)	86	2.60
6-31G*	[3s2p1d/2s] (6d)	72	2.85
6-31G	[3s2p/2s]	48	6.81

[a] Of the form [1/2] where the first term refers to the contracted set on carbon and oxygen and the second refers to the contracted set on hydrogen. The numbers in parentheses refer to the number of functions used for each type of polarization function.

[b] Number of basis functions.

[c] Mean absolute deviation relative to cc-pVQZ.

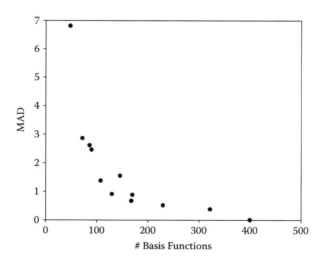

FIGURE 6.45   B3PW91 basis set dependence of methyloxirane rotational strengths.

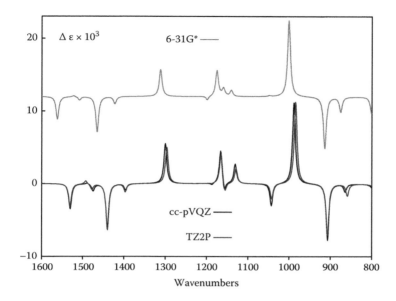

FIGURE 6.46   Comparison of the B3PW91/TZ2P and B3PW91/6-31G* VCD spectra of methyloxirane to the B3PW91/cc-pVQZ VCD spectrum.

CONCLUSIONS

The studies of the accuracies of 19 density functionals in predicting the VCD rotational strengths of camphor, α-pinene, and methyloxirane lead to conclusions regarding the optimum functionals to be used in predicting VCD spectra. For all three molecules, hybrid functionals are statistically more accurate than nonhybrid pure functionals. For all three molecules, the most accurate hybrid functionals include B3PW91 and B3P86. For camphor and α-pinene the hybrid B3LYP functional is less accurate than B3PW91, but predicts VCD spectra that are qualitatively similar. In predicting VCD spectra of other molecules, for optimum accuracy hybrid functionals should be used. It is important to note that the relative ordering of the different functionals, in terms of accuracy, varies for each molecule. Therefore, it is important to use more than one functional when predicting the VCD spectrum of a molecule.

The studies of the accuracies of 12 basis sets in predicting the VCD rotational strengths of camphor, α-pinene, and methyloxirane also lead to conclusions regarding the optimum basis sets to be used in predicting VCD spectra. For all three molecules, the four most accurate basis sets, relative to the large reference cc-pVQZ, are aug-cc-pVTZ, cc-pVTZ, 6-311++G(2d,2p), and TZ2P. For these molecules, the cc-pVTZ and TZ2P VCD spectra are quantitatively identical to the cc-pVQZ VCD spectra. Qualitatively, the 6-31G* VCD spectra are similar to the cc-pVTZ VCD spectra.

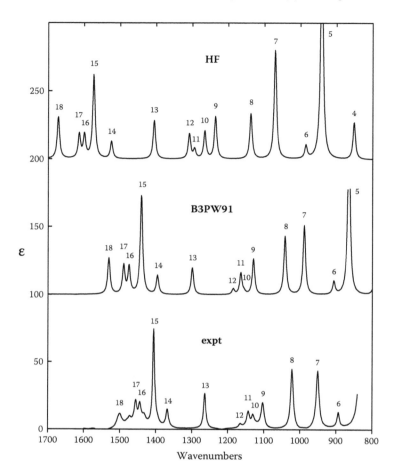

FIGURE 6.47 Comparison of the HF/cc-pVTZ and B3PW91/cc-pVTZ IR spectra of methyloxirane to the experimental IR spectrum of (–)-methyloxirane.

The comparisons of HF/cc-pVTZ and DFT, B3PW91/cc-pVTZ VCD rotational strengths of camphor, α-pinene, and methyloxirane, confirm that VCD rotational strengths calculated using DFT are more accurate than VCD rotational strengths calculated using HF theory.

DETERMINATION OF ABSOLUTE CONFIGURATION

To determine the absolute configuration (AC) of a chiral molecule, using VCD, the VCD spectra of the two enantiomers are calculated, using optimum density functionals and basis sets, and compared to the experimental VCD spectrum of one of the enantiomers, (+) or (–). The calculated VCD spectrum in best agreement

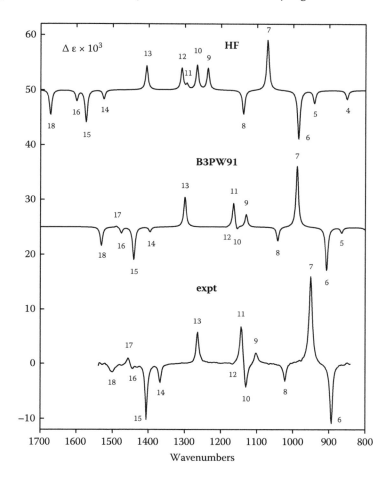

FIGURE 6.48 Comparison of the HF/cc-pVTZ and B3PW91/cc-pVTZ VCD spectra of (*S*)-methyloxirane to the experimental VCD spectrum of (–)-methyloxirane.

with the experimental VCD spectrum defines the optimum functional and basis set combination used, and also the AC of the experimental enantiomer. To further evaluate the reliability of the AC determined by comparison of the calculated and experimental VCD spectra, the experimental VCD spectrum is assigned, based on the optimum calculated VCD spectrum, and the experimental rotational strengths determined using Lorentzian fitting. The calculated rotational strengths of the two enantiomers are then compared to the experimental rotational strengths, as shown in Figure 6.50 for camphor. The enantiomer giving superior quantitative agreement with the experimental rotational strengths has the AC of the experimental enantiomer. The VCD of camphor proves that the AC of (+)-camphor is 1*R*,4*R*, and not 1*S*,4*S*.

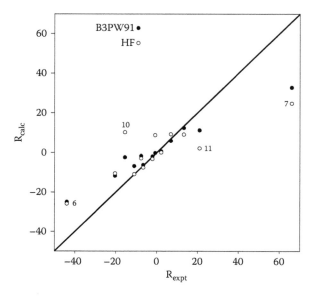

FIGURE 6.49 Comparison of the HF/cc-pVTZ and B3PW91/cc-pVTZ rotational strengths to the experimental rotational strengths of modes 6–18. Rotational strengths R are in 10^{-44} esu^2 cm^2.

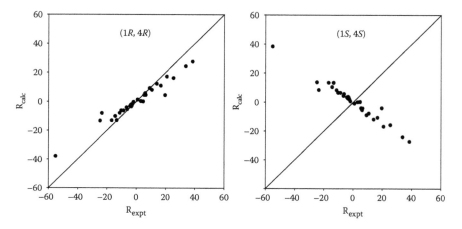

FIGURE 6.50 Comparison of the B3PW91/cc-pVTZ calculated rotational strengths for (1R,4R)- and (1S,4S)-camphor to the experimental rotational strengths of modes 22–59 of (+)-camphor. Rotational strengths R are in 10^{-44} esu^2 cm^2.

REFERENCES

1. P.J. Stephens and M.A. Lowe, Vibrational Circular Dichroism, *Ann. Rev. Phys. Chem.*, 36, 213–241, 1985.
2. M.A. Lowe, P.J. Stephens, and G.A. Segal, The Theory of Vibrational Circular Dichroism: Trans 1,2-Dideuteriocyclobutane and Propylene Oxide, *Chem. Phys. Lett.*, 123, 108–116, 1986.
3. P.J. Stephens, The Theory of Vibrational Optical Activity, in *Understanding Molecular Properties*, ed. J. Avery, J.P. Dahl, and A.E. Hansen, D. Reidel, 1987, pp. 333–342.
4. R.W. Kawiecki, F. Devlin, P.J. Stephens, R.D. Amos, and N.C. Handy, Vibrational Circular Dichroism of Propylene Oxide, *Chem. Phys. Lett.*, 145, 411–417, 1988.
5. P.J. Stephens, The *A Priori* Prediction of Vibrational Circular Dichroism Spectra: A New Approach to the Study of the Stereochemistry of Chiral Molecules, *Croat. Chem. Acta.*, 62, 429–440, 1989.
6. P.J. Stephens, Vibronic Interactions in the Electronic Ground State: Vibrational Circular Dichroism Spectroscopy, in *Vibronic Processes in Inorganic Chemistry*, ed. C.D. Flint, Kluwer, 1989, pp. 371–384.
7. F.J. Devlin, P.J. Stephens, and R.D. Amos, Vibrational Circular Dichroism of Propylene Oxide, R. W. Kawiecki, *J. Phys. Chem.*, 95, 9817–9831, 1991.
8. F.J. Devlin and P.J. Stephens, *Ab Initio* Calculation of Vibrational Circular Dichroism Spectra of Chiral Natural Products Using MP2 Force Fields: Camphor, *J. Am. Chem. Soc.*, 116, 5003–5004, 1994.
9. P.J. Stephens, C.F. Chabalowski, F.J. Devlin, and K.J. Jalkanen, *Ab Initio* Calculation of Vibrational Circular Dichroism Spectra Using Large Basis Set MP2 Force Fields, *Chem. Phys. Lett.*, 225, 247–257, 1994.
10. L. Alagna, S. Di Fonzo, T. Prosperi, S. Turchini, P. Lazzeretti, M. Malagoli, R. Zanasi, C.R. Natoli, and P.J. Stephens, Random Phase Approximation Calculations of K-Edge Rotational Strengths of Chiral Molecules: Propylene Oxide, *Chem. Phys. Lett.*, 223, 402–410, 1994.
11. P.J. Stephens, F.J. Devlin, C.S. Ashvar, C.F. Chabalowski, and M.J. Frisch, Theoretical Calculation of Vibrational Circular Dichroism Spectra, *Faraday Discuss.*, 99, 103–119, 1994.
12. F.J. Devlin, J.W. Finley, P.J. Stephens, and M.J. Frisch, *Ab Initio* Calculation of Vibrational Absorption and Circular Dichroism Spectra Using Density Functional Force Fields: A Comparison of Local, Non-Local and Hybrid Density Functionals, *J. Phys. Chem.*, 99, 16883–16902, 1995.
13. F.J. Devlin, P.J. Stephens, J.R. Cheeseman, and M.J. Frisch, Prediction of Vibrational Circular Dichroism Spectra Using Density Functional Theory: Camphor and Fenchone, *J. Am. Chem. Soc.*, 118, 6327–6328, 1996.
14. F.J. Devlin, P.J. Stephens, J.R. Cheeseman, and M.J. Frisch, *Ab Initio* Prediction of Vibrational Absorption and Circular Dichroism Spectra of Chiral Natural Products Using Density Functional Theory: Camphor and Fenchone, *J. Phys. Chem. A*, 101, 6322–6333, 1997.
15. F.J. Devlin, P.J. Stephens, J.R. Cheeseman, and M.J. Frisch, *Ab Initio* Prediction of Vibrational Absorption and Circular Dichroism Spectra of Chiral Natural Products Using Density Functional Theory: α-Pinene, *J. Phys. Chem. A*, 101, 9912–9924, 1997.
16. P.J. Stephens, Vibrational Circular Dichroism, Theory, in *Encyclopedia of Spectroscopy and Spectrometry*, Academic Press, London, 1999, pp. 2415–2421.
17. P.J. Stephens and F.J. Devlin, Determination of the Structure of Chiral Molecules Using *Ab Initio* Vibrational Circular Dichroism Spectroscopy, *Chirality*, 12, 172–179, 2000.

18. P.J. Stephens, F.J. Devlin, and A. Aamouche, Determination of the Structures of Chiral Molecules Using Vibrational Circular Dichroism Spectroscopy, *Chirality: Physical Chemistry*, ed. J.M. Hicks, Vol. 810, ACS Symposium Series, 2002, chap. 2, pp. 18–33.

19. P.J. Stephens, Vibrational Circular Dichroism Spectroscopy: A New Tool for the Stereochemical Characterization of Chiral Molecules, in *Computational Medicinal Chemistry for Drug Discovery*, ed. P. Bultinck, H. de Winter, W. Langenaecker, and J.P. Tollenaere, Marcel Dekker, New York, 2004, chap. 26, pp. 699–725.

20. PeakFit, 4th ed., Jandel Scientific Software, San Rafael, CA, 1995. (IBM, Armonk, NY, acquired SPSS Inc. in 2009.)

21. (a) D.W. Marquardt, An Algorithm for Least-Squares Estimation of Nonlinear Parameters, *J. Soc. Indust. Appl. Math.*, 11, 431–441, 1963; (b) K. Levenberg, A Method for the Solution of Certain Non-Linear Problems in Least Squares, *Quart. Appl. Math.*, 2, 164–168, 1944.

22. S.F. Sousa, P.A. Fernandes, and M.J. Ramos, General Performance of Density Functionals, *J. Phys. Chem. A*, 111, 10439–10452, 2007.

23. (a) P.J. Stephens and M.A. Lowe, Vibrational Circular Dichroism, *Ann. Rev. Phys. Chem.*, 36, 213–241, 1985; (b) M.A. Lowe, P.J. Stephens, and G.A. Segal, The Theory of Vibrational Circular Dichroism: Trans 1,2-Dideuteriocyclobutane and Propylene Oxide, *Chem. Phys. Lett.*, 123, 108–116, 1986; (c) P.J. Stephens, The Theory of Vibrational Optical Activity, in *Understanding Molecular Properties*, ed. J. Avery, J.P. Dahl, and A.E. Hansen, D. Reidel, 1987, pp. 333–342; (d) R.W. Kawiecki, F. Devlin, P.J. Stephens, R.D. Amos, and N.C. Handy, Vibrational Circular Dichroism of Propylene Oxide, *Chem. Phys. Lett.*, 145, 411–417, 1988; (e) P.J. Stephens, The *A Priori* Prediction of Vibrational Circular Dichroism Spectra: A New Approach to the Study of the Stereochemistry of Chiral Molecules, *Croat. Chem. Acta.*, 62, 429–440, 1989; (f) P.J. Stephens, Vibronic Interactions in the Electronic Ground State: Vibrational Circular Dichroism Spectroscopy, in *Vibronic Processes in Inorganic Chemistry*, ed. C.D. Flint, Kluwer, 1989, pp. 371–384; (g) R.W. Kawiecki, F.J. Devlin, P.J. Stephens, and R.D. Amos, Vibrational Circular Dichroism of Propylene Oxide, *J. Phys. Chem.*, 95, 9817–9831, 1991; (h) P.J. Stephens, C.F. Chabalowski, F.J. Devlin, and K.J. Jalkanen, *Ab Initio* Calculation of Vibrational Circular Dichroism Spectra Using Large Basis Set MP2 Force Fields, *Chem. Phys. Lett.*, 225, 247–257, 1994; (i) L. Alagna, S. Di Fonzo, T. Prosperi, S. Turchini, P. Lazzeretti, M. Malagoli, R. Zanasi, C.R. Natoli, and P.J. Stephens, Random Phase Approximation Calculations of K-Edge Rotational Strengths of Chiral Molecules: Propylene Oxide, *Chem. Phys. Lett.*, 223, 402–410, 1994; (j) F.J. Devlin, J.W. Finley, P.J. Stephens, and M.J. Frisch, *Ab Initio* Calculation of Vibrational Absorption and Circular Dichroism Spectra Using Density Functional Force Fields: A Comparison of Local, Non-Local and Hybrid Density Functionals, *J. Phys. Chem.*, 99, 16883–16902, 1995; (k) P.J. Stephens and F.J. Devlin, Determination of the Structure of Chiral Molecules Using *Ab Initio* Vibrational Circular Dichroism Spectroscopy, *Chirality*, 12, 172–179, 2000; (l) P.J. Stephens, F.J. Devlin, and A. Aamouche, Determination of the Structures of Chiral Molecules Using Vibrational Circular Dichroism Spectroscopy, *Chirality: Physical Chemistry*, ed. J.M. Hicks, Vol. 810, ACS Symposium Series, American Chemical Society, 2002, chap. 2, pp. 18–33; (m) P.J. Stephens, Vibrational Circular Dichroism Spectroscopy: A New Tool for the Stereochemical Characterization of Chiral Molecules, in *Computational Medicinal Chemistry for Drug Discovery*, ed. P. Bultinck, H. de Winter, W. Langenaecker, and J.P. Tollenaere, Marcel Dekker, New York, 2004, chap. 26, pp. 699–725.

7 Applications of VCD Spectroscopy to Organic Chemistry

INTRODUCTION

Since the calculation of vibrational rotational strengths, using *ab initio* density functional theory (DFT), became possible in the 1990s, we have applied vibrational circular dichroism (VCD) spectroscopy to the determinations of the stereochemistries of many chiral molecules. Thirty-four molecules studied are shown in Figure 7.1.

Most frequently, the purposes of VCD studies were the determinations of ACs. When the AC of a molecule had previously been assigned, VCD was used to determine the reliability of the assignment. In many cases, specifically molecules **1**, **2**, **3**, **4**, **5**, **7**, **9**, **10**, **11**, **12**, **14**, **16**, **21**, **26**, **28**, **29**, and **32**, the previously assigned ACs were confirmed to be correct. However, in several cases, specifically molecules **6**, **15**, **24**, and **25**, our VCD studies proved the previously assigned ACs to be incorrect. In several cases, molecules **8**, **13**, **18**, **19**, **20**, **22**, **23**, **27**, **30**, **31**, **33**, and **34**, the ACs of the molecules had not been determined, so our studies provided the first AC assignments.

In the cases of molecules that were obtained by asymmetric syntheses using reactions whose stereochemistries were not predictable, molecules **8**, **13**, **18**, **19**, **20**, **22**, **33**, and **34**, the purposes of the AC determinations were to assist the understanding of the reaction stereochemistries. In one case, molecule **14**, VCD was used to determine the structure of the product of a Baeyer-Villiger reaction. Another focus of our studies has been the determination of the ACs of natural products, molecules **1**, **21**, **23**, **24**, **25**, **27**, **30**, **31**, and **33**, and pharmacologically active synthetic organic molecules, **13** and **22**. In addition, VCD has been applied to the determination of the ACs of a chiral organometallic molecule, **29**, and an isotopically chiral molecule, **32**. In the case of molecule **17**, VCD was used to determine its supramolecular structure in solution. The majority of the molecules studied are conformationally flexible. For conformationally flexible molecules, VCD analysis determines the structures and relative populations of the conformers.

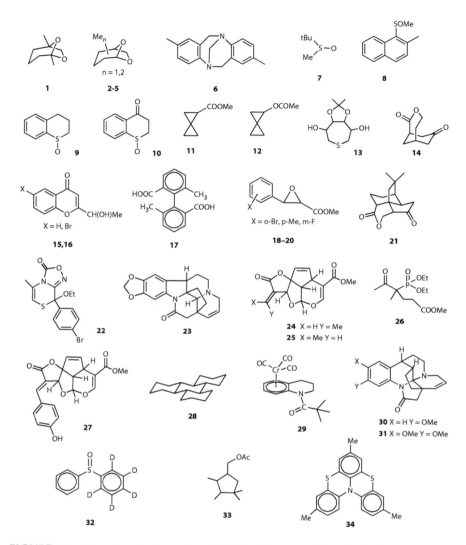

FIGURE 7.1 Chiral molecules whose VCD has been studied using DFT. **1**. Frontalin: 1,5-Dimethyl-6,8-dioxabicyclo [3.2.1] octane [1,2]. **2**. Exo-7-methyl-6,8-dioxa-bicyclo[3.2.1] octane [2]. **3**. Endo-7-methyl-6,8-dioxa-bicyclo[3.2.1]octane [2]. **4**. Exo-5,7-dimethyl-6,8-dioxa-bicyclo[3.2.1]octane [2]. **5**. Endo-5,7-dimethyl-6,8-dioxa-bicyclo[3.2.1]octane [2]. **6**. Tröger's base [3,4]. **7**. Tertiarybutyl-methyl-sulfoxide [5]. **8**. 1-(2-Methyl-naphthyl)-methyl-sulfoxide [6]. **9**. 1-Thiochroman-sulfoxide [7]. **10**. 1-Thiochromanone-sulfoxide [8]. **11**. Spiropentylcarboxylic acid methyl ester [9]. **12**. Spiropentylacetate [9]. **13**. 3,6-Dihydroxy-4,5-O-isopropylidene-thiepane [10]. **14**. 3-Oxabicyclo[4.3.1]decane-2,8-dione [11]. **15**. 2-(1-Hydroxyethyl)-chromen-4-one [12]. **16**. 6-Bromo-2-(1-hydroxyethyl)-chromen-4-one [12]. **17**. 2,2′-Dimethyl-biphenyl-6,6′-dicarboxylic acid [13]. **18**. Trans-o-bromo-phenyl-glycidic acid methyl ester [14]. **19**. Trans-p-methyl-phenyl-glycidic acid methyl ester [14]. **20**. Trans-m-fluoro-phenyl-glycidic acid methyl ester [14]. **21**. Quadrone [15]. **22**. 8-(4-Bromo-phenyl)-8-ethoxy-5-methyl-8H-[1,4]-thiazino-[3,4-c]

To further illustrate the power of VCD spectroscopy for determining the stereochemistries of chiral molecules, thirteen of these studies are now summarized in the sections that follow.

REFERENCES

1. C.S. Ashvar, P.J. Stephens, T. Eggimann, and H. Wieser, Vibrational Circular Dichroism Spectroscopy of Chiral Pheromones: Frontalin (1,5-Dimethyl-6,8- Dioxabicyclo [3.2.1] Octane), *Tet. Asymm.*, 9, 1107–1110, 1998.
2. C.S. Ashvar, F.J. Devlin, P.J. Stephens, K.L. Bak, T. Eggimann, and H. Wieser, Vibrational Absorption and Circular Dichroism of Mono- and Di-Methyl Derivatives of 6,8-Dioxabicyclo [3.2.1] Octane, *J. Phys. Chem. A*, 102, 6842–6857, 1998.
3. A. Aamouche, F.J. Devlin, and P.J. Stephens, Determination of Absolute Configuration Using Circular Dichroism: Tröger's Base Revisited Using Vibrational Circular Dichroism, *J. Chem. Soc. Chem. Comm.*, 361–362, 1999.
4. F.J. Devlin, and P.J. Stephens, Structure, Vibrational Absorption and Circular Dichroism Spectra and Absolute Configuration of Tröger's Base, A. Aamouche, *J. Am. Chem. Soc.*, 122, 2346–2354, 2000.
5. A. Aamouche, F.J. Devlin, P.J. Stephens, J. Drabowicz, B. Bujnicki, and M. Mikolajczyk, Vibrational Circular Dichroism and Absolute Configuration of Chiral Sulfoxides: tert-Butyl Methyl Sulfoxide, *Chem. Eur. J.*, 6, 4479–4486, 2000.
6. P.J. Stephens, A. Aamouche, F.J. Devlin, S. Superchi, M.I. Donnoli, and C. Rosini, Determination of Absolute Configuration Using Vibrational Circular Dichroism Spectroscopy: The Chiral Sulfoxide 1-(2-Methylnaphthyl) Methyl Sulfoxide, *J. Org. Chem.*, 66, 3671–3677, 2001.
7. F.J. Devlin, P.J. Stephens, P. Scafato, S. Superchi, and C. Rosini, Determination of Absolute Configuration Using Vibrational Circular Dichroism Spectroscopy: The Chiral Sulfoxide 1-Thiochroman S-Oxide, *Tet. Asymm.*, 12, 1551–1558, 2001.
8. F.J. Devlin, P.J. Stephens, P. Scafato, S. Superchi and C. Rosini, Determination of Absolute Configuration Using Vibrational Circular Dichroism Spectroscopy: The Chiral Sulfoxide 1-Thiochromanone S-Oxide, *Chirality*, 14, 400–406, 2002.
9. F.J. Devlin, P.J. Stephens, C. Österle, K.B. Wiberg, J.R. Cheeseman, and M.J. Frisch, Configurational and Conformational Analysis of Chiral Molecules Using IR and VCD Spectroscopies: Spiropentylcarboxylic Acid Methyl Ester and Spiropentyl Acetate, *J. Org. Chem.*, 67, 8090–8096, 2002.
10. V. Cerè, F. Peri, S. Pollicino, A. Ricci, F.J. Devlin, P.J. Stephens, F. Gasparrini, R. Rompietti, and C. Villani, Synthesis, Chromatographic Separation, VCD Spectroscopy and *Ab Initio* DFT Studies of Chiral Thiepane Tetraol Derivatives, *J. Org. Chem.*, 70, 664–669, 2005.
11. P.J. Stephens, D.M. McCann, F.J. Devlin, T.C. Flood, E. Butkus, S. Stoncius, and J.R. Cheeseman, Determination of Molecular Structure Using Vibrational Circular Dichroism (VCD) Spectroscopy: The Keto-Lactone Product of Baeyer-Villiger Oxidation of (+)-(1*R*,5*S*)-Bicyclo[3.3.1]Nonane-2,7-Dione, *J. Org. Chem.*, 70, 3903–3913, 2005.

FIGURE 7.1 (Opposite) [1,2,4]oxadiazol-3-one [16,17]. **23**. Schizozygine [18]. **24**. Plumericin [19]. **25**. Iso-plumericin [19]. **26**. Diethyl [2-(2methoxycarbonyl-ethyl)-3-oxo-but-2-yl]-phosphonate [20]. **27**. Prismatomerin [21,22]. **28**. Anti-trans-anti-trans-anti-trans-perhydrotriphenylene [23]. **29**. Tricarbonyl-η⁶-N-pivaloyl-tetrahydroquinoline-chromium(0) [24]. **30**. Isoschizogaline [25]. **31**. Isoschizogamine [25]. **32**. Phenyl, perdeuteriophenyl, sulfoxide [26]. **33**. Mealybug sex pheromone [27]. **34**. Heterohelicene [28].

12. F.J. Devlin, P.J. Stephens, and P. Besse, Are The Absolute Configurations of 2-(1-Hydroxyethyl)-Chromen-4-One and Its 6-Bromo Derivative Determined by X-Ray Crystallography Correct? A Vibrational Circular Dichroism Study of Their Acetate Derivatives, *Tet. Asymm.*, 16, 1557–1566, 2005.

13. M. Urbanová, V. Setnička, F.J. Devlin, and P.J. Stephens, Determination of Molecular Structure in Solution Using Vibrational Circular Dichroism Spectroscopy: The Supramolecular Tetramer of S-2,2'-Dimethyl-Biphenyl-6,6'-Dicarboxylic Acid, *J. Am. Chem. Soc.*, 127, 6700–6711, 2005.

14. F.J. Devlin, P.J. Stephens, and O. Bortolini, Determination of Absolute Configuration Using Vibrational Circular Dichroism Spectroscopy: Phenyl Glycidic Acid Derivatives Obtained via Asymmetric Epoxidation Using Oxone and a Keto Bile Acid, *Tet. Asymm.*, 16, 2653–2663, 2005.

15. P.J. Stephens, D.M. McCann, F.J. Devlin, and A.B. Smith III, Determination of the Absolute Configurations of Natural Products Via Density Functional Theory Calculations of Optical Rotation, Electronic Circular Dichroism and Vibrational Circular Dichroism: The Cytotoxic Sesquiterpene Natural Products Quadrone, Suberosenone, Suberosanone and Suberosenol A Acetate, *J. Nat. Prod.*, 69, 1055–1064, 2006.

16. E. Carosati, G. Cruciani, A. Chiarini, R. Budriesi, P. Ioan, R. Spisani, D. Spinelli, B. Cosimelli, F. Fusi, M. Frosini, R. Matucci, F. Gasparrini, A. Ciogli, P.J. Stephens, and F.J. Devlin, Calcium Channel Antagonists Discovered by a Multidisciplinary Approach, *J. Med. Chem.*, 49, 5206–5216, 2006.

17. P.J. Stephens, F.J. Devlin, F. Gasparrini, A. Ciogli, D. Spinelli, and B. Cosimelli, Determination of the Absolute Configuration of a Chiral Oxadiazol-3-One Calcium Channel Blocker, Resolved Using Chiral Chromatography, via Concerted Density Functional Theory Calculations of Its Vibrational Circular Dichroism, Electronic Circular Dichroism and Optical Rotation, *J. Org. Chem.*, 72, 4707–4715, 2007.

18. P.J. Stephens, J.J. Pan, F.J. Devlin, M. Urbanová, and J. Hájíček, Determination of the Absolute Configurations of Natural Products via Density Functional Theory Calculations of Vibrational Circular Dichroism, Electronic Circular Dichroism and Optical Rotation: The Schizozygane Alkaloid Schizozygine, *J. Org. Chem.*, 72, 2508–2524, 2007.

19. P.J. Stephens, J.J. Pan, F.J. Devlin, K. Krohn, and T. Kurtán, Determination of the Absolute Configurations of Natural Products via Density Functional Theory Calculations of Vibrational Circular Dichroism, Electronic Circular Dichroism and Optical Rotation: The Iridoids Plumericin and Iso-Plumericin, *J. Org. Chem.*, 72, 3521–3536, 2007.

20. S. Delarue-Cochin, J.J. Pan, A. Dauteloup, F. Hendra, R.G. Angoh, D. Joseph, P.J. Stephens, and C. Cavé, Asymmetric Michael Reaction: novel Efficient Access to Chiral β-Ketophosphonates, *Tet. Asymm.*, 18, 685–691, 2007.

21. K. Krohn, D. Gehle, S.K. Dey, N. Nahar, M. Mosihuzzaman, N. Sultana, M.H. Sohrab, P.J. Stephens, J.J. Pan, and F. Sasse, Prismatomerin, a New Iridoid from *Prismatomeris tetrandra*. Structure Elucidation, Determination of Absolute Configuration and Cytotoxicity, *J. Nat. Prod.*, 70, 1339–1343, 2007.

22. P.J. Stephens, J.J. Pan, and K. Krohn, Determination of the Absolute Configurations of Pharmacological Natural Products via Density Functional Theory Calculations of Vibrational Circular Dichroism: The New, Cytotoxic, Iridoid Prismatomerin, *J. Org. Chem.*, 72, 7641–7649, 2007.

23. P.J. Stephens, F.J. Devlin, S. Schürch, and J. Hulliger, Determination of the Absolute Configuration of Chiral Molecules via Density Functional Theory Calculations of Vibrational Circular Dichroism and Optical Rotation: The Chiral Alkane D_3-anti-trans-anti-trans-anti-trans-Perhydro Triphenylene, *Theor. Chem. Acc.*, 119, 19–28, 2008.

24. P.J. Stephens, F.J. Devlin, C. Villani, F. Gasparrini, and S.L. Mortera, Determination of the Absolute Configurations of Chiral Organometallic Complexes via Density Functional Theory Calculations of Their Vibrational Circular Dichroism Spectra: The Chiral Chromium Tricarbonyl Complex of N-Pivaloyl-Tetrahydroquinoline, *Inorg. Chim. Acta*, 361, 987–999, 2008.
25. P.J. Stephens, J.J. Pan, F.J. Devlin, M. Urbanova, O. Julinek, and J. Hájíček, Determination of the Absolute Configurations of Natural Products via Density Functional Theory Calculations of Vibrational Circular Dichroism, Electronic Circular Dichroism and Optical Rotation: The Isoschizozygane Alkaloids Isoschizogaline and Isoschizogamine, *Chirality*, 20, 454–470, 2008.
26. J. Drabowicz, A. Zajac, P. Lyzwa, P.J. Stephens, J.J. Pan, and F.J. Devlin, Determination of the Absolute Configurations of Isotopically Chiral Molecules Using Vibrational Circular Dichroism (VCD) Spectroscopy: The Isotopically Chiral Sulfoxide Perdeuteriophenyl-Phenyl-Sulfoxide, *Tet. Asymm.*, 19, 288–294, 2008.
27. B. Figadère, F.J. Devlin, J.G. Millar, and P.J. Stephens, Determination of the Absolute Configuration of the Sex Pheromone of the Obscure Mealybug by Vibrational Circular Dichroism Analysis, *J. Chem. Soc. Chem. Comm.*, 1106–1108, 2008.
28. G. Lamanna, C. Faggi, F. Gasparrini, A. Ciogli, C. Villani, P.J. Stephens, F.J. Devlin, and S. Menichetti, Efficient Thia-Bridged Triarylamine Heterohelicenes: Synthesis, Resolution and Absolute Configuration Determination, *Chem. Eur. J.*, 14, 5747–5750, 2008.

TRÖGER'S BASE

The heterocyclic amine, **1**, was first synthesized by Tröger in 1887 [1] and subsequently named Tröger's base. **1** contains two stereogenic N atoms of identical chirality and therefore has two enantiomers:

1

(R,R)-**1**   (S,S)-**1**

In 1944, **1** was resolved for the first time by Prelog and Wieland [2] using chromatography on a lactose column. In 1967, the AC of **1** was assigned for the first time by Mason et al. [3], using the electronic CD (ECD) spectrum, together with the coupled oscillator theory (also known as the exciton coupling theory [4]).

In 1991, Wilen et al. crystallized a salt of **1** containing monoprotonated (+)-**1** and the monoanion of (–)-1,1′-binaphthalene-2,2′-diyl hydrogen phosphate, **2**, and determined its structure using x-ray crystallography [5]. Given the AC $R(-)$ of **2**, the crystal structure of the salt led to the conclusion that the AC of (+)-**1** is S,S. This AC was opposite to that assigned by Mason et al. [3]. In 1999, given the commercial availability of (+)-**1**, (–)-**1**, and (±)-**1**, we redetermined the AC of **1** using VCD, in order to resolve the uncertainty then existing regarding its AC [6]. We confirmed that the AC of **1** is $R,R(-)/S,S(+)$, the AC assigned by Wilen et al. [5]. Here, we present new experimental data and new calculations, which further confirm that the AC of **1** is $R,R(-)/S,S(+)$.

The enantiomers and racemate of **1** were obtained from Aldrich. The enantiomeric excesses of both enantiomers were ≥99.5, determined by HPLC. The IR and VCD spectra of **1** were measured using 0.085 M solutions in CCl_4, and cells of pathlengths 546, 236, and 113 μ. The IR spectra of (+), (–), and (±)-**1** were quantitatively identical, proving that the samples were of identical purities. The baseline of the VCD spectra of (+) and (–)-**1** was the VCD spectrum of (±)-**1**. The IR and VCD spectra of (+)-**1** over the frequency range 817–1,650 cm^{-1} are shown in Figure 7.2. The VCD spectrum is the half-difference spectrum, $\frac{1}{2}[\Delta\varepsilon(+) - \Delta\varepsilon(-)]$.

As discussed in Chapter 5, Tröger's base is predicted to be conformationally rigid. The B3PW91/cc-pVTZ and B3LYP/cc-pVTZ equilibrium geometries were obtained by reoptimization of the B3LYP/6-31G* geometry obtained in Chapter 5. The B3PW91/cc-pVTZ and B3LYP/cc-pVTZ harmonic frequencies, dipole strengths, and rotational strengths obtained for the fundamental transitions of modes 34–87 of (S,S)-**1** are given in Table 7.1. The B3PW91/cc-pVTZ and B3LYP/cc-pVTZ IR spectra, calculated using Lorentzian bandshapes ($\gamma = 4.0$ cm^{-1}), are compared to the experimental IR spectrum in Figure 7.3. The B3PW91 and B3LYP IR spectra are qualitatively extremely similar; the largest difference is for modes 40–44. Comparison to the experimental IR spectrum shows that the B3PW91 spectrum of modes 40–44 is in better agreement with the experimental spectrum. The assignment of the experimental IR spectrum, shown in Figure 7.3, is therefore based on the B3PW91/cc-pVTZ spectrum.

The Lorentzian fit of the experimental IR spectrum is shown in Figure 7.4. The experimental frequencies and dipole strengths obtained are given in Table 7.1, and compared to the B3PW91/cc-pVTZ and B3LYP/cc-pVTZ frequencies and dipole strengths in Figures 7.5 and 7.6. As usual, the calculated frequencies are 0–4% greater than the experimental frequencies, due to the absence of anharmonicity. The B3PW91 dipole strengths and experimental dipole strengths are in good agreement, supporting the reliability of the assignment of the experimental IR spectrum.

The B3PW91/cc-pVTZ and B3LYP/cc-pVTZ VCD spectra, calculated using Lorentzian bandshapes ($\gamma = 4.0$ cm^{-1}), are compared to the experimental VCD spectrum in Figure 7.7. The B3PW91 and B3LYP VCD spectra are qualitatively extremely similar. However, the VCD of mode 50 is positive and negative for B3PW91 and B3LYP; in the experimental VCD spectrum mode 50 is positive. The assignment of the experimental VCD spectrum, shown in Figure 7.7, is therefore based on the B3PW91/cc-pVTZ spectrum.

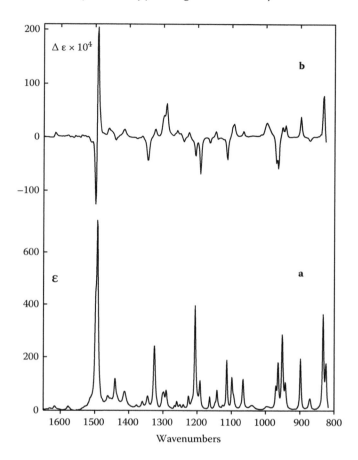

FIGURE 7.2   (a) IR spectrum of (+)-**1**, pathlength 236 μ. (b) VCD spectrum of (+)-**1**, 546 μ pathlength for ranges 1,800–1,575 cm⁻¹, 1,481–1,213 cm⁻¹, 1,186–975 cm⁻¹, and 960–840 cm⁻¹; 236 μ pathlength for ranges 1,575–1,504 cm⁻¹, 1,213–1,186 cm⁻¹, and 975–960 cm⁻¹; 113 μ pathlength for ranges 1,504–1,481 cm⁻¹ and 840–825 cm⁻¹.

The Lorentzian fit of the experimental VCD spectrum is shown in Figure 7.8. The experimental frequencies and rotational strengths obtained are given in Table 7.1. The experimental rotational strengths are compared to the B3PW91/cc-pVTZ and B3LYP/cc-pVTZ rotational strengths in Figure 7.9. The agreement of the calculated and experimental rotational strengths is very good. The B3PW91 rotational strengths are in better agreement than the B3LYP rotational strengths.

The VCD experimental frequencies are in excellent agreement with the IR experimental frequencies, proving that the assignments of the experimental IR and VCD spectra are consistent. The excellent agreement of the B3PW91/cc-pVTZ VCD spectrum of (*S,S*)-**1** and the experimental VCD spectrum of (+)-**1**, and of the B3PW91/cc-pVTZ rotational strengths of (*S,S*)-**1** and the experimental rotational strengths of (+)-**1**, proves that the AC of **1** is (*S,S*)-(+), and that the AC determined by Mason et al. [3] using ECD was incorrect.

TABLE 7.1

Calculated and Experimental Frequencies, Dipole Strengths, and Rotational Strengths for 1[a]

	Calculation[b]						Experiment[c]					
	B3LYP			B3PW91			IR			VCD		
Mode	ν	D	R	ν	D	R	ν	D	γ	ν	R	γ
87	1,657	0.1	−3.5	1,670	0.3	−6.8	1,617	8.4	3.0	1,615	10.5	2.8
86	1,655	6.3	15.6	1,668	8.1	19.1						
85	1,610	1.4	8.4	1,624	2.1	9.7	1,578	10.3	4.1	1,579	3.7	2.8
84	1,608	9.9	−8.9	1,621	12.5	−8.9						
83	1,535	127.9	−143.0	1,540	139.4	−169.0	1,498	129.7	2.5	1,498	−219.1	2.5
82	1,529	346.9	192.4	1,533	388.9	226.9	1,492	390.9	3.0	1,492	339.2	2.9
81	1,507	3.3	−3.5	1,497	4.6	−17.9	1,461	29.9	3.9	1,458	32.5	4.3
80	1,501	49.6	40.2	1,494	53.9	43.7						
79	1,500	0.2	5.4	1,490	5.9	22.1	1,452	20.1	4.1	1,448	15.4	4.3
78	1,488	17.7	9.2	1,478	18.8	8.4	1,440	77.8	3.7			
77	1,488	10.8	−10.4	1478	13.1	−9.6				1,439	−12.2	2.7
76	1,483	38.1	−9.9	1471	41.9	−11.5	1,431	21.8	4.0			
75	1,482	2.4	−2.3	1,469	2.7	−2.2						
74	1,447	8.6	−1.4	1,449	9.5	3.6	1,415	32.2	4.1	1,415	21.6	3.4
73	1,443	80.7	33.0	1,445	84.5	22.9	1,409	39.9	4.4	1,411	10.0	3.2
72	1,417	0.6	−0.1	1,406	0.9	0.1	1,378	13.2	4.1	1,377	−6.5	3.9
71	1,417	0.3	−0.2	1,406	0.3	−0.4						
70	1,392	16.3	5.0	1,388	9.4	4.7	1,361	20.6	3.6	1,361	3.6	1.9
69	1,375	23.7	−47.4	1,383	26.0	−52.5	1,345	32.5	3.4	1,345	−96.2	3.8
68	1,358	131.2	−5.0	1,363	128.6	14.6	1,325	165.7	3.1	1,324	28.2	3.3
67	1,356	4.7	12.4	1,356	0.5	3.9						
66	1,332	48.6	40.8	1,344	84.3	23.6	1,302	18.9	2.5	1,299	71.8	3.7
65	1,327	3.3	5.9	1,336	8.4	12.4	1,298	32.7	3.2			
64	1,320	41.3	68.3	1,321	43.5	81.5	1,291	42.8	3.1	1,290	115.5	3.5
63	1,296	1.1	2.5	1,295	4.3	10.1	1,266	4.5	2.4	1,265	5.5	2.7
62	1,290	10.2	10.0	1,288	4.4	4.9	1,259	18.4	3.1	1,259	18.4	3.0
61	1,274	9.2	10.0	1,279	7.2	2.1	1,250	11.2	3.1	1,250	10.2	3.0
60	1,257	12.9	−13.2	1,265	1.2	−4.8	1,240	11.6	3.1	1,240	−16.0	2.8
59	1,246	10.2	4.3	1,250	12.1	10.8	1,225	23.9	2.1	1,225	15.5	2.6
58	1,230	317.8	−36.6	1,236	280.6	−74.2	1,206	225.7	2.4	1,206	−53.2	2.5

TABLE 7.1 (CONTINUED)
Calculated and Experimental Frequencies, Dipole Strengths, and Rotational Strengths for 1[a]

	Calculation[b]						Experiment[c]					
	B3LYP			B3PW91			IR			VCD		
Mode	ν	D	R	ν	D	R	ν	D	γ	ν	R	γ
57	1,215	37.4	−88.2	1,219	33.9	−86.9	1,192	69.1	3.0	1,192	−120.5	2.9
56	1,184	19.8	−1.8	1,187	22.3	−1.6	1,164	27.0	2.5	1,164	−16.6	2.3
55	1,173	1.0	6.7	1,172	5.7	14.2	1,148	11.9	3.2	1,147	23.4	3.0
54	1,169	24.6	−1.1	1,167	34.8	−6.6	1,142	45.9	2.6	1,141	−9.9	2.7
53	1,133	84.6	−76.4	1,137	88.6	−62.9	1,114	90.8	1.8	1,113	−72.5	2.5
52	1,114	47.1	40.9	1,126	5.0	22.0	1,099	61.6	2.0	1,098	22.4	2.1
51	1,112	16.5	36.3	1,124	58.3	42.5	1,093	28.1	2.3	1,093	50.1	3.2
50	1,083	59.2	−11.3	1,087	41.6	9.7	1,066	81.0	2.6	1,066	19.6	2.7
49	1,066	3.9	1.9	1,058	6.0	2.0	1,044	16.4	6.6 ⎤			
									⎬	1,039	10.6	6.9
48	1,065	29.4	20.1	1,058	33.6	12.8	1,037	21.4	6.4 ⎦			
47	1,031	0.5	2.0	1,034	1.4	0.7				1,014	7.1	5.6
46	1,023	1.1	−10.4	1,019	0.0	0.7	997	13.2	3.9	998	95.8	5.2
45	1,017	29.5	85.5	1,013	27.5	85.8	990	10.7	4.3	989	49.2	5.8
44	988	44.1	−44.0	988	39.8	−51.7	970	45.4	2.1	970	−84.2	2.6
43	979	81.5	−57.3	984	74.0	−55.0	964	104.6	2.0	964	−105.3	2.5
42	976	8.2	−0.6	977	158.9	−6.8 ⎤						
						⎬	952	172.5	2.0	952	43.8	2.8
41	975	60.1	−18.2	976	1.0	1.2 ⎦						
40	959	207.6	84.2	966	114.5	67.6	943	78.0	2.8	943	47.8	2.9
39	931	0.0	0.0	935	0.1	2.1						
38	911	151.3	65.0	920	101.1	53.9	899	120.9	1.9	898	78.5	2.5
37	896	4.7	−4.4	894	4.3	−4.7 ⎤						
						⎬	871	52.8	3.7	872	−24.3	3.6
36	893	10.5	−13.7	891	11.3	−12.8 ⎦						
35	852	212.4	119.0	851	229.9	129.0	832	265.5	2.1	832	176.9	2.4
34	848	68.5	−12.2	846	72.5	−16.7	824	120.2	2.1			

[a] Frequencies ν and bandwidths γ in cm^{-1}; dipole strengths D in 10^{-40} esu^2 cm^2; rotational strengths R in 10^{-44} esu^2 cm^2. Experimental rotational strengths are for (+)-1; calculated rotational strengths are for (S,S)-1.

[b] Using the cc-pVTZ basis set.

[c] From Lorentzian fitting of the IR and VCD spectra.

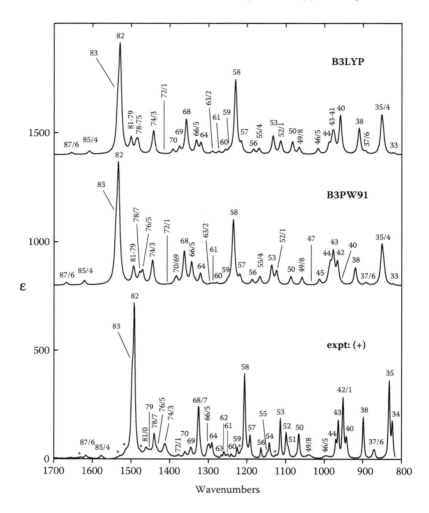

FIGURE 7.3   Comparison of the B3PW91/cc-pVTZ and B3LYP/cc-pVTZ IR spectra of **1** to the experimental IR spectrum of (+)-**1**. Assignment of the latter is based on the B3PW91/cc-pVTZ spectrum. Asterisks denote bands not assigned as fundamentals of **1**.

In order to extend the studies of the functional dependence of the accuracy of DFT rotational strengths, discussed in Chapter 6, we have compared the mean absolute deviations of rotational strengths, calculated for (*S,S*)-**1** using the 19 functionals used in Chapter 6 and the cc-pVTZ basis set, and the experimental rotational strengths of (+)-**1**. The results are given in Table 7.2 and Figure 7.10. Two hybrid functionals, B3P86 and HSE1PBE, give slightly more accurate rotational strengths than B3PW91. Most of the functionals, except for BLYP, BMK, M05-2X, CAMB3LYP, OLYP, and HCTH, are more accurate than B3LYP. The number of modes for which the calculated rotational strengths are opposite in sign to the experimental rotational strengths

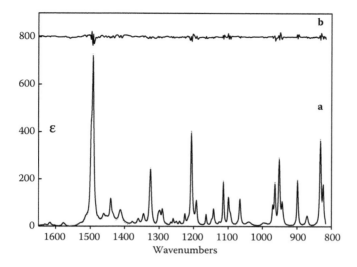

FIGURE 7.4 (a) Lorentzian fit of the experimental IR spectrum of (+)-**1** over the range 1,650–816 cm^{-1}. The solid line is the experimental spectrum; the dotted line is the Lorentzian fit. (b) The difference spectrum: the experimental spectrum minus the Lorentzian fit.

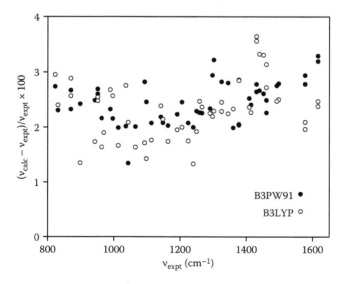

FIGURE 7.5 Comparison of the B3PW91/cc-pVTZ and B3LYP/cc-pVTZ frequencies to the experimental IR frequencies of modes 34–87.

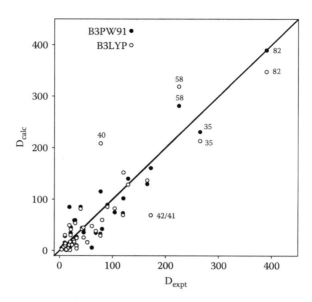

FIGURE 7.6   Comparison of the B3PW91/cc-pVTZ and B3LYP/cc-pVTZ dipole strengths to the experimental dipole strengths of modes 34–87. For bands assigned to multiple vibrational modes, calculated dipole strengths are the sums of the dipole strengths of contributing modes. Dipole strengths D are in 10^{-40} esu^2 cm^2.

varies from 0 to 9. For B3PW91 and B3LYP the numbers of incorrect signs are 1 and 5. The accuracies of the pure functionals BP86, BPW91, OLYP, and BLYP are substantially lower than the accuracies of the related hybrid functionals B3P86, B3PW91, O3LYP, and B3LYP, supporting the expectation that hybrid functionals are more accurate than pure functionals.

To evaluate the accuracies of other basis sets, relative to cc-pVTZ and cc-pVQZ, calculations have been repeated using the 10 additional basis sets listed in Table 7.3, and the B3PW91 functional. The values of the mean absolute deviations of the calculated rotational strengths from the cc-pVQZ rotational strengths of modes 35–87 are given in Table 7.3. In Figure 7.11 the dependence of the mean absolute deviation on the number of basis functions is plotted. As the number of basis functions diminishes, the mean absolute deviation increases substantially.

The basis sets TZ2P and 6-31G* are frequently used in calculating VCD spectra. The B3PW91/TZ2P and B3PW91/6-31G* VCD spectra of (S,S)-**1** are compared to the B3PW91/cc-pVQZ VCD spectrum in Figure 7.12. The TZ2P VCD spectrum is quantitatively very nearly identical to the cc-pVQZ VCD spectrum, proving that the basis set error of TZ2P is not significantly greater than the basis set error of cc-pVTZ. The 6-31G* VCD spectrum is both quantitatively and qualitatively very different from the cc-pVQZ and TZ2P VCD spectra. The qualitative difference is greatest in the frequency range 1,300–1,400 cm^{-1}. The

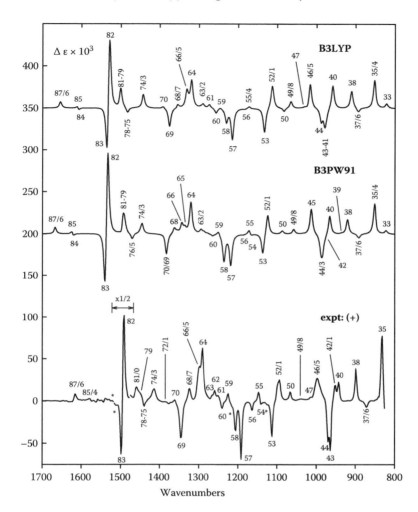

FIGURE 7.7 Comparison of the B3PW91/cc-pVTZ and B3LYP/cc-pVTZ VCD spectra of (*S,S*)-**1** to the experimental VCD spectrum of (+)-**1**. The assignment of the experimental spectrum is based on the B3PW91/cc-pVTZ spectrum. Asterisks denote bands not assigned as fundamentals of **1**.

6-31G* basis set error is therefore much greater than the basis set errors of TZ2P and cc-pVTZ.

To confirm that DFT is more accurate than the Hartree-Fock (HF) theory, the IR and VCD spectra have been calculated using HF theory and the cc-pVTZ basis set. The spectra obtained are compared to the experimental spectra in Figures 7.13 and 7.14. The experimental rotational strengths are compared to the HF/cc-pVTZ and B3PW91/cc-pVTZ rotational strengths in Figure 7.15. The B3PW91/cc-pVTZ

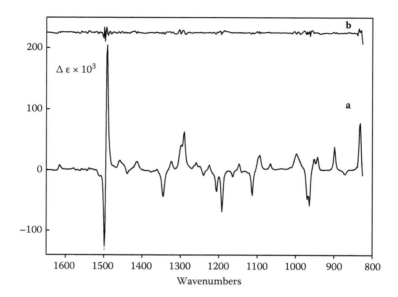

FIGURE 7.8 (a) Lorentzian fit of the experimental VCD spectrum of (+)-**1** over the range 1,650–825 cm⁻¹. The solid line is the experimental spectrum; the dotted line is the Lorentzian fit. (b) The difference spectrum: the experimental spectrum minus the Lorentzian fit.

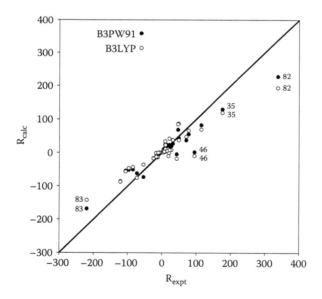

FIGURE 7.9 Comparison of the B3PW91/cc-pVTZ and B3LYP/cc-pVTZ rotational strengths to the experimental rotational strengths of modes 35–87. For bands assigned to multiple vibrational modes, calculated rotational strengths are the sums of the rotational strengths of contributing modes. Rotational strengths R are in 10^{-44} esu^{2}cm^2.

TABLE 7.2
The Functional Dependence of the Mean Absolute Deviations of cc-pVTZ-Calculated and Experimental Rotational Strengths of Modes 35–87 of 1

Functional	Type[a]	MAD[b]	No. Wrong Signs
BLYP	GGA	33.20	9
BMK	HM-GGA	32.68	8
M05-2X	HM-GGA	29.91	6
CAMB3LYP	H-X	27.45	3
OLYP	GGA	26.80	4
HCTH	GGA	26.56	6
B3LYP	H-GGA	26.14	5
B1B95	HM-GGA	26.06	8
BB95	M-GGA	24.85	9
PBE1PBE	H-GGA	23.55	6
BPW91	GGA	23.46	4
BP86	GGA	23.39	4
τHCTHHYB	HM-GGA	22.99	2
τHCTH	M-GGA	22.43	5
O3LYP	H-GGA	22.19	1
MPW1PW91	H-GGA	22.06	2
B3PW91	H-GGA	21.75	1
B3P86	H-GGA	20.65	0
HSE1PBE	H-S	20.12	2

[a] GGA, generalized gradient approximation; M-GGA, meta-generalized gradient approximation; H-GGA, hybrid generalized gradient approximation; HM-GGA, hybrid meta-generalized gradient approximation; H-X, range-separated hybrid (with short-range DFT exchange and long-range HF exchange); H-S, range-separated hybrid (short-range HF exchange and long-range DFT exchange).

[b] Mean absolute deviation relative to experiment in 10^{-44} esu^2 cm^2.

rotational strengths are clearly in better agreement with the experimental rotational strengths than are the HF/cc-pVTZ rotational strengths, especially for modes 42/41, 43, 57, 58, 59, and 82. The mean absolute deviations between calculated and experimental rotational strengths of modes 35–87 are 21.8 and 47.8 $\times$ 10^{-44} esu^{2}cm^2 for B3PW91/cc-pVTZ and HF/cc-pVTZ, respectively, confirming this conclusion.

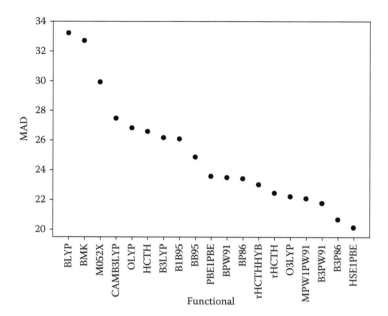

FIGURE 7.10 Functional dependence of the mean absolute deviation (MAD) of the rotational strengths of (S,S)-**1** calculated using the cc-pVTZ basis set.

TABLE 7.3
The Basis Set Dependence of the Mean Absolute Deviations of B3PW91-Calculated and B3PW91/cc-pVQZ Rotational Strengths of Modes 35–87 of (S,S)-**1**

Basis Set	Contracted Set[a]	No. Fns.[b]	MAD[c]
cc-pVQZ	[5s4p3d2f1g/4s3p2d1f] (5d,7f,9g)	1,585	0.00
cc-pVTZ	[4s3p2d1f/3s2p1d] (5d,7f)	822	1.11
aug-cc-pVTZ	[5s4p3d2f/4s3p2d] (5d,7f)	1,288	2.86
6-311++G(2d,2p)	[5s4p2d/4s2p] (5d)	693	2.88
TZ2P	[5s4p2d/3s2p] (6d)	713	4.61
6-311++G**	[5s4p1d/4s1p] (5d)	544	5.75
6-311G**	[4s3p1d/3s1p] (5d)	450	6.21
6-31G**	[3s2p1d/2s1p] (6d)	375	9.27
6-31G*	[3s2p1d/2s] (6d)	321	9.59
aug-cc-pVDZ	[4s3p2d/3s2p] (5d)	599	11.53
cc-pVDZ	[3s2p1d/2s1p] (5d)	356	11.80
6-31G	[3s2p/2s]	207	16.11

[a] Of the form [1/2] where the first term refers to the contracted set on carbon and nitrogen and the second refers to the contracted set on hydrogen. The numbers in parentheses refer to the number of functions used for each type of polarization function.

[b] Number of basis functions.

[c] Mean absolute deviation relative to cc-pVQZ for modes 35–87.

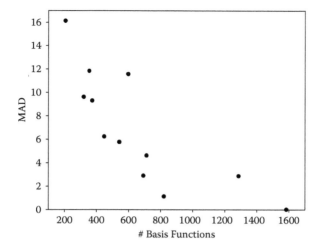

FIGURE 7.11   B3PW91 basis set dependence of (S,S)-**1** rotational strengths.

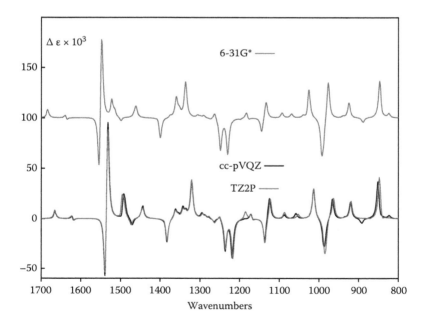

FIGURE 7.12   Comparison of the B3PW91/TZ2P and B3PW91/6-31G* VCD spectra of (S,S)-**1** to the B3PW91/cc-pVQZ VCD spectrum.

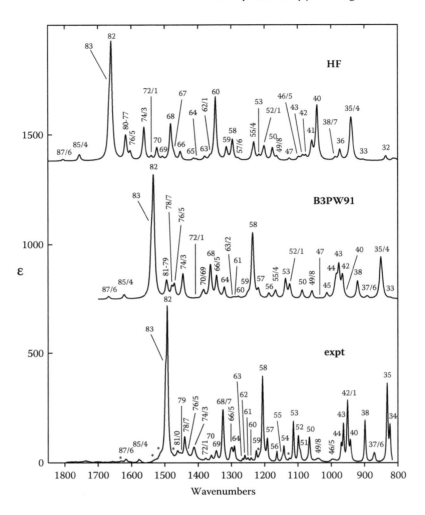

FIGURE 7.13   Comparison of the HF/cc-pVTZ and B3PW91/cc-pVTZ IR spectra of (S,S)-**1** to the experimental IR spectrum of (+)-**1**.

The AC of **1** obtained from the new experimental VCD spectrum and the new DFT calculations using the large cc-pVTZ basis set confirms that the AC of **1**, previously determined [6] using a less reliable FT-VCD instrument and DFT calculations using the smaller basis set 6-31G*, was correct.

The determination of the AC of **1** using VCD strongly supports the utility of VCD spectroscopy in determining the ACs of chiral organic molecules.

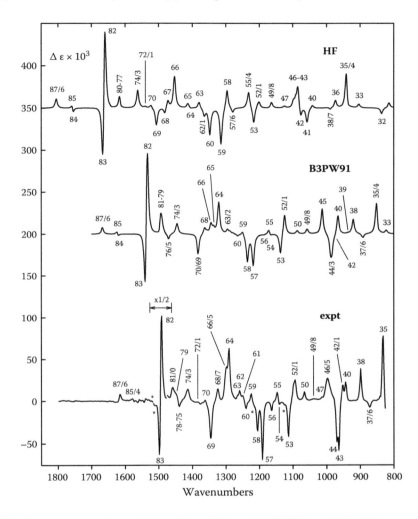

FIGURE 7.14 Comparison of the HF/cc-pVTZ and B3PW91/cc-pVTZ VCD spectra of (S,S)-**1** to the experimental VCD spectrum of (+)-**1**.

REFERENCES

1. J. Tröger, Über einige mittelst nascirenden formaldehydes entstehende basen, *J. Prakt. Chem.*, 36, 225–245, 1887.
2. V. Prelog, P. Wieland, Über die Spaltung der Tröger'schen Base in optische Antipoden, ein Beitrag zur Stereochemie des dreiwertigen Stickstoffs, *Helv. Chim. Acta*, 27, 1127–1134, 1944.
3. S.F. Mason, G.W. Vane, K. Schofield, R.J. Wells, J.S. Whitehurst, The circular dichroism and absolute configuration of Tröger's base, *J. Chem. Soc. B*, 553–556, 1967.

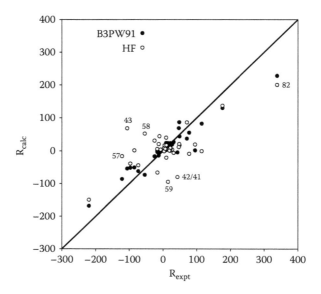

FIGURE 7.15 Comparison of the HF/cc-pVTZ and B3PW91/cc-pVTZ rotational strengths to the experimental rotational strengths of modes 35–87. Rotational strengths R are in 10^{-44} $esu^2 cm^2$.

4. N. Harada, K. Nakanishi, *Circular Dichroic Spectroscopy: Exciton Coupling in Organic Stereochemistry*, University Science Books, Mill Valley, CA, 1983.
5. S.H. Wilen, J.Z. Qi, P.G. Williard, Resolution, Asymmetric Transformation, and Configuration of Tröger's Base. Application of Tröger's Base as a Chiral Solvating Agent, *J. Org. Chem.*, 56, 485–487, 1991.
6. A. Aamouche, F.J. Devlin, P.J. Stephens, Structure, Vibrational Absorption and Circular Dichroism Spectra and Absolute Configuration of Tröger's Base, *J. Am. Chem. Soc.*, 122, 2346–2354, 2000.

2-(1-HYDROXYETHYL)-CHROMEN-4-ONE AND ITS 6-Br DERIVATIVE

Enantiomers of the two chromenones, 2-(1-hydroxyethyl)-chromen-4-one, **1**, and 6-Br-2-(1-hydroxyethyl)-chromen-4-one, **2**, were first prepared by Besse et al. in 1999 [1], with the expectation that these compounds would exhibit antiasthmatic properties. The ACs of (–)-**1** and (–)-**2** were determined using x-ray crystallography. The stereogenic C atoms, C_9, in (–)-**1** and (–)-**2** were concluded to be R and S, respectively. Since 6-Br substitution of **1** seems unlikely to change the sign of the optical rotation, it seems unlikely that (–)-**1** and (–)-**2** have opposite ACs. Since (–)-**2** has a Br heavy atom, its x-ray-determined AC is expected to be more reliable than that of (–)-**1** which does not.

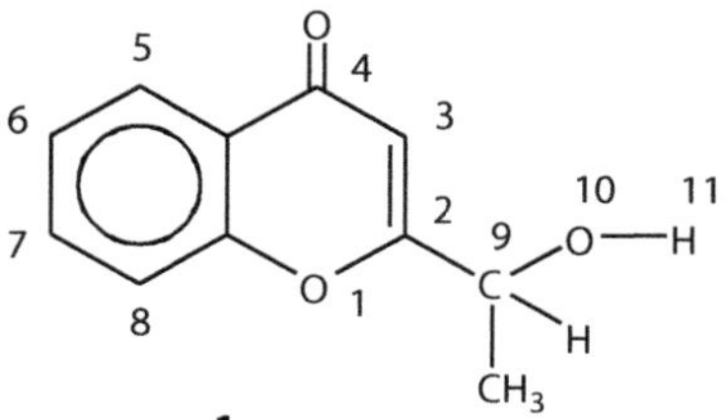

1 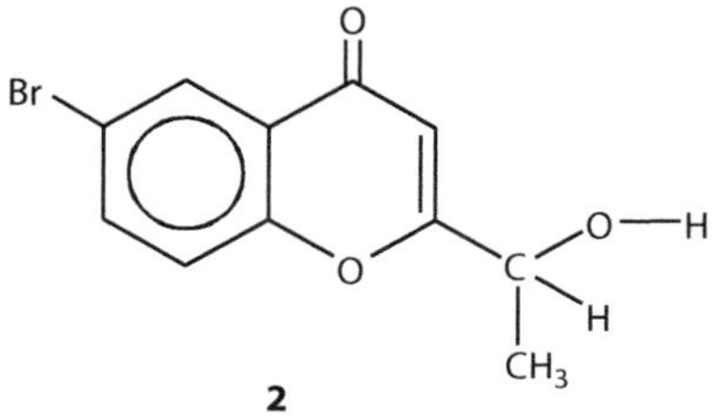   **2**

The surprising switch of AC from R(–) for **1** to S(–) for **2**, as suggested by the x-ray analysis, led us to reinvestigate the ACs of these two molecules using VCD spectroscopy [2]. In order to eliminate the intermolecular aggregation of **1** and **2** due to intermolecular hydrogen bonding of their OH groups, and to reduce the conformational flexibilities of **1** and **2**, their acetate derivatives, **3** and **4**, were studied.

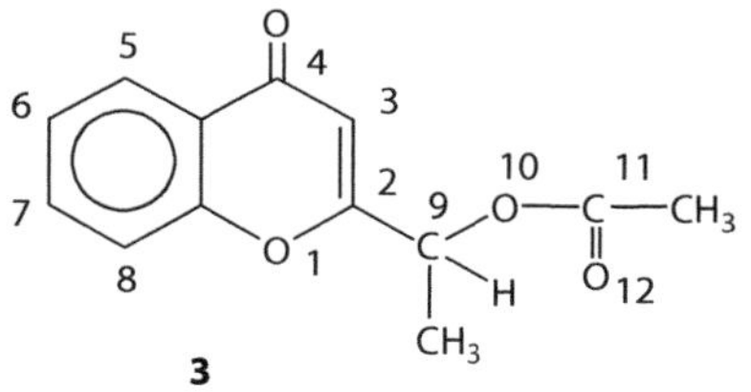

3 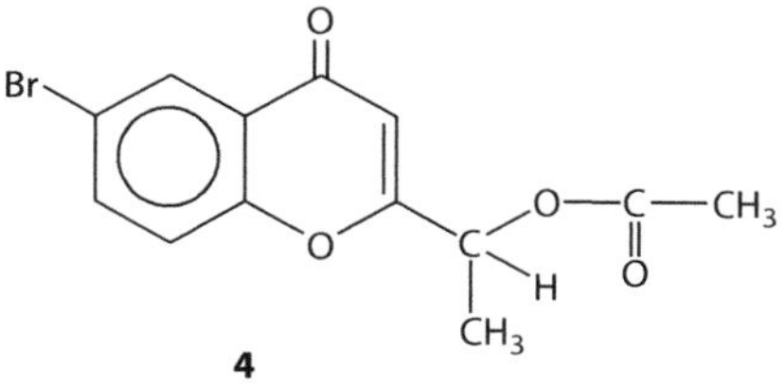   **4**

EXPERIMENTAL IR AND VCD SPECTRA

Samples of (±)-**3**, (+)-**3**, (–)-**3**, (±)-**4**, (+)-**4**, and (–)-**4** were synthesized by Dr. Besse, Université Blaise Pascal, Aubière, France [2]. The IR and VCD spectra of (±)-**3**, (+)-**3**, and (–)-**3**, and (±)-**4**, (+)-**4**, and (–)-**4** were measured using CCl_4 solutions of concentrations 0.06 and 0.07 M for **3** and **4**, respectively, and a 109 μ cell. The $[\alpha]_D$ values of the CCl_4 solutions were +90.7 for (+)-**3**, –100.5 for (–)-**3**, +84.8 for (+)-**4**, and –95.5 for (–)-**4**. The VCD spectra of (+)-**3**, (–)-**3**, (+)-**4**, and (–)-**4** were obtained using the VCD spectra of (±)-**3** and (±)-**4** as baselines. Given the different $[\alpha]_D$ values of (+)-**3** and (–)-**3**, the ratio of their ee's is the ratio of the magnitudes of the $[\alpha]_D$ values. In order to obtain the half-difference VCD spectrum, ½$[\Delta\varepsilon(+) - \Delta\varepsilon(-)]$ for **3**, the VCD of (+)-**3** was normalized to the ee of (–)-**3**. The half-difference VCD spectrum of **4** was obtained in the same way.

In Figure 7.16, the IR and VCD spectra of **3** and **4** are shown for the frequency range 1,500–1,900 cm⁻¹. In this region, for both **3** and **4**, there are two strong IR bands, of virtually identical frequencies in **3** and **4**, which are assigned as the C=O stretching modes of **3** and **4**. In both (+)-**3** and (+)-**4** the lower-frequency C=O stretching bands exhibit strong positive VCD, while the higher-frequency C=O stretching bands exhibit similar bisignate VCD. Empirically, this suggests that (+)-**3** and (+)-**4** have the same ACs.

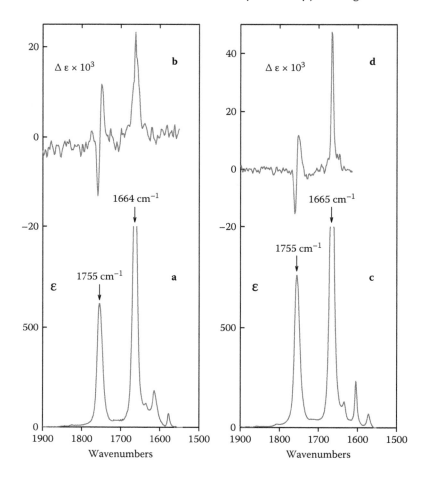

FIGURE 7.16 Experimental IR and VCD spectra of **3** and **4** in CCl$_4$ solution in the C=O stretching region. (a) IR: (−)-**3**. (b) VCD: (+)-**3**. (c) IR: (−)-**4**. (d) VCD: (+)-**4**.

CONFORMATIONAL ANALYSIS

The bicyclic chromen-4-one moieties of **3** and **4** are conformationally rigid. The conformational flexibilities of **3** and **4** are due to the CH(CH$_3$)(OAc) substituents. The *cis* conformation of the acetate group is always much lower in energy than the *trans* conformation. For the *cis* acetate conformation, rotation can occur about the C$_2$–C$_9$ and C$_9$–O$_{10}$ bonds. To identify the most stable conformations of **3**, a two-dimensional (2D) DFT potential energy surface (PES) scan was carried out for *R*-**3** at the B3LYP/6-31G* level, varying the dihedral angles O$_1$C$_2$C$_9$O$_{10}$ and C$_2$C$_9$O$_{10}$C$_{11}$, with the results shown in Figure 7.17. Three stable conformations, **a**, **b**, and **c**, with energies within 3 kcal/mol, were located. B3LYP/6-31G* optimization of the lowest-energy structures in the 3 valleys of the PES, followed by reoptimization at the B3LYP/TZ2P and B3PW91/TZ2P levels, and harmonic frequency calculations at

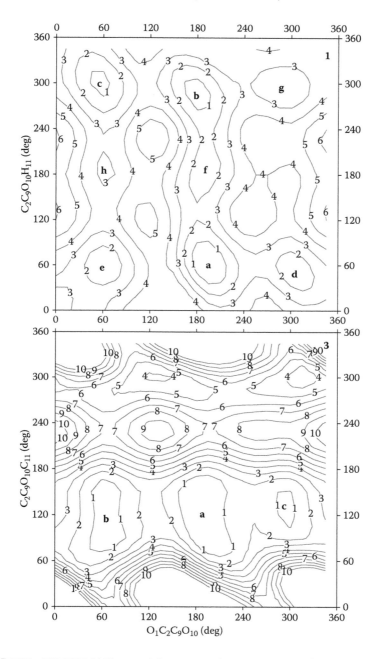

FIGURE 7.17 B3LYP/6-31G* potential energy surfaces of *R*-**1** and *R*-**3**. Contour spacing is 1 kcal/mol.

the optimized geometries, led to the values of the dihedral angles $O_1C_2C_9O_{10}$ and $C_2C_9O_{10}C_{11}$, the relative energies, the relative free energies, and the room temperature populations of the conformations **3a**, **3b**, and **3c**, given in Tables 7.4 and 7.5. The B3LYP and B3PW91 structures of **3a**, **3b**, and **3c** are very similar. The B3LYP/TZ2P structures of **3a**, **3b**, and **3c** are shown in Figure 7.18.

In order to evaluate the degree of conformational rigidification of **1** due to the conversion to the acetate **3**, a PES scan identical to that for *R*-**3** was also carried out for *R*-**1**, with the results also shown in Figure 7.17. For **1**, eight stable conformations with relative energies within 3 kcal/mol are identified. The conversion of the OH group of **1** to the OAc group of **3** thus does cause substantial conformational rigidification, as expected on the basis of the earlier study of endo-borneol [3].

TABLE 7.4
TZ2P Dihedral Angles of the
Conformations of *R*-3 and *R*-4

	$O_1C_2C_9O_{10}{}^a$		$C_2C_9O_{10}C_{11}{}^a$	
	B3LYP	**B3PW91**	**B3LYP**	**B3PW91**
3a	−172.2	−172.4	144.5	143.2
3b	65.2	64.6	146.0	145.2
3c	−66.3	−65.1	139.0	137.9
4a	−172.4	−172.5	144.1	142.8
4b	65.0	64.4	145.9	145.1
4c	−66.4	−65.2	139.0	137.5

a Angles are in degrees.

TABLE 7.5
TZ2P Relative Energies and Free Energies, and Populations
of the Conformations of *R*-3 and *R*-4

	ΔE^a		ΔG^a		$P(\%)^b$	
	B3LYP	**B3PW91**	**B3LYP**	**B3PW91**	**B3LYP**	**B3PW91**
3a	0.00	0.00	0.00	0.00	57.4	55.0
3b	0.27	0.34	0.49	0.48	24.6	24.1
3c	0.61	0.60	0.68	0.57	18.0	20.9
4a	0.00	0.00	0.00	0.00	52.9	50.1
4b	0.32	0.38	0.33	0.28	30.0	30.8
4c	0.60	0.59	0.66	0.56	17.0	19.1

a ΔE and ΔG are in kcal/mol.
b Populations are based on ΔG values; T = 293 K.

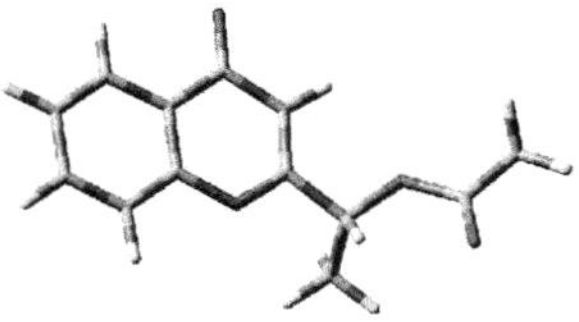

3a 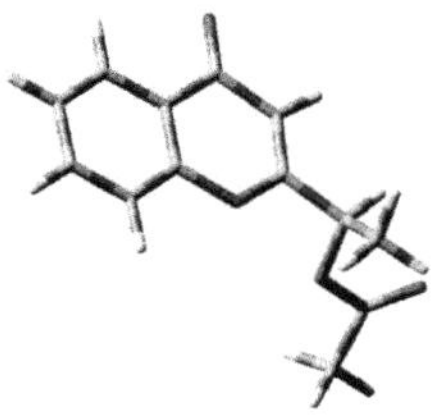   3b

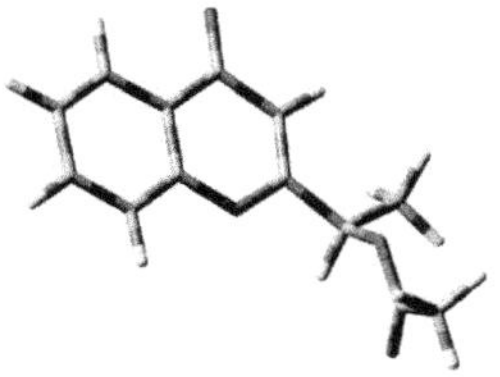

3c

FIGURE 7.18 B3LYP/TZ2P geometries of the conformations **a**, **b** and **c** of *R*-**3**.

In the case of **4**, the conformations **4a**, **4b**, and **4c** were obtained by replacement of the 6-H atoms of **3a**, **3b**, and **3c** by Br and reoptimization. The dihedral angles, relative energies, relative free energies, and populations of **4a**, **4b**, and **4c** obtained at the B3LYP/TZ2P and B3PW91/TZ2P levels, also given in Tables 7.4 and 7.5, are very close to those of **3a**, **3b**, and **3c**, demonstrating that the Br-substitution of **3** is a minor perturbation to its conformational structures and energies.

ANALYSIS OF THE IR AND VCD SPECTRA

The B3LYP/TZ2P and B3PW91/TZ2P conformationally averaged IR and VCD spectra of *R*-**3** are compared to the experimental IR and VCD spectra of (–)-**3** and (+)-**3** over the range 1,550–1,900 cm^{-1} in Figure 7.19. The B3LYP and B3PW91 IR spectra are identical, and unambiguously assign the experimental IR spectrum, as shown in Figure 7.19. The intense experimental bands at 1,664 and 1,755 cm^{-1} are assigned as the C=O stretching modes 68 and 69. The ratio of the intensities of these two bands is correctly predicted.

The B3LYP and B3PW91 VCD spectra are also identical. The conformationally averaged VCD of mode 68 is positive and intense. The conformationally averaged VCD of mode 69 is bisignate, due to the opposite signs of the rotational strengths of conformation **3a** and conformations **3b** and **3c**. The predicted VCD spectra of modes 68 and 69 are in excellent agreement with the experimental VCD spectrum, as shown in Figure 7.19, leading to the conclusion that the AC of **3** is *R*(+).

The B3LYP/TZ2P and B3PW91/TZ2P conformationally averaged IR and VCD spectra of *R*-**4** are compared to the experimental IR and VCD spectra of (–)-**4** and (+)-**4** over the range 1,550–1,900 cm^{-1} in Figure 7.20. Modes 69 and 70 of **4** are the same C=O stretching modes as modes 68 and 69 of **3**. The predicted spectra of

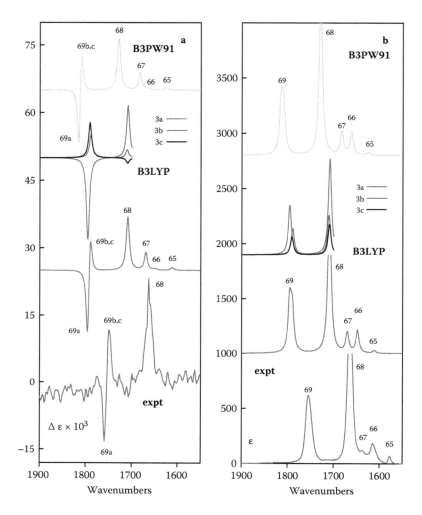

FIGURE 7.19 (a) Comparison of the conformationally averaged B3LYP/TZ2P (second from bottom) and the B3PW91/TZ2P (top) VCD spectra of modes 65–69 of *R*-**3** to the experimental VCD spectrum of (+)-**3** (bottom) (from Figure 7.16). (b) Comparison of the conformationally averaged B3LYP/TZ2P (second from bottom) and the B3PW91/TZ2P (top) IR spectra of modes 65–69 of **3** to the experimental IR spectrum of (−)-**3** (bottom) (from Figure 7.16). The population-weighted B3LYP/TZ2P VCD and IR spectra for each conformer, **3a–3c**, are also shown (second from top). Bandshapes in calculated spectra are Lorentzian (γ = 4.0 cm^{-1}).

R-**4** are identical to the spectra of *R*-**3**. The assignments of the experimental IR and VCD spectra of **4**, shown in Figure 7.20, are identical to the assignments for **3**. As a result, the AC of **4** is also *R*(+).

Since acetylations of (+)-**1** and (+)-**2** lead to (+)-**3** and (+)-**4** [2], the ACs of both **1** and **2** are also *R*(+)/*S*(−). For **2**, the AC is identical to the AC obtained by x-ray

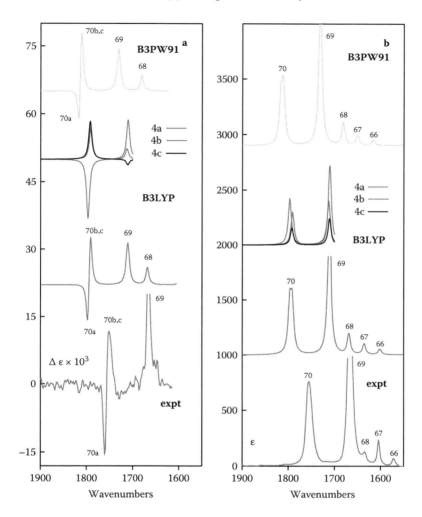

FIGURE 7.20 (a) Comparison of the conformationally averaged B3LYP/TZ2P (second from bottom) and the B3PW91/TZ2P (top) VCD spectra of modes 68–70 of *R*-**4** to the experimental VCD spectrum of (+)-**4** (bottom) (from Figure 7.16). (b) Comparison of the conformationally averaged B3LYP/TZ2P (second from bottom) and the B3PW91/TZ2P (top) IR spectra of modes 66–70 of **4** to the experimental IR spectrum of (–)-**4** (bottom) (from Figure 7.16). The population-weighted B3LYP/TZ2P VCD and IR spectra for each conformer, **4a**–**4c**, are also shown (second from top). Bandshapes in calculated spectra are Lorentzian ($\gamma = 4.0$ cm^{-1}).

crystallography. For **1**, the ACs are opposite, proving that the analysis of the x-ray crystallography of **1** was incorrect. This result confirms that the determination of the AC of a molecule using the Bijvoet x-ray crystallography method requires the molecule to contain a heavy atom.

This study was the first utilization of conformational rigidification via derivatization in determining ACs using VCD. The determination of the ACs of **1** and **2**, using the

VCD of **3** and **4**, makes it clear that derivatization resulting in conformational rigidifi-cation greatly facilitates AC determination.

The VCD study of the ACs of **1** and **2** illustrates the utility of VCD in determin-ing the reliability of previously assigned ACs. That the AC of **1** determined by x-ray crystallography was incorrect was the first demonstration of an incorrect x-ray AC, using VCD spectroscopy.

REFERENCES

1. P. Besse, G. Baziard-Mouysset, K. Boubekeur, P. Palvadeau, H. Veschambre, M. Payard, G. Mousset, Microbiological Reductions of Chromen-4-One Derivatives, *Tet. Asymm.*, 10, 4745–4754, 1999.
2. F.J. Devlin, P.J. Stephens, and P. Besse, Are The Absolute Configurations of 2-(1-Hydroxyethyl)-Chromen-4-One and Its 6-Bromo Derivative Determined by X-ray Crystallography Correct? A Vibrational Circular Dichroism Study of Their Acetate Derivatives, *Tet. Asymm.*, 16, 1557–1566, 2005.
3. F.J. Devlin. P.J. Stephens, and P. Besse, Conformational Rigidification Via Derivatization Facilitates the Determination of Absolute Configuration Using Chiroptical Spectroscopy: A Case Study of the Chiral Alcohol *endo*-Borneol, *J. Org. Chem.*, 70, 2980–2993, 2005.

THE SUPRAMOLECULAR TETRAMER OF 2,2′-DIMETHYL-BIPHENYL-6,6′-DICARBOXYLIC ACID

2,2′-Dimethyl-biphenyl-6,6′-dicarboxylic acid, **1**, is chiral. The enantiomers are:

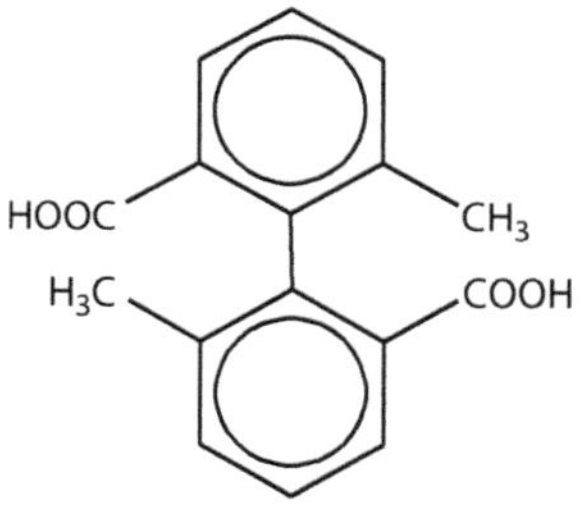

1

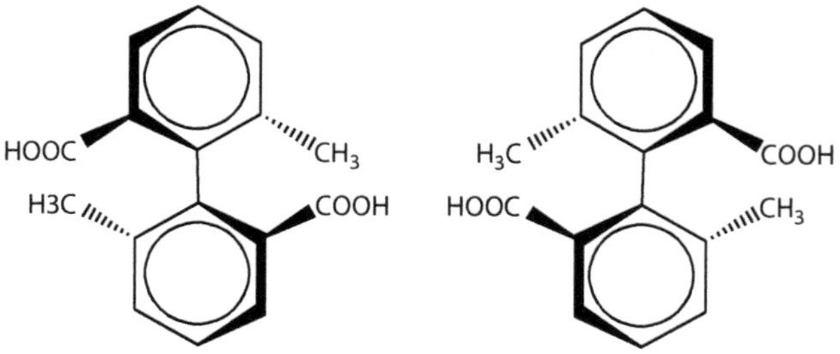

R-1   **S-1**

The x-ray crystal structure of *R*-**1** contained tetramers of **1**, with intermolecular hydrogen bonding of COOH groups [1].

Solutions of **1** in $CDCl_3$ exhibit concentration-dependent IR and VCD spectra [2], demonstrating that in solution aggregation of **1** also occurs.

The question raised by Professor Marie Urbanová and colleagues (Institute of Chemical Technology, Prague, Czech Republic) was: Is the tetramer of **1** formed in solution?

In order to answer this question, we predicted the IR and VCD spectra of the tetramer of *S*-**1** and compared the spectra to the experimental IR and VCD spectra of optically pure (+)-*S*-**1** [2]. The most intense IR absorption and VCD in the frequency range 1,100–1,800 cm^{-1} occurred at ~1,700 cm^{-1}, due to the carbonyl stretching vibrational modes, so the evaluation of the structure of **1** concentrated on these transitions.

The tetramer of **1** contains four hydrogen-bonded $(COOH)_2$ groups. To assess the conformational flexibility of a $(COOH)_2$ group, a 2D PES scan of the dimer of benzoic acid was carried out, using B3LYP/6-31G*, with the results shown in Figure 7.21. The two minima in the PES correspond to the two coplanar tautomers of the $(COOH)_2$ group:

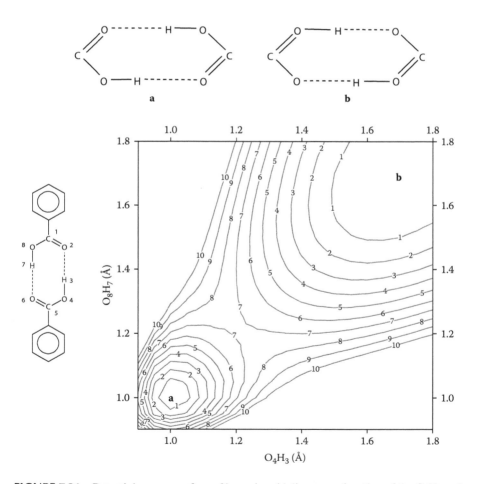

FIGURE 7.21 Potential energy surface of benzoic acid dimer as a function of the O_4H_3 and O_8H_7 distances (Å). The two minima are shown. Contour spacing is 1 kcal/mol.

Conversion of tautomer **a** to **b** involves the oscillation of both H atoms between both pairs of O atoms; the PES scan predicts that the barrier to the conversion is 7–8 kcal/mol, not a high energy. It is therefore predicted that the tetramer of **1** has six inequivalent conformations: **aaaa**, **aaab**, **aabb**, **abab**, **abbb**, and **bbbb** [2]. The B3LYP/6-31G* relative energies and free energies of these six conformations are given in Table 7.6.

The B3LYP/6-31G* IR spectra of the C=O stretching modes of the six conformations of *S*-**1** are compared to the experimental IR spectrum of *S*-**1** in Figure 7.22. The closest agreement of the calculated and experimental spectra is for the **aaab** conformation. The

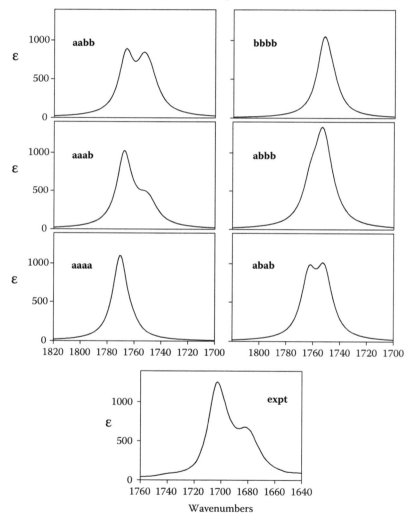

FIGURE 7.22 B3LYP/6-31G* IR spectra for the C=O stretching modes of the six conformations of (*S*-**1**)$_4$ and the experimental IR spectrum of *S*(+)-**1** (CDCl$_3$/0.16 M). Bandshapes in calculated spectra are Lorentzian (γ-values obtained from the fit to the experimental absorption spectrum).

B3LYP/6-31G* VCD spectra of *S*-**1** are compared to the experimental VCD spectrum of *S*-**1** in Figure 7.23. Again, the closest agreement is for the **aaab** conformation.

The calculated IR and VCD spectra of **aaab** of *S*-**1** over the frequency range 1,100–1,700 cm^{-1} are compared to the experimental IR and VCD spectra of *S*-**1** in Figure 7.24. The calculated and experimental IR and VCD spectra are moderately similar.

As a result, it was concluded that in a 0.16 M CDCl$_3$ solution, **1** is tetrameric with the conformation **aaab**, the structure shown in Figure 7.25.

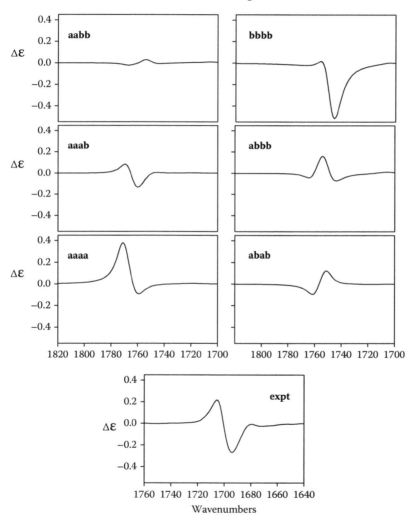

FIGURE 7.23 B3LYP/6-31G* VCD spectra for the C=O stretching modes of the six conformations of (*S*-**1**)$_4$ and the experimental VCD spectrum of 100% ee *S*(+)-**1** (CDCl$_3$/0.16 M). Bandshapes in calculated spectra are Lorentzian (γ-values obtained from the fit to the experimental absorption spectrum).

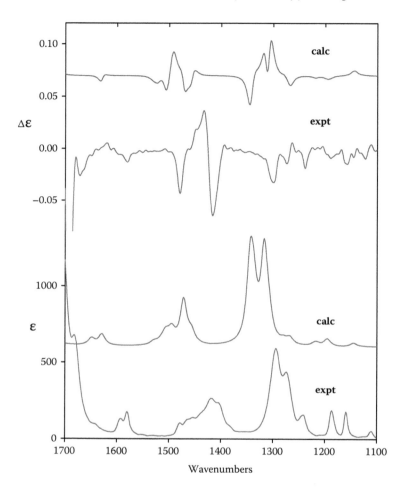

FIGURE 7.24 B3LYP/6-31G* IR and VCD spectra (upper) of the **aaab** conformer of $(S\text{-}1)_4$ and the experimental IR and VCD spectra (lower) of $S(+)\text{-}1$ ($CDCl_3$/0.16 M) in the region 1,100–1,700 cm^{-1}. Bandshapes in calculated spectra are Lorentzian ($\gamma = 4.0$ cm^{-1}).

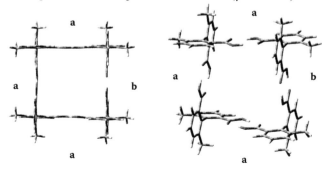

FIGURE 7.25 The B3LYP/6-31G* structure of the **aaab** conformer of $(S\text{-}1)_4$.

TABLE 7.6
B3LYP/6-31G* Relative Energies and Free
Energies of the Conformations of $(S-1)_4$[a]

Conformer	ΔE	ΔG
aaaa (D_4)	0.00	1.81
aaab (C_2)	1.12	0.00
aabb (C_1)	1.96	2.86
abab (D_2)	2.10	3.67
abbb (C_2)	3.11	2.41
bbbb (D_4)	3.94	6.69

[a] ΔE and ΔG in kcal/mol.

This study was the first to use VCD to determine the structure of a supramolecular species. Since supramolecular chemistry has recently become a very important field of chemistry [3], the study of **1** demonstrates the potential utility of VCD in this field.

REFERENCES

1. M. Tichy, T. Kraus, J. Zavada, I. Cisarova, and J. Podlaha, Self-assembly of Enantiopure 2,2′-Dimethylbiphenyl-6,6′-Dicarboxylic Acid. Formation of Chiral Squares and Their Columnar Stacking, *Tet. Asymm.*, 10, 3277–3280, 1999.
2. M. Urbanová, V. Setnička, F.J. Devlin, and P.J. Stephens, Determination of Molecular Structure in Solution Using Vibrational Circular Dichroism Spectroscopy: The Supramolecular Tetramer of *S*-2,2′-Dimethyl-Biphenyl-6,6′-Dicarboxylic Acid, *J. Am. Chem. Soc.*, 127, 6700–6711, 2005.
3. (a) J.-M. Lehn, Toward Complex Matter: Supramolecular Chemistry and Self-organization, *Proc. Natl. Acad. Sci. U.S.A.*, 99, 4763–4768, 2002. (b) J.-M. Lehn, Supramolecular Polymer Chemistry—Scope and Perspectives, *Polym. Int.*, 51, 825–839, 2002. (c) J.-M. Lehn, Toward Self-Organization and Complex Matter, *Science*, 295, 2400–2403, 2002. (d) J.-M. Lehn, Supramolecular Chemistry: From Molecular Information towards Self-organization and Complex Matter, *Rep. Prog. Phys.*, 67, 249–265, 2004.

TRANS-PARA-METHYL-PHENYL-GLYCIDIC ACID

In 2001 and 2002, Professor Bortolini and coworkers (Universita di Ferrara, Italy) developed an asymmetric epoxidation technique, using a bile acid and oxone, which converted derivatives of trans-cinnamic acid, **1**, to enantiomers of derivatives of trans-phenyl-glycidic acid, **2** [1].

Given the complexity of the mechanism of the epoxidation reaction, the ACs of the trans-phenyl-glycidic acid derivatives synthesized were not predictable and were not determined.

In 2002, Professor Bortolini supplied us with methyl esters of the (+) and (±) isomers of the p-CH$_3$, o-Br, and m-F derivatives of **2**, in order to determine their ACs using VCD [2]. The methyl esters were prepared in order to eliminate intermolecular hydrogen bonding due to the COOH group.

Here, we discuss the determination of the AC of the p-CH$_3$ derivative of **2**, **3**, using its methyl ester, **4**. The (+) enantiomer of **4** resulted from the (+) enantiomer of **3**, so the ACs of (+)-**3** and (+)-**4** are identical [2]. The ee of (+)-**4** was determined, using gas chromatography, to be 79% [2].

3   **4**

The IR and VCD spectra of (+)-**4** were measured using a 0.23 M CCl$_4$ solution and cells of pathlengths 109, 239, and 597 μ. The VCD baselines were the VCD spectra of a 0.23 M CCl$_4$ solution of (±)-**4**. The experimental IR and VCD spectra of (+)-**4** are shown in Figure 7.26, over the range 820–1,550 cm^{-1}. The VCD spectrum is normalized to 100% ee.

CONFORMATIONAL ANALYSIS

The COOCH$_3$ group can rotate around the C$_2$-C$_7$ bond of **4** and the p-CH$_3$-phenyl group can rotate around the C$_3$-C$_4$ bond. Conformational analysis of **4** was based on a 2D B3LYP/6-31G* PES scan, varying the dihedral angles O$_8$C$_7$C$_2$H$_6$ and C$_9$C$_4$C$_3$H$_5$. The COOCH$_3$ group can be *cis* or *trans*, but the *trans* structure is always much higher in energy so the *cis* geometry was used for the PES scan. The PES scan of (2S,3R)-**4**, the AC of (+)-**4** concluded from the comparison of time-dependent density functional theory (TDDFT) calculation of the specific rotation [α]$_D$ of (2S,3R)-**4** to the experimental specific rotation, [α]$_D$, of (+)-**4** [3], is shown in Figure 7.27. Two low-energy stable conformations, **a** and **b**, are predicted. The two other minima of the PES are much higher in energy, and are therefore ignored. Reoptimization of the structures of conformations **a** and **b** obtained from the PES scan, at the B3LYP/TZ2P and B3PW91/TZ2P levels, led to the O$_8$C$_7$C$_2$H$_6$ and C$_9$C$_4$C$_3$H$_5$ dihedral angles and relative energies given in Table 7.7. The dihedral angles O$_8$C$_7$C$_2$H$_6$ of **a** and **b** differ by ~180°; the angles C$_9$C$_4$C$_3$H$_5$ are identical. The two conformations of **4**

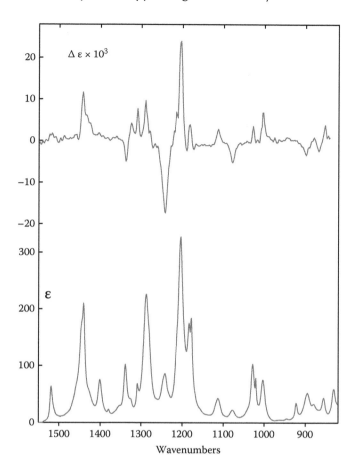

FIGURE 7.26 Experimental IR (bottom) and VCD (top) spectra of CCl_4 solutions of **4** in the mid-IR spectral region. IR: (+)-**4**, 0.23 M, pathlength 239 μ; VCD: (+)-**4**, 0.23 M, pathlength 597 μ (1,167–841 cm⁻¹), pathlength 239 μ (1,600–1,184 cm⁻¹), pathlength 109 μ (1,182–1,169 cm⁻¹).

thus result from rotation of the $COOCH_3$ group about the C_2-C_7 bond. Harmonic frequency calculations proved the stability of the conformations **a** and **b** and led to the relative free energies and room temperature populations given in Table 7.7. The B3PW91/TZ2P structures of conformations **a** and **b** of (2S,3R)-**4** are:

4a   **4b**

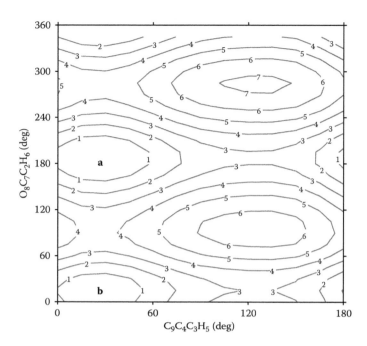

FIGURE 7.27 B3LYP/6-31G* potential energy surface of (2S,3R)-**4**. Dihedral angles were varied in 15° steps. Contour spacing is 1 kcal/mol.

TABLE 7.7

TZ2P Dihedral Angles, Relative Energies and Free Energies, and Populations of the Conformations of (2S,3R)-4

	$O_8C_7C_2H_6{}^a$		$C_9C_4C_3H_5{}^a$		ΔE^b		ΔG^b		$P(\%)^c$	
	B3LYP	**B3PW91**	**B3LYP**	**B3PW91**	**B3LYP**	**B3PW91**	**B3LYP**	**B3PW91**	**B3LYP**	**B3PW91**
4a	−171.1	−170.7	23.6	23.7	0.00	0.00	0.00	0.00	63.1	70.5
4b	15.0	15.9	23.3	23.5	0.24	0.26	0.31	0.51	36.9	29.5

ᵃ   Angles are in degrees. AC is (2S,3R).
ᵇ   ΔE and ΔG are in kcal/mol.
ᶜ   Populations are based on ΔG values; T = 293 K.

ANALYSIS OF THE IR AND VCD SPECTRA

Comparisons of the B3LYP/TZ2P and B3PW91/TZ2P conformationally averaged IR and VCD spectra of (2S,3R)-**4** to the experimental IR and VCD spectra of (+)-**4** are shown in Figure 7.28. The B3PW91 spectra are in better agreement with the experimental spectra than are the B3LYP spectra. For example, the relative intensities of the intense bands of the experimental IR spectrum at ~1,200 cm⁻¹ are correctly and incorrectly predicted by B3PW91 and B3LYP, respectively. In Figure 7.29, the contributions of conformations **a** and **b** to the conformationally

TABLE 7.8
Calculated and Experimental Frequencies, Dipole Strengths, and Rotational Strengths for 4[a]

Mode	Calculation[b] 4a			Calculation[b] 4b			Experiment[c]		
	ν	D	R	ν	D	R	ν	D	R
57	1,555	42.5	2.2	1,555	45.5	1.1	1,519	28.9	
56	1,495	21.3	25.0	1,494	39.0	−51.2			
							1,459	35.7	
55	1,493	69.2	−27.8	1,494	34.6	69.5			
54	1,483	18.4	−3.6	1,482	19.4	1.4			
							1,448	77.8	
53	1,482	27.2	0.9	1,481	26.0	−0.3			
52	1,474	229.1	20.5	1,470	127.7	15.7	1,441	176.8	
51	1,461	41.6	16.7	1,461	57.6	7.8	1,425	30.0	
50	1,435	88.9	0.4	1,432	69.3	5.9	1,401	61.9	
49	1,407	2.9	−0.1	1,407	4.0	−0.4	1,379	17.8	
48	1,371	146.5	−12.6				1,338	90.7	−7.4
				1,360	37.6	13.4	1,326	21.8	7.7
47	1,343	19.6	5.8	1,342	12.6	1.2			
46	1,331	46.4	35.6				1,308	47.1	12.5
				1,323	17.6	−11.0	1,295	24.7	
45				1,315	1,220.1	28.3	1,288	257.9	20.1
	1,274	189.7	−103.7				1,243	98.7	−65.7
44				1,263	57.6	−13.5	1,232	24.2	
	1,238	11.0	1.5						
							1,213	70.2	8.1
43				1,238	6.2	−0.4			
	1,228	604.4	82.8				1,204	378.6	61.6
42	1,221	503.7	−4.5	1,223	32.7	1.1	1,196	35.9	
41	1,205	195.4	4.8	1,207	17.4	−2.9	1,184	113.3	
40	1,200	95.9	−6.4	1,202	19.2	0.4	1,177	134.9	
39	1,172	2.9	−2.0	1,171	2.5	0.9	1,155	12.8	
38	1,139	25.8	6.5	1,139	39.5	2.1	1,121	10.7	
37	1,135	31.8	21.4	1,132	9.6	−2.4	1,112	58.6	7.5
36				1,104	51.5	−31.2	1,078	27.4	−19.5
	1,099	11.9	−7.6				1,070	8.1	
35				1,063	246.6	17.4	1,028	119.7	6.8
	1,057	31.9	−1.8						
							1,020	43.4	
34				1,058	28.7	−4.5			
	1,041	3.5	−3.5						
33	1,034	100.0	19.4	1,039	22.2	−2.0	1,003	98.6	19.2

[a] Frequencies in cm^{-1}; dipole strengths in 10^{-40} esu^2 cm^2; rotational strengths in 10^{-44} esu^2 cm^2.

[b] B3PW91/TZ2P-calculated rotational strengths are for (2S,3R)-4.

[c] From Lorentzian fitting of the IR and VCD spectra of (+)-4.

averaged B3PW91/TZ2P IR and VCD spectra are shown. Conformations **a** and **b** give very different contributions to both the IR and VCD spectra. As a result, eight bands of the experimental IR spectrum are assigned to conformation **a**, and four bands are assigned to conformation **b**. Six bands of the experimental VCD spectrum are assigned to conformation **a**, and four bands to conformation **b**. The agreements of the B3PW91/TZ2P and experimental IR and VCD spectra provide substantial confirmation of the conformational analysis of **4**. The agreement of the calculated VCD spectrum of (2*S*,3*R*)-**4** with the experimental VCD spectrum of (+)-**4** proved that the AC of **4** is (2*S*,3*R*)-(+).

In Table 7.8, the B3PW91/TZ2P frequencies, dipole strengths, and rotational strengths of conformations **a** and **b** of (2*S*,3*R*)-**4** and the experimental frequencies, dipole strengths, and rotational strengths obtained by Lorentzian fitting of the experimental IR and VCD spectra of (+)-**4** are given. The excellent agreement of the calculated and experimental rotational strengths, compared in Figure 7.30, supports the reliabilities of the conformational analysis of **4** and the assignment of the AC as (2*S*,3*R*)-(+).

Since the ACs of (+)-**4** and (+)-**3** are identical, the VCD analysis of the AC of (+)-**4** proves that the AC of (+)-**3** is 2*S*,3*R*. The determination of the AC of **3** using the VCD of **4** led to the understanding of the mechanism of the asymmetric epoxidation reaction, developed by Professor Bortolini. This study illustrates the utility of VCD in analyzing the mechanisms of asymmetric organic reactions.

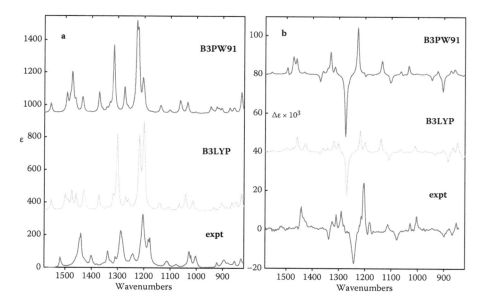

FIGURE 7.28 Comparison of the B3PW91/TZ2P and B3LYP/TZ2P IR (a) and VCD (b) spectra of (2*S*,3*R*)-**4** to the experimental IR (a) and VCD (b) spectra of (+)-**4**. The conformationally averaged calculated spectra were obtained using equilibrium populations calculated from relative free energies (Table 7.7).

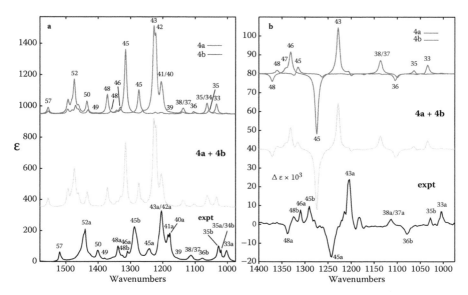

FIGURE 7.29 (SEE COLOR INSERT.) (a) Comparison of the conformationally averaged B3PW91/TZ2P IR spectrum of **4** (green) to the experimental IR spectrum of (+)-**4** (black). The population-weighted B3PW91/TZ2P IR spectra for each conformer (**4a** (red) and **4b** (blue)) are also shown. (b) Comparison of the conformationally averaged B3PW91/TZ2P VCD spectrum of (2*S*,3*R*)-**4** (green) to the experimental VCD spectrum of (+)-**4** (black). The population-weighted B3PW91/TZ2P VCD spectra for each conformer (**4a** (red) and **4b** (blue)) are also shown. Bands assigned to the fundamentals of single conformations **4a** and **4b** are numbered in red and blue, respectively; bands assigned to both conformations are numbered in black.

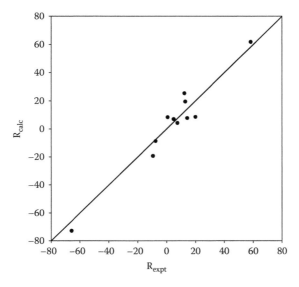

FIGURE 7.30 Comparison of the B3PW91/TZ2P-calculated rotational strengths of (2*S*,3*R*)-**4** (in 10^{-44} esu^2 cm^2) to the experimental values for (+)-**4**.

REFERENCES

1. (a) O. Bortolini, M. Fogagnolo, G. Fantin, and S. Maietti, A. Medici, Asymmetric Epoxidation of Cinnamic Acid Derivatives Using Dioxiranes Generated *In Situ* from Dehydrocholic Acid, *Tet. Asymm.*, 12, 1113–1115, 2001; (b) O. Bortolini, G. Fantin, M. Fogagnolo, R. Forlani, S. Maietti, and P. Pedrini, Improved Enantioselectivity in the Epoxidation of Cinnamic Acid Derivatives with Dioxiranes from Keto Bile Acids, *J. Org. Chem.*, 67, 5802–5806, 2002.
2. F.J. Devlin, P.J. Stephens, and O. Bortolini, Determination of absolute configuration Using Vibrational Circular Dichroism Spectroscopy: Phenylglycidic Acid Derivatives Obtained via Asymmetric Epoxidation Using Oxone and a Keto Bile Acid, *Tet. Asymm.*, 16, 2653–2663, 2005.
3. P.J. Stephens, F.J. Devlin, J.R. Cheeseman, M.J. Frisch, O. Bortolini, and P. Besse, Determination of Absolute Configuration Using *Ab Initio* Calculation of Optical Rotation, *Chirality*, 15, S57–S64, 2003.

THE CYTOTOXIC SESQUITERPENE NATURAL PRODUCT, QUADRONE

The sesquiterpene natural product quadrone, **1**, was first isolated in 1978 from the fungus *Aspergillus terreus* and shown to be cytotoxic [1]. The natural product exhibited optical rotation, $[\alpha]_D = -52.7$ (c 0.59, EtOH) and is therefore chiral. X-ray crystallography showed the structure of **1** to be:

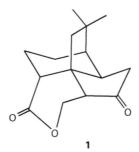

1

but did not define its AC [1]. The two enantiomers of **1** are:

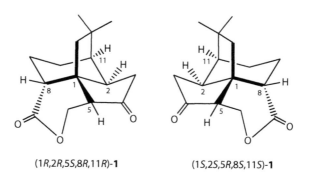

(1*R*,2*R*,5*S*,8*R*,11*R*)-**1**   (1*S*,2*S*,5*R*,8*S*,11*S*)-**1**

Considerable effort was subsequently devoted to the total synthesis of **1**. In 1984, Smith and Konopelski reported the first synthesis of (+)-**1** [2]. Based on the AC of the compound from which (+)-**1** was synthesized, the AC of (+)-**1** was assigned to be $1S,2S,5R,8S,11S$. The AC of the natural product, (−)-**1**, is therefore $1R,2R,5S,8R,11R$. Subsequently, in 1991, Smith et al. reported syntheses of both enantiomers of **1** [3], providing further support for the AC of **1**.

To demonstrate the utility of VCD spectroscopy in determining the ACs of natural products, a study of **1** in collaboration with Professor Amos B. Smith III (University of Pennsylvania) was initiated in 2005 [4], using 2.4 mg of synthetic (−)-**1**. Solutions of **1** in $CDCl_3$ were used to measure its IR and VCD spectra and specific rotations. Given the unavailability of (+)-**1** and (±)-**1**, the baselines for the VCD spectra of (−)-**1** were the VCD spectra of the $CDCl_3$ solvent. The IR spectrum of a 0.24 M solution of (−)-**1** in a 99 μ pathlength cell is shown in Figure 7.31 over the range 800–1,550 cm^{-1}, except for the region ~900 cm^{-1}, where the $CDCl_3$ absorption is very intense. The maximum absorbance is 0.52 at 1,141 cm^{-1}. The VCD spectra of the solution of (−)-**1** and of the $CDCl_3$ solvent in the same cell are shown in Figure 7.32 over the range 800–1,550 cm^{-1}. The large artefact at ~900 cm^{-1} is due to the $CDCl_3$ absorption. Subtraction of the solvent baseline leads to the VCD spectrum of (−)-**1** shown in Figure 7.31. The absorbance of the 1,742 cm^{-1} C=O stretching absorption band of the 0.24 M solution in the 99 μ pathlength cell is ~1.6, much too high to permit reliable measurement of its VCD. The 0.24 M solution was therefore diluted to 0.14 M and the IR and VCD spectra remeasured in the 99 μ cell, with the results over the range 1,650–1,850 cm^{-1} shown in Figures 7.31 and 7.32. The optical rotations of **1** were measured after dilution of the 0.14 M solution; $[\alpha]_D$ was −44.6 (c 0.12, $CDCl_3$), slightly lower in magnitude than the value reported for naturally occurring **1** [1]. The difference in $[\alpha]_D$ could be due to the differences in the concentrations and solvents used, or to the ee of the synthetic (−)-**1** being slightly lower than the ee of the natural product.

CONFORMATIONAL ANALYSIS

Conformational analysis of quadrone using the MMFF94 force field surprisingly predicts only three conformations within a 20 kcal/mol window. The relative energies of these conformations, **a**, **b**, and **c**, are given in Table 7.9. In **a**, the cyclohexane ring has a chair conformation, the lactone ring has a boat conformation, and the cyclopentanone ring is planar except for C_2. In **b**, both the cyclohexane and lactone rings have chair conformations. In **c**, the cyclohexane ring has a twisted-boat conformation and the lactone ring has a half-chair conformation.

Optimization of the MMFF94 conformations **a**–**c** using DFT at the B3LYP/6-31G* level leads to the relative energies given in Table 7.9. The DFT relative energies of **a**–**c** are similar to the MMFF94 values. B3LYP/6-31G* harmonic frequency calculations for **a**–**c** confirm their stabilities and lead to the relative free energies and room temperature equilibrium populations given in Table 7.9. The relative free energies are similar to the relative energies. The population of **a** is predicted to be >99.9%. Surprisingly, quadrone is therefore predicted to be a conformationally rigid molecule.

In order to evaluate this conclusion, we have carried out a 2D PES scan of $(1R,2R,5S,8R,11R)$-**1** at the B3LYP/6-31G* level, varying the two dihedral angles

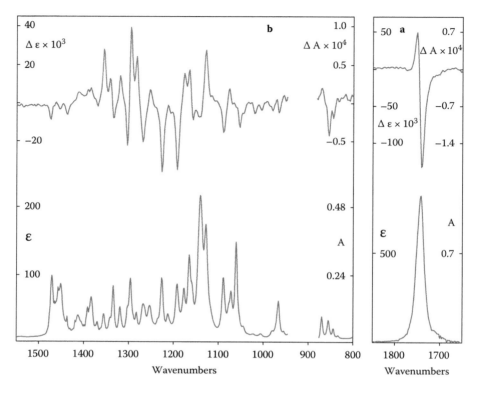

FIGURE 7.31 Experimental IR and VCD spectra of CDCl$_3$ solutions of (–)-**1**, 0.14 M (a) and 0.24 M (b). The baselines are the CDCl$_3$ IR and VCD spectra.

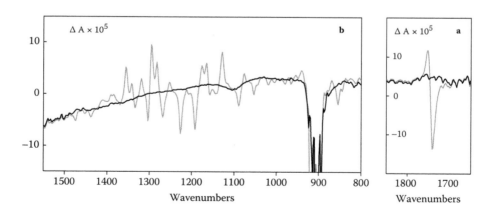

FIGURE 7.32 Experimental VCD spectra (light) of CDCl$_3$ solutions of (–)-**1**, 0.14 M (a) and 0.24 M (b) and the CDCl$_3$ solvent (dark).

TABLE 7.9

Conformational Analysis of 1

Conf.	$\Delta E^{a,b}$	$\Delta E^{a,c}$	$\Delta G^{a,c}$	$P(\%)^d$
a	0.00	0.00	0.00	99.98
b	4.93	5.43	5.09	0.02
c	5.54	5.68	5.76	0.01

[a] kcal/mol.
[b] MMFF94.
[c] B3LYP/6-31G*.
[d] Populations calculated using ΔG values at 298 K.

$C_1C_2C_3C_4$ and $C_8C_1C_5C_6$ from $-60°$ to $+60°$ in steps of $10°$, starting from conformation **a**. The resulting PES, shown in Figure 7.33, possesses only one valley. Optimization of the lowest-energy structure in the scan leads to conformation **a**, confirming that this is indeed the minimum of the valley.

Conformational analysis of quadrone thus leads to the conclusion that only conformation **a** is significantly populated at room temperature. Further support for this conclusion is provided by comparison of the B3LYP/6-31G* structure of **a** to the structure of quadrone determined using x-ray crystallography [1a,3]. Key dihedral angles of the two independent molecules identified in the unit cell of crystalline quadrone and the B3LYP/6-31G* dihedral angles of $(1R,2R,5S,8R,11R)$-**1** are given in Table 7.10. The agreement of the B3LYP/6-31G* dihedral angles of conformation **a** and the x-ray dihedral angles is excellent. Crystalline quadrone thus consists of molecules having conformation **a**, consistent with the prediction that **a** is the lowest-energy conformation.

Analysis of the experimental IR and VCD spectra of **1** is now carried out by comparison to the spectra of conformation **a** calculated at the B3LYP/cc-pVTZ and B3PW91/cc-pVTZ levels. The dihedral angles of the reoptimized geometries are given in Table 7.10, and are very similar to the B3LYP/6-31G* dihedral angles. The B3PW91/cc-pVTZ structure of $(1R,2R,5S,8R,11R)$-**1a** is shown in Figure 7.34.

ANALYSIS OF THE IR AND VCD SPECTRA

The frequencies, dipole strengths, and rotational strengths of modes 33–88 of $(1R,2R,5S,8R,11R)$-**1a**, calculated using the functionals B3LYP and B3PW91 and the basis set cc-pVTZ, are listed in Table 7.11. The B3LYP/cc-pVTZ and B3PW91/cc-pVTZ IR spectra of **1a** are compared to the experimental IR spectrum in Figure 7.35. The B3PW91 spectrum is in better agreement with the experimental spectrum than

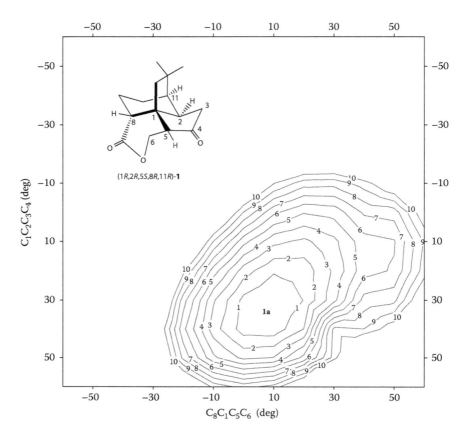

FIGURE 7.33 The B3LYP/6-31G* 2D PES scan of (1*R*,2*R*,5*S*,8*R*,11*R*)-**1**. All contours greater than 10 kcal have been omitted for clarity.

is the B3LYP spectrum, especially for modes 53–56, and is therefore used to assign the experimental spectrum, as detailed in Figure 7.35. Fundamental transitions of modes 33–36, 40, 43–46, 48–58, 61–63, 66, 69–74, 77, 78, and 83 are clearly resolved in the experimental spectrum. Modes 41/42, 59/60, 64/65, 67/68, 75/76, 79/80, 81/82, 84/86, and 87/88 are not resolved, as is also the case in the B3PW91 spectrum. Lorentzian fitting of the experimental IR spectrum, shown in Figure 7.36, leads to the frequencies, dipole strengths, and bandwidths given in Table 7.11. Comparison of the experimental frequencies and dipole strengths to the B3PW91/ cc-pVTZ frequencies and dipole strengths is shown in Figure 7.37 for modes 33–86. The agreement of calculated and experimental dipole strengths is good, except for modes 53, 54, and 56.

The B3LYP/cc-pVTZ and B3PW91/cc-pVTZ VCD spectra of (1*R*,2*R*,5*S*,8*R*,-11*R*)-**1a** are compared to the experimental VCD spectrum of (–)-**1** in Figure 7.38. The B3PW91 spectrum is in better agreement with the experimental spectrum

TABLE 7.10
Calculated and Experimental Dihedral Angles for 1[a]

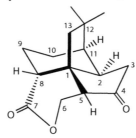

Dihedral	x-Ray[b]		B3LYP[c]	B3LYP[d]	B3PW91[d]
C_1-C_2-C_3-C_4	39.7	38.6	33.8	33.5	34.0
C_2-C_3-C_4-C_5	−34.4	−33.5	−26.0	−25.6	−26.0
C_3-C_4-C_5-C_1	15.4	14.9	7.9	7.4	7.6
C_4-C_5-C_1-C_2	9.9	9.9	13.3	13.6	13.8
C_5-C_1-C_2-C_3	−30.9	−30.6	−29.8	−29.7	−30.1
C_1-C_5-C_6-O	44.7	46.4	42.4	42.0	42.6
C_5-C_6-O-C_7	−51.6	−57.4	−52.7	−52.0	−52.5
C_6-O-C_7-C_8	4.6	8.9	7.1	6.8	6.6
O-C_7-C_8-C_1	45.5	45.4	44.4	44.0	44.8
C_7-C_8-C_1-C_5	−47.8	−51.1	−49.7	−49.1	−49.7
C_8-C_1-C_5-C_6	4.2	6.3	7.8	7.3	7.5
C_1-C_2-C_{11}-C_{10}	−74.2	−74.0	−75.3	−75.3	−75.0
C_2-C_{11}-C_{10}-C_9	56.2	57.9	57.5	57.5	57.7
C_{11}-C_{10}-C_9-C_8	−36.1	−39.7	−37.6	−37.7	−38.4
C_{10}-C_9-C_8-C_1	35.4	38.7	36.5	36.6	37.4
C_9-C_8-C_1-C_2	−56.6	−58.8	−57.4	−57.4	−58.0
C_8-C_1-C_2-C_{11}	77.6	77.4	77.8	77.7	77.7
C_9-C_8-C_1-C_5	−172.9	−175.5	−173.7	−173.7	−174.1
C_{10}-C_{11}-C_2-C_3	169.9	169.3	166.7	166.9	167.3
C_5-C_4-O-C_3	178.4	176.3	179.5	179.8	179.9

[a] Dihedral angles in degrees.
[b] x-ray crystallography: refs. 1a, 3. The two sets of dihedral angles are for the two independent molecules in the unit cell.
[c] Calculated 6-31G* dihedral angles for the (1R,2R,5S,8R,11R)-AC of **1a**.
[d] Calculated cc-pVTZ dihedral angles for the (1R,2R,5S,8R,11R)-AC of **1a**.

than is the B3LYP spectrum, especially for modes 47–56. The assignment of the experimental VCD spectrum, based on the B3PW91/cc-pVTZ spectrum, is detailed in Figure 7.38. Lorentzian fitting of the VCD, based on this assignment, shown in Figure 7.39, leads to the frequencies, rotational strengths, and bandwidths given in Table 7.11. The VCD frequencies are in excellent agreement with the IR

FIGURE 7.34   The B3PW91/cc-pVTZ structure of (1*R*,2*R*,5*S*,8*R*,11*R*)-**1a**. The hydrogen atoms have been removed for clarity.

frequencies, confirming the consistency of the assignments of the experimental IR and VCD spectra. Comparison of the B3PW91/cc-pVTZ and B3LYP/cc-pVTZ rotational strengths of (1*R*,2*R*,5*S*,8*R*,11*R*)-**1** to the experimental rotational strengths of (–)-**1** is shown in Figure 7.40 for modes 34–86. The quantitative agreement of the B3PW91 rotational strengths is excellent. In contrast, the agreement of the B3PW91/cc-pVTZ rotational strengths of (1*S*,2*S*,5*R*,8*S*,11*S*)-**1**, also shown in Figure 7.40, is atrociously bad. Quantitative documentation of this is provided by the deviations of the rotational strengths calculated for the two enantiomers of **1** from the experimental rotational strengths, given in Table 7.12. It follows that the AC of **1** is unambiguously (1*R*,2*R*,5*S*,8*R*,11*R*)-(–), confirming the AC obtained via total synthesis by Smith et al. [2,3]. The poorer agreement of the B3LYP/cc-pVTZ and experimental rotational strengths confirms the superior accuracy of the B3PW91/cc-pVTZ VCD spectrum. The excellent agreement of the B3PW91/cc-pVTZ and experimental rotational strengths confirms that the ee of the sample of (–)-**1** was close to 100%.

   The B3LYP/cc-pVTZ and B3PW91/cc-pVTZ IR and VCD spectra of the C=O stretching modes, 87 and 88, are in excellent qualitative agreement with the experimental IR and VCD spectra (Figures 7.35 and 7.38), further confirming the AC of **1** obtained from modes 34–86. However, the quantitative agreements of the calculated and experimental dipole and rotational strengths are much poorer than for modes 34–86. It is likely that the poorer agreement of the rotational strengths for modes 87 and 88 is due to greater artefacts in the experimental VCD, commonly observed in measurements of the VCD of C=O stretching modes, as was the case in our measurement of the VCD of camphor, presented in Chapter 2. The magnitudes of the artefacts in the VCD of **1** could be larger than in camphor, due to the use of a solvent baseline, instead of a racemate baseline.

   In reference 4, the B3LYP and B3PW91 calculations of the IR and VCD spectra of **1** used the basis set TZ2P. Comparison of the B3PW91/TZ2P and B3PW91/cc-pVTZ rotational strengths of (1*R*,2*R*,5*S*,8*R*,11*R*)-**1**, shown in Figure 7.41 for

TABLE 7.11

Calculated and Experimental Frequencies, Dipole Strengths, and Rotational Strengths for 1[a]

Mode	Calculation[b]						Experiment[c]					
	B3LYP			B3PW91			IR			VCD		
	ν	D	R	ν	D	R	ν	D	γ	ν	R	γ
88	1,819	445.0	193.6	1,836	448.4	202.1	1,750	298.2	6.7	1,751	237.5	5.8
87	1,812	771.8	−282.5	1,829	774.6	−296.6	1,742	931.2	8.1	1,741	−445.5	6.8
86	1,520	5.3	−10.8	1,509	5.1	−8.8⎤						
85	1,516	7.2	−6.6	1,505	22.7	−3.2 ⎬	1,471	61.5	3.7	1,474	−13.3	3.0
84	1,514	38.5	−0.4	1,505	32.2	−1.7⎦						
83	1,508	17.4	0.7	1,495	21.2	−0.9	1,458	40.3	4.4			
82	1,501	38.8	−6.9	1,488	40.7	−7.0⎤						
						⎬	1,451	51.8	4.2	1,454	−4.6	2.8
81	1,497	29.8	−1.9	1,484	33.9	0.1⎦						
80	1,491	2.0	−1.7	1,478	4.3	−2.1⎤						
						⎬	1,438	10.8	3.6	1,437	−10.2	3.0
79	1,485	4.7	−9.6	1,470	8.7	−14.8⎦						
78	1,460	21.3	14.5	1,445	24.8	17.0	1,412	37.2	6.2	1,411	17.4	5.8
77	1,431	21.5	1.4	1,419	22.7	3.1	1,391	23.8	3.9	1,392	9.9	3.5
76	1,416	8.4	−3.3	1,408	40.2	−6.2⎤						
						⎬	1,383	40.9	3.7	1,384	10.7	3.2
75	1,412	18.1	0.6	1,404	2.8	9.4⎦						
74	1,407	7.4	−2.9	1,395	7.1	−4.4	1,370	8.1	2.7	1,369	−4.3	2.0
73	1,390	10.2	22.4	1,383	17.9	32.7	1,355	20.3	3.1	1,355	53.4	3.5
72	1,378	2.9	−1.1	1,371	12.0	12.4	1,342	9.6	2.6	1,341	18.5	2.5
71	1,364	27.7	−3.4	1,361	45.3	−13.1	1,334	41.0	2.6	1,334	−19.1	3.1
70	1,349	43.4	24.5	1,347	48.2	20.8	1,319	25.2	2.9	1,319	24.7	3.0
69	1,329	33.6	−28.4	1,328	34.9	−44.5	1,303	7.7	1.7	1,303	−54.3	2.8
68	1,321	38.3	30.9	1,320	56.2	77.4⎤						
						⎬	1,296	75.3	4.3	1,295	96.0	4.0
67	1,318	10.9	28.9	1,316	8.9	−1.4⎦						
66	1,310	14.6	26.0	1,307	5.5	15.4	1,283	13.5	2.6	1,282	37.8	2.7
65	1,291	5.6	10.7	1,290	22.0	7.6⎤						
						⎬	1,267	49.8	5.1	1,269	−64.9	5.1
64	1,289	35.4	−29.6	1,285	37.7	−15.9⎦						
63	1,281	43.2	34.4	1,280	38.2	15.4	1,253	49.2	5.4	1,253	24.0	4.3
62	1,263	2.4	1.4	1,260	11.7	0.1	1,236	7.8	3.7			
61	1,249	74.7	−60.2	1,248	59.9	−64.1	1,226	69.0	3.6	1,226	−80.4	3.7
60	1,236	12.4	−5.7	1,235	19.2	10.1⎤				1,213	3.8	1.8
						⎬	1,212	24.1	4.3			
59	1,230	10.5	13.2	1,232	4.0	−4.3⎦				1,208	−5.3	4.2
58	1,210	80.8	−100.3	1,213	83.2	−95.5	1,192	68.8	4.0	1,192	−88.2	4.0
57	1,200	13.0	13.4	1,198	26.6	24.4	1,177	50.7	3.9	1,176	43.4	4.1

(Continued)

TABLE 7.11 (CONTINUED)
Calculated and Experimental Frequencies, Dipole Strengths, and Rotational Strengths for 1[a]

Mode	B3LYP ν	B3LYP D	B3LYP R	B3PW91 ν	B3PW91 D	B3PW91 R	IR ν	IR D	IR γ	VCD ν	VCD R	VCD γ
56	1,187	52.8	25.9	1,186	166.5	77.7	1,165	86.0	3.6	1,165	39.5	3.3
55	1,180	87.9	1.3	1,179	90.4	8.4	1,159	27.3	3.2	1,158	−25.0	3.5
54	1,151	250.9	104.6	1,161	389.8	−53.9	1,141	267.7	5.4	1,142	−38.5	5.5
53	1,145	374.2	24.9	1,152	76.2	99.7	1,128	162.0	4.8	1,129	83.3	4.3
52	1,130	7.3	1.5	1,139	5.3	−8.4	1,116	13.4	4.3	1,113	3.3	3.0
51	1,102	92.7	−44.6	1,108	74.9	−65.7	1,089	79.2	3.6	1,089	−41.6	3.6
50	1,092	17.8	17.6	1,099	12.9	16.7	1,078	18.6	3.4	1,077	19.5	2.9
49	1,081	62.6	9.5	1,092	84.2	34.9	1,072	43.5	3.3			
48	1,077	92.3	−52.1	1,088	67.2	3.2	1,061	121.8	3.4			
47	1,073	32.0	49.4	1,073	1.2	−1.9				1,053	−29.2	3.4
46	1,056	11.2	−9.8	1,063	9.8	−12.8	1,045	13.6	5.1	1,044	−6.6	2.4
45	1,036	3.7	−27.2	1,037	1.3	−17.4	1,023	5.8	6.0	1,019	−11.1	2.8
44	1,024	4.4	−8.3	1,020	4.8	−6.4	1,006	5.0	3.9	1,005	−6.2	2.4
43	988	16.2	−20.0	996	4.3	−6.2	979	10.4	5.0	980	−9.0	3.1
42	982	16.5	24.0	985	30.7	−12.1				971	8.4	2.6
							966	57.1	3.5			
41	977	22.1	−5.0	981	10.0	29.9				966	−14.7	3.0
40	963	6.8	6.7	969	10.1	11.7	953	9.3	5.4	955	15.0	4.3
39	949	3.3	17.3	945	3.0	16.6						
38	932	17.7	−33.4	938	14.6	−44.6						
37	903	9.5	12.9	910	9.3	16.6						
36	879	25.4	12.6	883	31.9	16.6	870	24.5	2.2	870	32.4	4.5
35	865	32.2	−29.6	869	24.2	−34.8	855	25.4	2.8	854	−50.0	3.5
34	852	8.6	−13.7	857	10.2	−10.2	844	9.8	2.2	844	−13.0	2.3
33	837	2.2	−5.7	845	3.3	−3.2	834	2.7	2.3			

[a] Frequencies ν and bandwidths γ in cm⁻¹; dipole strengths D in 10^{-40} esu² cm²; rotational strengths R in 10^{-44} esu² cm². Experimental rotational strengths are for (−)-1; calculated rotational strengths are for (1R,2R,5S,8R,11R)-1a.

[b] Using the cc-pVTZ basis set.

[c] From Lorentzian fitting of the IR and the VCD spectra.

modes 33–86, demonstrates that the basis set error of the TZ2P basis set is not significantly greater than the basis set error of the larger cc-pVTZ basis set, as demonstrated in Chapter 6.

In reference 4, the determination of the AC of (−)-1 using VCD was supported by comparison of TDDFT calculations of the optical rotation (OR) and electronic circular dichroism (ECD) of (1R,2R,5S,8R,11R)-1 to the experimental OR and ECD of (−)-1.

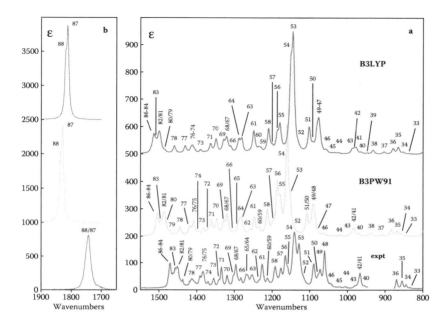

FIGURE 7.35 Calculated, B3LYP/cc-pVTZ, B3PW91/cc-pVTZ, and experimental IR spectra of (–)-**1**. The calculated spectra have Lorentzian bandshapes, (a) (γ = 4.0 cm^{-1}) and (b) (γ = 6.0 cm^{-1}). The assignment of the experimental spectrum is based on the B3PW91/cc-pVTZ spectrum.

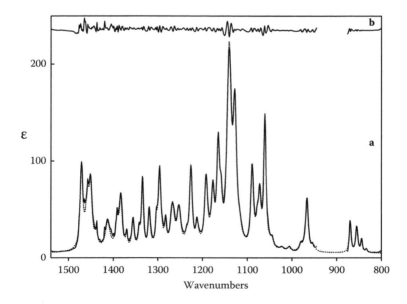

FIGURE 7.36 (a) Lorentzian fit of the experimental IR spectrum of (–)-**1** over the range 1,550–800 cm^{-1}. The solid line is the experimental spectrum; the dotted line is the Lorentzian fit. (b) The difference spectrum: the experimental spectrum minus the Lorentzian fit.

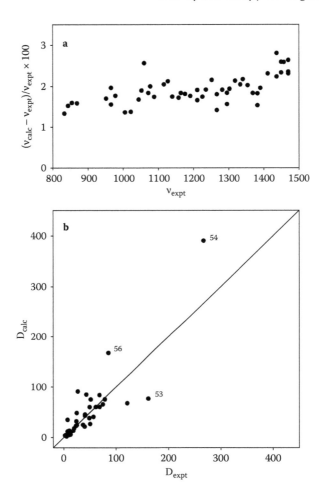

FIGURE 7.37 Comparison of B3PW91/cc-pVTZ-calculated and experimental frequencies (a) and dipole strengths (b) of modes 33–86 of (–)-**1**. The line in b is of slope +1.

The determination of the AC of the natural product quadrone using VCD demonstrates the utility of VCD in determining the ACs of natural products. Further applications of VCD to the determinations of the ACs of the natural products, schizozygine, plumericin, and the sex pheromone of the obscure mealybug will be discussed later in this chapter.

REFERENCES

1. (a) R.L. Ranieri, G.J. Calton, Quadrone, A New Antitumor Agent from *Aspergillus terreus*, *Tetra. Lett.*, 499–502, 1978; (b) G.J. Calton, R.L. Ranieri, M.A. Espenshade, Quadrone, a New Antitumor Substance Produced by *Aspergillus terreus*. Production, Isolation and Properties.*J. Antibiot.*, 31, 38–42, 1978.

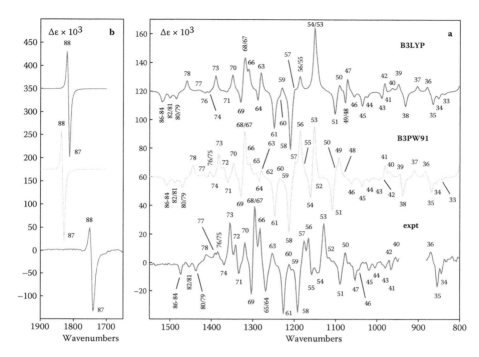

FIGURE 7.38 Calculated, B3LYP/cc-pVTZ, and B3PW91/cc-pVTZ VCD spectra of (1*R*,2*R*,5*S*,8*R*,11*R*)-**1**, and experimental VCD spectra of (−)-**1**. The calculated spectra have Lorentzian bandshapes, (a) (γ = 4.0 cm⁻¹) and (b) (γ = 6.0 cm⁻¹). The assignment of the experimental spectrum is based on the B3PW91/cc-pVTZ spectrum.

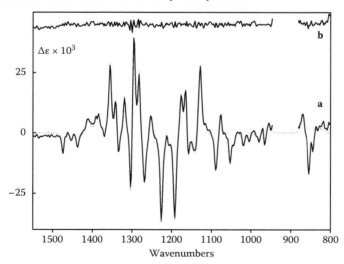

FIGURE 7.39 (a) Lorentzian fit of the experimental VCD spectrum of (−)-**1** over the range 1,550–800 cm⁻¹. The solid line is the experimental spectrum; the dotted line is the Lorentzian fit. (b) The difference spectrum: the experimental spectrum minus the Lorentzian fit.

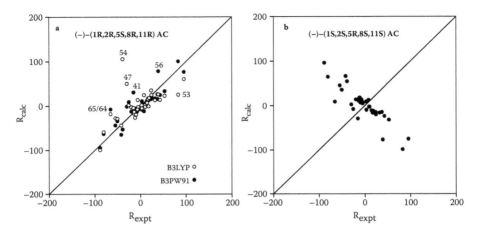

FIGURE 7.40 Comparison of experimental rotational strengths of modes 34–86 for (–)-**1** to the B3LYP/cc-pVTZ and B3PW91/cc-pVTZ rotational strengths for (1*R*,2*R*,5*S*,8*R*,11*R*)-**1** (a) and the B3PW91/cc-pVTZ rotational strengths for (1*S*,2*S*,5*R*,8*S*,11*S*)-**1** (b). The lines have slopes of +1.

TABLE 7.12
Deviations of the B3PW91/cc-pVTZ Rotational Strengths Calculated for Both Enantiomers of 1 from the Experimental Rotational Strengths of Modes 34–86

Enantiomer	Mean Absolute Deviation[a]	RMS Deviation[b]
(1*R*,2*R*,5*S*,8*R*,11*R*)	13.2	18.4
(1*S*,2*S*,5*R*,8*S*,11*S*)	52.7	72.6

[a] Average of the absolute deviations of calculated rotational strengths for (1*R*,2*R*,5*S*,8*R*,11*R*)-**1** and (1*S*,2*S*,5*R*,8*S*,11*S*)-**1** from the experimental rotational strengths of (–)-**1**.

[b] RMS deviation of calculated and experimental rotational strengths.

2. A.B. Smith, J.P. Konopelski, Total Synthesis of (+)-Quadrone: Assignment of Absolute Stereochemistry, *J. Org. Chem.*, 49, 4094–4095, 1984.

3. A.B. Smith, J.P. Konopelski, B.A. Wexler, P.A. Sprengeler, Quadrone Structural and Synthetic Studies. Total Synthesis of Natural (–)-Quadrone, the (+)-Enantiomer and the Racemate. Conformational Analysis, Circular Dichroism and Determination of Absolute Stereochemistry, *J. Am. Chem. Soc.*, 113, 3533–3542, 1991.

4. P.J. Stephens, D.M. McCann, F.J. Devlin, A.B. Smith III, Determination of the Absolute Configurations of Natural Products via Density Functional Theory Calculations of Optical Rotation, Electronic Circular Dichroism and Vibrational Circular Dichroism: The Cytotoxic Sesquiterpene Natural Products Quadrone, Suberosenone, Suberosanone and Suberosenol A Acetate, *J. Nat. Prod.*, 69, 1055–1064, 2006.

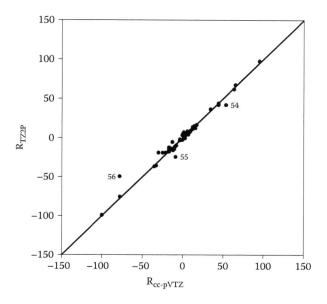

FIGURE 7.41 Comparison of the B3PW91/TZ2P and B3PW91/cc-pVTZ rotational strengths of modes 33–86 for (1R,2R,5S,8R,11R)-**1**. The line has a slope of +1.

A CHIRAL OXADIAZOL-3-ONE CALCIUM CHANNEL BLOCKER

The drug diltiazem, **1**, is an L-type calcium channel blocker, used to treat hypertension and angina pectoris. Budriesi et al. synthesized a large number of derivatives of thiazino-oxadiazolone, **2**, and assayed their activities as calcium channel blockers [1]. The chiral derivative, 8-(4-bromophenyl)-8-ethoxy-5-methyl-8H-[1,4] thiazino-[3,4-c][1,2,4]oxadiazol-3-one, **3**, was the most active, the racemate having an activity 20 times greater than diltiazem. In order to determine the enantioselectivity of the calcium channel blocker activity of **3**, chiral chromatography was used to resolve the enantiomers of **3** by Professor Gasparrini (University of Rome, Italy), [1,2], and the activities of the two enantiomers were determined; the activity of (−)-**3** was shown to be greater than that of (+)-**3** [1].

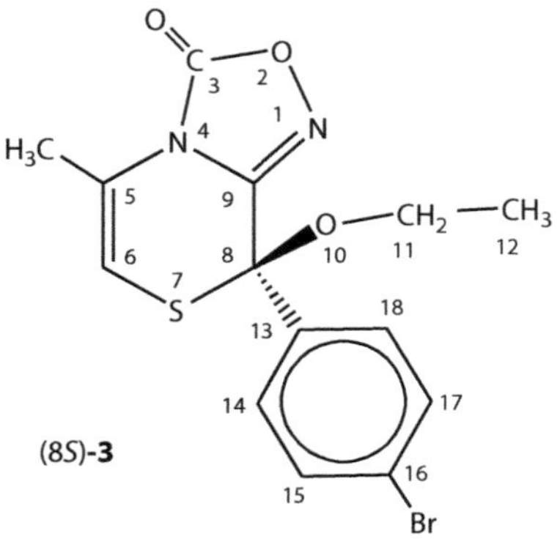

(8S)-**3**

Since chromatography does not determine the ACs of enantiomers, we determined the AC of **3** using VCD in 2005 [1,2]. The VCD spectra of (+), (–), and (±)-**3** were measured using 0.05 M CDCl$_3$ solutions and a cell pathlength of 597 μ. The ee's of the (+) and (–) samples were determined by chromatography to be 94 and >99%, respectively. The baselines of the (+) and (–) VCD spectra were the (±) VCD spectrum. Normalization of the (+) and (–) VCD spectra to 100% ee led to the half-difference VCD spectrum, ½[Δε(+) – Δε(–)] shown in Figure 7.42, together with the IR spectrum of (±)-**3**, over the range 800–1,700 cm^{-1}.

Molecule **3** is conformationally flexible due to the rotations of the ethoxide group around the C$_8$-O$_{10}$ and O$_{10}$-C$_{11}$ bonds, and the p-Br-phenyl group around the C$_8$-C$_{13}$ bond, and puckering of the thiazino ring. Conformational analysis using MMFF94 led to 16 conformations, within 10 kcal/mol [2]. Reoptimizations using B3LYP/6-31G* led to 13 independent conformations [2]. Calculations of the relative free energies and room temperature populations predicted that only the three lowest free energy conformations have populations of >2% [2]. The B3LYP/6-31G* structures of these three conformations, **a-c**, of S-**3**, are:

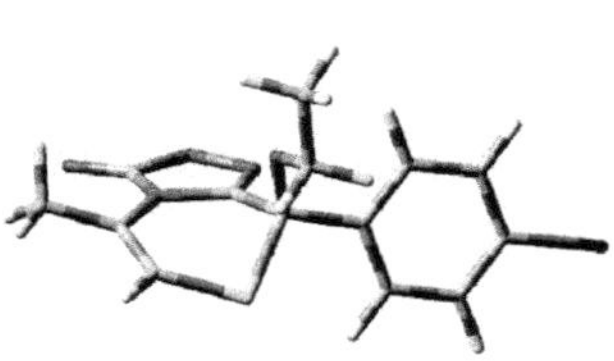

a

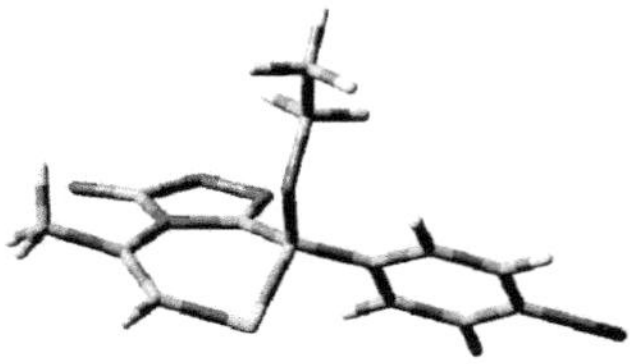

b

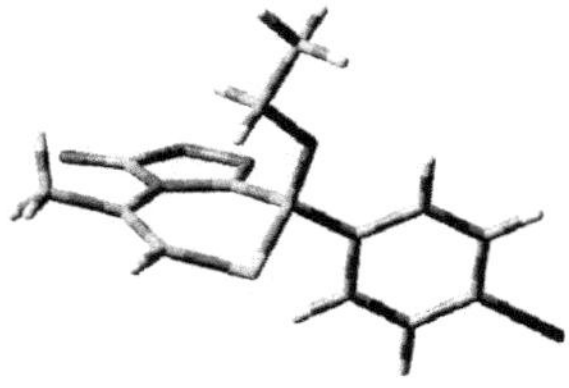

c

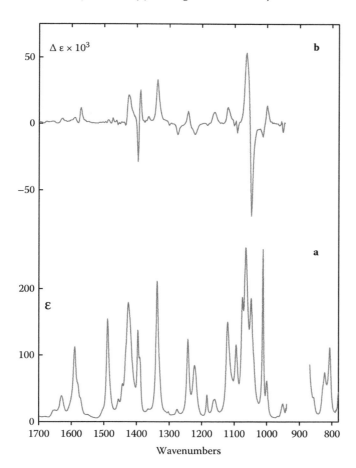

FIGURE 7.42   (a) IR spectrum of (±)-**3**; (b) VCD spectrum of (+)-**3**. The gap at ~900 cm⁻¹ is
due to strong absorption of the CDCl₃ solvent.

The B3LYP/TZ2P and B3PW91/TZ2P conformationally averaged IR and VCD
spectra of *S*-**3** are compared to the experimental IR and VCD spectra of (±)-**3** and (+)-**3**
in Figures 7.43 and 7.44. The B3PW91 IR and VCD spectra are in better agreement
with the experimental IR and VCD spectra than the B3LYP IR and VCD spectra.
As a result, the assignments of the experimental IR and VCD spectra shown in
Figures 7.43 and 7.44 are based on the B3PW91/TZ2P spectra. The agreements of
the calculated and experimental IR and VCD spectra support the reliability of the
conformational analysis of **3**.

Comparisons of the B3PW91/TZ2P dipole strengths and rotational strengths
of *S*-**3** to the experimental dipole and rotational strengths, obtained by Lorentzian

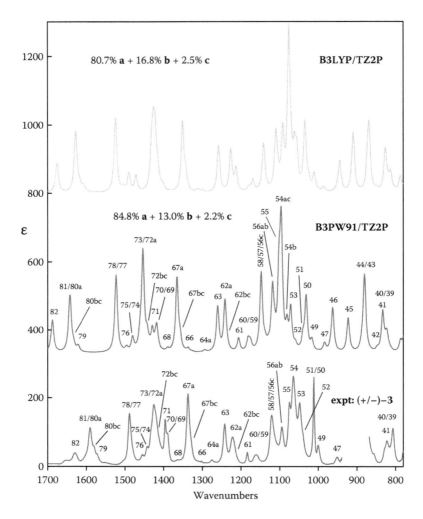

FIGURE 7.43 Comparison of the conformationally averaged B3LYP/TZ2P and B3PW91/TZ2P IR spectra of **3** to the experimental IR spectrum of (±)-**3**. The calculated spectra have Lorentzian bandshapes ($\gamma = 4.0$ cm^{-1}). Assignment of the experimental spectrum is based on the B3PW91/TZ2P spectrum. When no conformation is specified the fundamentals of **a**, **b**, and **c** are unresolved.

fitting of the experimental IR and VCD spectra of (±)-**3** and (+)-**3**, are shown in Figures 7.45 and 7.46.

The agreements of the calculated and experimental dipole and rotational strengths support the assignments of the experimental IR and VCD spectra. The agreements of the calculated and experimental VCD spectra and rotational

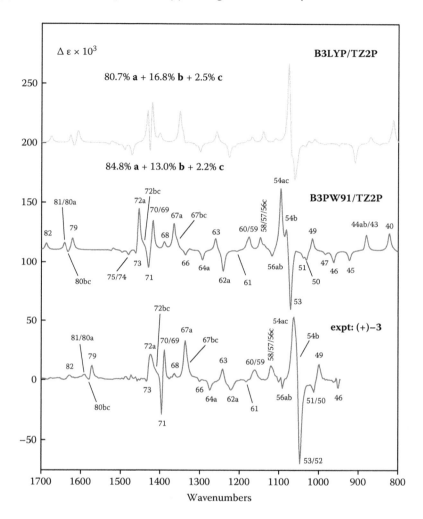

FIGURE 7.44 Comparison of the conformationally averaged B3LYP/TZ2P and B3PW91/TZ2P VCD spectra of (*S*)-**3** to the experimental VCD spectrum of (+)-**3**. Assignment of the latter is based on the B3PW91/TZ2P VCD spectrum. When no conformation is specified the fundamentals of **a**, **b**, and **c** are unresolved. The calculated spectra have Lorentzian band-shapes ($\gamma = 4.0$ cm^{-1}).

strengths prove that the AC of **3** is *S*(+)/*R*(−). Since (−)-**3** is more active as a calcium channel blocker than (+)-**3**, we conclude that the *R* enantiomer of **3** is more active.

This study illustrates the utility of VCD in determining the AC of a chiral molecule exhibiting pharmaceutical activity.

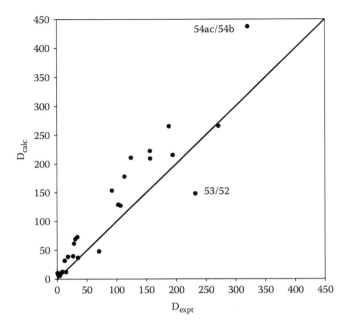

FIGURE 7.45 Comparison of B3PW91/TZ2P and experimental vibrational dipole strengths of modes 82–49. Dipole strengths are in 10^{-40} esu^2 cm^2.

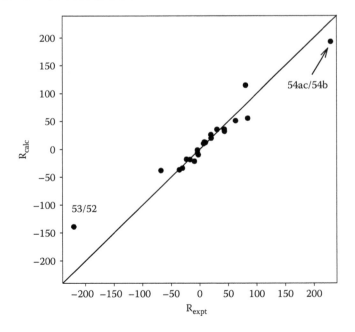

FIGURE 7.46 Comparison of the B3PW91/TZ2P rotational strengths for (S)-**3** to the experimental rotational strengths of modes 82–49 of (+)-**3**. Rotational strengths are in 10^{-44} esu^2 cm^2.

REFERENCES

1. E. Carosati, G. Cruciani, A. Chiarini, R. Budriesi, P. Ioan, R. Spisani, D. Spinelli, B.Cosimelli, F. Fusi, M. Frosini, R. Matucci, F. Gasparrini, A. Ciogli, P.J. Stephens, and F.J. Devlin, Calcium Channel Antagonists Discovered by a Multidisciplinary Approach, *J. Med. Chem.*, 49, 5206–5216, 2006.
2. P.J. Stephens, F.J. Devlin, F. Gasparrini, A. Ciogli, D. Spinelli, and B. Cosimelli, Determination of the Absolute Configuration of a Chiral Oxadiazol-3-One Calcium Channel Blocker, Resolved Using Chiral Chromatography, via Concerted Density Functional Theory Calculations of Its Vibrational Circular Dichroism, Electronic Circular Dichroism and Optical Rotation, *J. Org. Chem.*, 72, 4707–4715, 2007.

THE ALKALOID NATURAL PRODUCT, SCHIZOZYGINE

The East African plant *Schizozygia caffaeoides*, belonging to the Apocynaceae family, has been used traditionally in Kenya for the treatment of skin diseases due to fungal infections [1]. In the 1960s, a group of alkaloids were isolated from this plant by Renner et al. [2]. One of these was named schizozygine. Schizozygine, **1**, exhibits optical rotation: $[\alpha]_D = +15.5$ (c 1, CHCl$_3$) [2a], and is therefore chiral. Renner et al. deduced the structure and relative configuration of schizozygine, using its chemical reactions and spectroscopic properties [2]. The AC was not determined, so there were two possibilities:

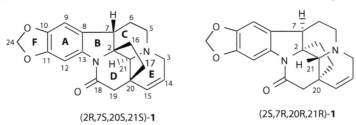

(2R,7S,20S,21S)-**1**   (2S,7R,20R,21R)-**1**

Surprisingly, until very recently, the AC of schizozygine has remained undetermined. In 2006, we therefore undertook the determination of the AC of schizozygine, in collaboration with Professors Hájíček and Urbanová (Prague, Czech Republic), using VCD [3].

EXPERIMENTAL IR AND VCD SPECTRA

The natural product, **1**, was isolated and purified by Professor Hájíček, the sample exhibiting $[\alpha]_D = +18.1$ (c 0.63, CHCl$_3$). Its IR and VCD spectra were measured using solutions in CDCl$_3$ of concentrations in the range 0.1–0.5 M, and a cell of pathlength 99 μ. Since (–)-**1** and (±)-**1** were not available, the baseline for the VCD spectra of (+)-**1** was the spectrum of the CDCl$_3$ solvent. In order to minimize artefacts in the composite VCD spectrum, experimental VCD spectra over frequency ranges with A > 1.0 were deleted. The resulting IR and VCD spectra over the range 800–1,750 cm^{-1} are shown in Figure 7.47.

CONFORMATIONAL ANALYSIS

MMFF94 Monte Carlo conformational searching, with a 20 kcal/mol window, surprisingly found only two conformations of **1**, **a** and **b**, differing in energy by 1.69 kcal/mol. These conformations differ principally with regard to the conformation

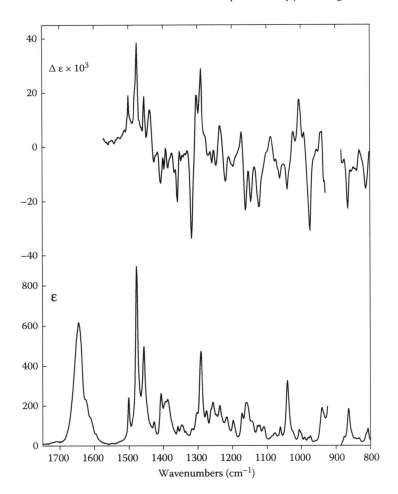

FIGURE 7.47 The experimental IR and VCD spectra of (+)-**1**.

of ring C, which is a boat in **a** and a chair in **b**. Reoptimization of **1a** and **1b** using DFT, at the B3LYP/6-31G* level, lowered the relative energy of **1b** and **1a** to 0.37 kcal/mol. At the same time, the conformation of ring F changed considerably: in the MMFF94 geometries of **1a** and **1b**, ring F is very close to planar; in contrast, in the B3LYP/6-31G* geometries of **1a** and **1b**, ring F is substantially puckered. In order to determine whether other conformations with oppositely puckered rings F exist, B3LYP/6-31G* PES scans of **1a** and **1b** were carried out as a function of the dihedral angle $C_9C_{10}OC_{24}$, with the results shown in Figure 7.48.

It is clear that additional stable conformations, **1a′** and **1b′**, with C_{24} oppositely puckered to **1a** and **1b**, exist. Further optimization led to the B3LYP/6-31G* geometries of **1a′** and **1b′**. The relative energies of **1a**, **1a′**, **1b**, and **1b′** are given in Table 7.13. Harmonic frequency calculations for **1a**, **1a′**, **1b**, and **1b′** confirmed

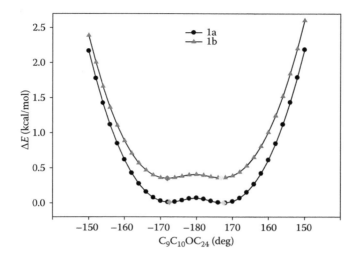

FIGURE 7.48 B3LYP/6-31G* PES scans for **1a** and **1b** as a function of the dihedral angle, $C_9C_{10}OC_{24}$. The dihedral angle was varied in 2° steps. The diamond symbols indicate the B3LYP/6-31G* optimized values of the dihedral angle for each conformation: **1a** = 172.5°, **1a′** = −172.5°, **1b** = −171.9°, **1b′** = 173.2°.

TABLE 7.13

Calculated Relative Energies (ΔE^a), Relative Free Energies (ΔG^a) and Room Temperature Percentage Populations (P^b) of the Conformers of 1

	MMFF94	B3LYP/6-31G*			B3LYP/TZ2P			B3PW91/TZ2P		
Conformer	ΔE	ΔE	ΔG	$P(\%)$	ΔE	ΔG	$P(\%)$	ΔE	ΔG	$P(\%)$
1a	0.00	0.000	0.038	36.20	0.000	0.029	35.13	0.000	0.020	35.59
1a′		0.011	0.000	38.62	0.023	0.000	36.89	0.030	0.000	36.82
1b	1.69	0.371	0.670	12.46	0.331	0.589	13.63	0.336	0.577	13.88
1b′		0.365	0.658	12.72	0.307	0.559	14.35	0.316	0.585	13.71

[a] In kcal/mol.
[b] Based on ΔG at T = 298 K.

that they are all stable conformations and gave the relative free energies and room temperature populations also given in Table 7.13.

Reoptimizations at the B3LYP/TZ2P and B3PW91/TZ2P levels led to the relative energies and free energies and populations also given in Table 7.13. There are only small changes from the B3LYP/6-31G* results. The final conclusion is that four conformations, **1a**, **1a′**, **1b**, and **1b′**, are significantly populated at room temperature and must all be included in predicting the IR and VCD spectra of **1**.

CALCULATED IR AND VCD SPECTRA

The IR and VCD spectra of the four conformations of **1**, **1a**, **1a′**, **1b**, and **1b′**, were calculated at the B3LYP/TZ2P and B3PW91/TZ2P levels. For both functionals, the IR spectra are identical for **1a** and **1a′** and for **1b** and **1b′**, but significantly different for **1a** and **1b** and for **1a′** and **1b′**, showing that the IR spectrum of **1** is insensitive to the puckering of ring F, but sensitive to the conformation of ring C. The B3LYP/TZ2P and B3PW91/TZ2P conformationally averaged IR spectra are compared to the experimental IR spectrum for the range 800–1,800 cm^{-1} in Figure 7.49. The B3PW91/TZ2P IR spectrum is in better agreement with the experimental spectrum than is the B3LYP/TZ2P spectrum; as a result, the experimental IR spectrum is assigned using the B3PW91/TZ2P spectrum, as detailed in Figure 7.49. For a number of modes, the calculated frequencies for **1a**, **1a′** and for **1b**, **1b′** differ significantly; these conformational splittings are observed in the experimental spectrum, providing support for the reliability of the conformational analysis of **1** and for the B3PW91/TZ2P geometries of the conformations **1a**, **1a′**, **1b**, and **1b′**. Lorentzian fitting of the experimental IR spectrum gives the frequencies and dipole strengths of the resolved bands [3].

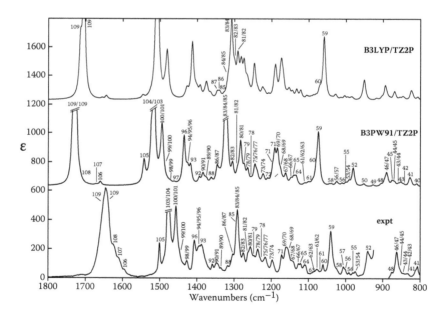

FIGURE 7.49 (SEE COLOR INSERT.) Comparison of the experimental and conformationally averaged B3LYP/TZ2P and B3PW91/TZ2P IR spectra of **1**. The assignment of the experimental spectrum is based on the B3PW91/TZ2P spectrum. The numbers define the fundamentals contributing to resolved bands. Red and green numbers indicate bands of **1a/1a′** and **1b/1b′**, respectively. Black numbers indicate superpositions of bands of **1a**, **1a′**, **1b**, and **1b′**. The bandshapes of the calculated spectra are Lorentzian ($\gamma = 4.0$ cm^{-1}).

The B3LYP/TZ2P and B3PW91/TZ2P conformationally averaged VCD spectra of (2*R*,7*S*,20*S*,21*S*)-**1** are in much better agreement with the experimental VCD spectrum of (+)-**1** than are the VCD spectra of (2*S*,7*R*,20*R*,21*R*)-**1**, leading to the expectation that the AC of (+)-**1** is (2*R*,7*S*,20*S*,21*S*). The VCD spectra of (2*R*,7*S*,20*S*,21*S*)-**1** are compared to the experimental VCD spectrum of (+)-**1** in Figure 7.50.

The resolved bands of the experimental VCD spectrum are first assigned on the basis of the assignment of the IR spectrum. Comparison of this assignment to the calculated spectra shows that, as with the IR spectra, the B3PW91/TZ2P VCD spectrum is in better agreement with the experimental VCD spectrum than is the B3LYP/TZ2P VCD spectrum. Comparison of the B3PW91/TZ2P and experimental VCD spectra confirms the assignment based on the IR spectrum assignment. The VCD assignment is detailed in Figure 7.50. Lorentzian fitting of the experimental VCD spectrum gives the rotational strengths of the resolved bands [3]. Comparison of these experimental rotational strengths to the conformationally averaged B3PW91/TZ2P rotational strengths of the assigned transitions of (2*R*,7*S*,20*S*,21*S*)-**1** is shown in Figure 7.51.

The agreement of the calculated and experimental rotational strengths is not perfect. However, comparison of the experimental rotational strengths to the rotational strengths of the same transitions for (2*S*,7*R*,20*R*,21*R*)-**1**, also shown in Figure 7.51, gives enormously worse agreement. As a result, the AC of (+)-**1** is unambiguously confirmed to be (2*R*,7*S*,20*S*,21*S*).

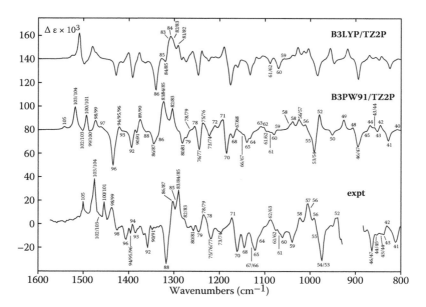

FIGURE 7.50 (SEE COLOR INSERT.) Comparison of the experimental VCD spectrum of (+)-**1** and the conformationally averaged B3LYP/TZ2P and B3PW91/TZ2P VCD spectra of (2*R*,7*S*,20*S*,21*S*)-**1**. Bandshapes of the calculated spectra are Lorentzian ($\gamma = 4.0$ cm^{-1}). The assignment of the experimental spectrum is based on the B3PW91/TZ2P-calculated spectrum. Fundamentals are numbered as in Figure 7.49.

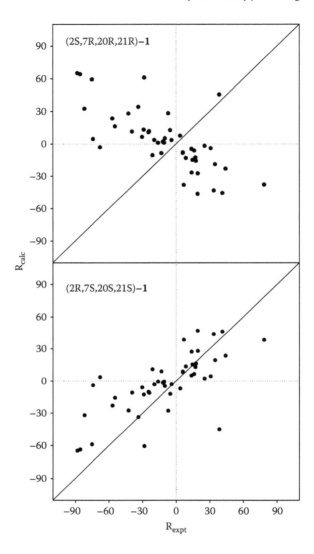

FIGURE 7.51 Comparison of the experimental and B3PW91/TZ2P-calculated rotational strengths of **1**. Experimental rotational strengths are for (+)-**1**. Calculated rotational strengths are for (2R,7S,20S,21S)-**1** and (2S,7R,20R,21R)-**1**. Calculated rotational strengths are population-weighted averages. The solid lines are of +1 slope. Rotational strengths R are in 10^{-44} esu^2 cm^2.

The deviations between the B3PW91/TZ2P rotational strengths of (2R,7S,-20S,21S)-**1** and the experimental rotational strengths of (+)-**1** are most likely due predominantly to errors in the experimental rotational strengths, resulting from significant artefacts in the experimental VCD spectrum, due to the availability of only a single enantiomer of **1**. Given the conformational flexibility of **1**, it is also possible that there are significant errors in the calculated rotational strengths due to errors in the predicted conformational populations.

Additional support for the AC of (+)-1, assigned on the basis of its VCD spectrum, was provided by comparison of time-dependent density functional theory (TDDFT) calculations of the optical rotatory dispersion and electronic circular dichroism of (2R,7S,20S,21S)-1 to the experimental ORD (589–436 nm) and ECD (360–220nm) [3].

Three other alkaloids, schizogaline, **2**, schizogamine, **3**, and 6,7-dehydro-19β-hydroxyschizozygine, **4**, have been isolated from *S. caffaeoides* and shown to have very similar structures to schizozygine [1,2]. Assuming a common biosynthetic pathway of the alkaloids **1–4** leads to the conclusion that the ACs of naturally occurring **2–4** are identical to that of (+)-**1**.

Additional studies of the ACs of alkaloid natural products using VCD have been carried out for iso-schizogaline and iso-schizogamine [4].

The determination of the ACs of schizozygine, iso-schizogaline and iso-schizogamine, using VCD, confirms the utility of VCD in determining the ACs of natural products, concluded from the preceding study of quadrone.

REFERENCES

1. R.M. Kariba, P.J. Houghton, A. Yenesew, Antimicrobial Activities of a New Schizozygane Indoline Alkaloid from Schizozygia caffaeoides and the Revised Structure of Isoschizogaline, *J. Nat. Prod.*, 65, 566–569, 2002.
2. (a) U. Renner, P. Kernweisz, Alkaloide aus Schizozygia caffaeoides (Boj.) Baill, *Experientia*, 19, 244–246, 1963; (b) U. Renner, Alkaloide aus Schizozygia caffaeoides. III. Strukturelle Beziehungen zwischen Schizozygin und einigen Nebenalkaloiden, *Lloydia*, 27, 406–415, 1964; (c) U. Renner, H. Fritz, Alkaloide aus Schizozygia caffaeoides (Boj.)Baill. II. Die Struktur des Schizozygins. *Helv. Chim. Acta*, 48, 308–317, 1965; (d) M. Hesse, U. Renner, Die Massenspektren von Schizozygin unddessen Derivaten, *Helv. Chim. Acta*, 49, 1875–1899, 1966.
3. P.J. Stephens, J.J. Pan, F.J. Devlin, M. Urbanová, and J. Hájíček, Determination of the Absolute Configurations of Natural Products via Density Functional Theory Calculations of Vibrational Circular Dichroism, Electronic Circular Dichroism and Optical Rotation: The Schizozygane Alkaloid Schizozygine, *J. Org. Chem.*, 72, 2508–2524, 2007.
4. P.J. Stephens, J.J. Pan, F.J. Devlin, M. Urbanová, O. Julinek, and J. Hájíček, Determination of the Absolute Configurations of Natural Products via Density Functional Theory Calculations of Vibrational Circular Dichroism, Electronic Circular Dichroism and Optical Rotation: The Isoschizozygane Alkaloids Isoschizogaline and Isoschizogamine. *Chirality*, 20, 454–470, 2008.

THE IRIDOID NATURAL PRODUCT, PLUMERICIN

In 1951 Little and Johnstone isolated the iridoid natural product, plumericin, **1**, from the roots of the plant *Plumeria multiflora*, and demonstrated that it possesses anti-fungal and antibacterial activity [1]. In 1960–61, Albers-Schönberg and Schmid iso-lated plumericin from the roots of *Plumeria rubra var. alba* and carried out both chemical reactions and spectroscopic studies of **1** [2]. On the basis of the results obtained, the structure of **1** was assigned as:

(1R,5S,8S,9S,10S)-**1**

Subsequently, plumericin was isolated from many other plants [3]. In 2005, Elsässer et al. determined the x-ray crystal structure of **1** and confirmed that the Albers-Schönberg and Schmid structure and relative configuration are both correct [4]. Given the absence of heavy atoms in **1**, the x-ray study could not confirm the AC of **1** assigned by Albers-Schönberg and Schmid. Elsässer et al. therefore calculated the ECD of the lowest-energy electronic transitions of **1**, using a semiempirical molecular-orbital theory methodology [5] and, on the basis of the comparison of the predicted ECD spectrum to the experimental ECD spectrum of naturally occurring **1**, concluded that the AC of the latter is in fact *1S,5R,8R,9R,10R*, opposite to the AC assigned by Albers-Schönberg and Schmid. This result also led to the conclusion that the ACs of many other iridoid natural products, which had been assigned by comparison to the Albers-Schönberg and Schmid AC of **1**, were incorrect.

In order to confirm the AC assigned by Elsässer et al., in 2006 we determined the AC of plumericin using VCD, in collaboration with Professors Krohn (Paderborn, Germany) and Kurtán (Debrecen, Hungary) [6].

EXPERIMENTAL IR AND VCD SPECTRA

The plumericin natural product was isolated from *Plumeria rubra* and purified as described by Elsässer et al. [4]. The $[\alpha]_D$ value was +178 (c 2.01, CDCl$_3$). The IR and VCD spectra of (+)-**1** were measured using solutions in both CHCl$_3$ and CDCl$_3$ of concentration 0.07 M and cells of pathlengths 109, 239, and 597 μ. The resulting IR and VCD spectra in the range 760–1,500 cm^{-1} are shown in Figure 7.52. Since (–)-**1**

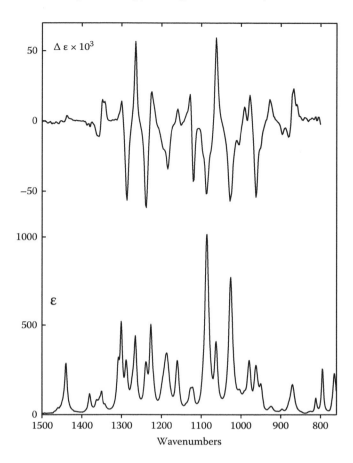

FIGURE 7.52 The mid-IR experimental IR and VCD spectra of (+)-**1**.

and (±)-**1** were not available, the baselines for the VCD spectra of (+)-**1** were the spectra of the solvents.

CONFORMATIONAL ANALYSIS

MMFF94 Monte Carlo conformational searching, with a 10 kcal/mol window, found four conformations of **1**, **a–d**, with the relative energies given in Table 7.14. DFT optimization of these conformations at the B3LYP/6-31G* level, and subsequently, at the B3LYP/TZ2P and B3PW91/TZ2P levels, led to the relative energies also given in Table 7.14. Harmonic frequency calculations confirmed the stabilities of the four conformations and gave the relative free energies and room temperature percentage populations also given in Table 7.14. The B3PW91/TZ2P geometries of **1a–1d** are shown in Figure 7.53. Conformations **1a** and **1b** differ in the orientation of

TABLE 7.14

Calculated Relative Energies (ΔE^a), Relative Free Energies (ΔG^a) and Room Temperature Percentage Populations (P^b) of the Conformations of 1

Conformer	MMFF94	B3LYP/6-31G*			B3LYP/TZ2P			B3PW91/TZ2P		
	ΔE	ΔE	ΔG	$P(\%)^b$	ΔE	ΔG	$P(\%)^b$	ΔE	ΔG	$P(\%)^b$
1a	0.00	0.00	0.00	59.26	0.00	0.00	70.68	0.00	0.00	68.89
1b	0.49	0.35	0.39	30.67	0.64	0.63	24.41	0.59	0.59	25.63
1c	1.07	0.94	1.22	7.55	1.56	1.69	4.06	1.44	1.62	4.44
1d	1.80	1.55	1.87	2.52	2.46	2.61	0.85	2.82	2.48	1.04

a In kcal/mol.

b Based on ΔG, at T = 298 K.

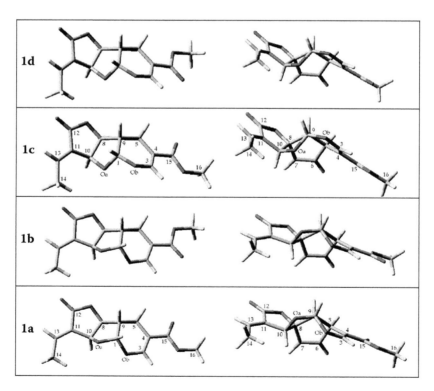

FIGURE 7.53 The B3PW91/TZ2P conformations of $(1R,5S,8S,9S,10S)$-**1**. The structures on the left are viewed with the plane of ring C horizontal, and C_9 pointing toward the viewer. The structures on the right are viewed along the C_1-C_9 bond with C_1 pointing toward the viewer.

the methoxycarbonyl substituent of ring D: in **1a** its C=O group is s-*trans* with respect to the $C_3=C_4$ group; in **1b** it is s-*cis*. In **1c** and **1d** the methoxycarbonyl group is s-*trans* and s-*cis*, respectively. In conformations **1a** and **1b**, the conformations of rings A–D are identical; in **1c** and **1d** rings B and D have different conformations; the O atoms of these rings are oppositely puckered.

Comparison of the predicted structures of conformations **1a–1d** to the x-ray structure of **1** of Elsässer et al. [4] demonstrates that the structure of **1a** is in excellent agreement with the x-ray structure, while the structures of **1b**, **1c**, and **1d** are in significantly worse agreement. Comparison of the heavy-atom dihedral angles for the B3PW91/TZ2P geometries to the x-ray dihedral angles are shown in Figure 7.54. Clearly, the structure of the lowest free energy conformation, **1a**, is in excellent agreement with the x-ray structure, supporting the reliability of the conformational analysis of **1** and of the B3PW91/TZ2P geometry for **1a**.

At room temperature, conformations **1a** and **1b** are predicted at all levels to have combined populations of 90–100%. As a result, the IR and VCD spectra are expected to be dominated by the contributions of **1a** and **1b**.

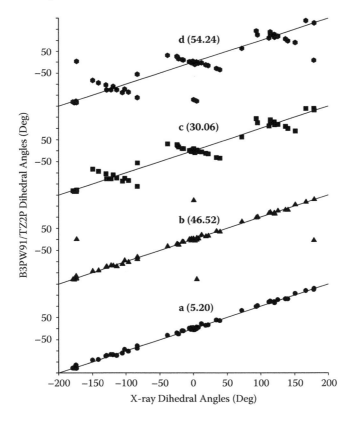

FIGURE 7.54 Comparison of the heavy-atom dihedral angles of the x-ray structure of **1** and the B3PW91/TZ2P-calculated dihedral angles for the conformations **a–d** of **1**. The numbers in parentheses are the RMS deviations between the calculated and x-ray dihedral angles.

CALCULATED IR AND VCD SPECTRA

The IR and VCD spectra of the four conformations of **1**, **1a–1d**, were calculated at the B3LYP/TZ2P and B3PW91/TZ2P levels. The conformationally averaged IR spectra are compared to the experimental IR spectrum over the range 750–1,550 cm⁻¹ in Figure 7.55.

The B3LYP/TZ2P and B3PW91/TZ2P spectra are not identical, the difference being greatest in the range 1,200–1,500 cm⁻¹. The B3PW91/TZ2P IR spectrum is in better agreement with the experimental IR spectrum, and is therefore used to assign the experimental spectrum, as detailed in Figure 7.55. As expected, the spectra of **1a** and **1b** dominate the calculated spectra. For a number of modes, the frequencies for **1a** and **1b** differ significantly; these conformational splittings are observed in the experimental spectrum, providing support for the reliability of the conformational analysis of **1** and for the B3PW91/TZ2P geometries of the conformations **1a** and **1b**.

The B3LYP/TZ2P and B3PW91/TZ2P conformationally averaged VCD spectra of (1*R*,5*S*,8*S*,9*S*,10*S*)-**1** are in much better agreement with the experimental VCD spectrum of (+)-**1** than are the VCD spectra of (1*S*,5*R*, 8*R*,9*R*,10*R*)-**1**, leading to the expectation that the AC of (+)-**1** is (1*R*,5*S*,8*S*,9*S*,10*S*). The VCD spectra of (1*R*,5*S*,8*S*,9*S*,10*S*)-**1** are compared to the experimental VCD spectrum of (+)-**1** over the range 800–1,420 cm⁻¹ in Figure 7.56. The resolved bands of the experimental VCD spectrum are first assigned on the basis of the assignment of the IR spectrum. Comparison of this assignment to the calculated spectra shows that, as with the IR spectra, the B3PW91/TZ2P VCD spectrum is in better agreement with the experimental VCD spectrum than is the B3LYP/TZ2P VCD spectrum. Comparison of the B3PW91/TZ2P and experimental VCD spectra confirms the assignment based on the IR spectrum assignment. The VCD assignment is detailed in Figure 7.56. Lorentzian fitting of the experimental VCD spectrum gives the rotational strengths of the resolved bands. Comparisons of these experimental rotational strengths to the conformationally averaged B3PW91/TZ2P rotational strengths of (1*R*,5*S*,8*S*,9*S*,10*S*)-**1** and (1*S*,5*R*,8*R*,9*R*,10*R*)-**1** are shown in Figure 7.57. The agreement is far superior for the AC 1*R*,5*S*,8*S*,9*S*,10*S*, leading to the unambiguous conclusion that this is the AC of naturally occurring plumericin, (+)-**1**. Our VCD study thus confirms the AC assigned by Albers-Schönberg and Schmid [2], and shows that the AC assigned by Elsässer et al. [4] is incorrect. The latter result leads to the conclusion that the semiempirical molecular-orbital method used for predicting the ECD of **1** is not reliable.

The deviations between the B3PW91/TZ2P rotational strengths of (1*R*,5*S*,-8*S*,9*S*,10*S*)-**1** and the experimental rotational strengths of (+)-**1** are most likely due predominantly to errors in the experimental rotational strengths, resulting from significant artefacts in the experimental VCD spectrum, due to the availability of only a single enantiomer of **1**. Given the conformational flexibility of **1**, it is also possible that there are significant errors in the calculated rotational strengths due to errors in the predicted conformational populations.

Additional support for the AC of (+)-**1**, assigned on the basis of its VCD spectrum, was provided by comparison of a TDDFT calculation of the ECD spectrum

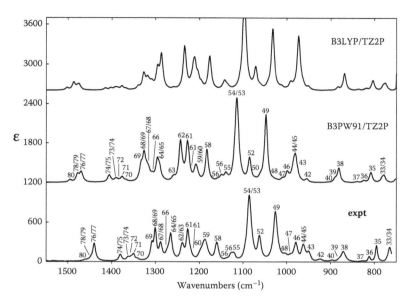

FIGURE 7.55 Comparison of the experimental and conformationally averaged B3LYP/TZ2P and B3PW91/TZ2P IR spectra of **1** for the range 750–1,550 cm⁻¹. Fundamentals are numbered as in Figure 7.56. Bandshapes of the calculated spectra are Lorentzian ($\gamma = 4.0$ cm⁻¹). The assignment of the experimental spectrum is based on the B3PW91/TZ2P-calculated spectrum.

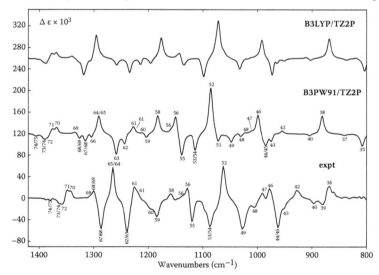

FIGURE 7.56 (SEE COLOR INSERT.) Comparison of the experimental and conformationally averaged B3LYP/TZ2P and B3PW91/TZ2P VCD spectra of $(1R,5S,8S,9S,10S)$-**1** for the range 800–1,420 cm⁻¹. Red, green, and cyan numbers indicate fundamentals of **1a**, **1b**, and **1c/1d**, respectively. Black numbers indicate superpositions of fundamentals of **1a** and **1b**. Bandshapes of the calculated spectra are Lorentzian ($\gamma = 4.0$ cm⁻¹). The assignment of the experimental spectrum is based on the B3PW91/TZ2P-calculated spectrum.

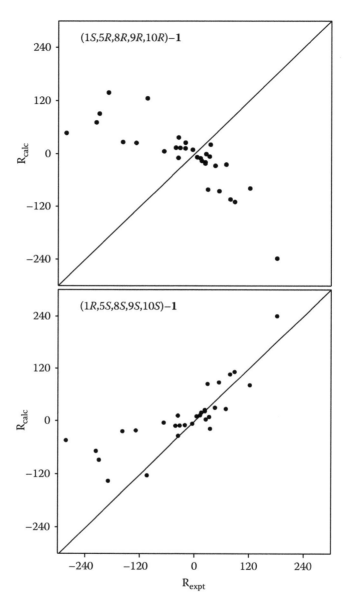

FIGURE 7.57 Comparison of the B3PW91/TZ2P-calculated and experimental rotational strengths of **1**. Experimental rotational strengths are for (+)-**1**. Calculated rotational strengths are for (1*R*,5*S*,8*S*,9*S*,10*S*)-**1** and (1*S*,5*R*,8*R*,9*R*,10*R*)-**1**. Calculated rotational strengths are population-weighted averages. The solid lines are of +1 slope. Rotational strengths R are in 10^{-44} esu² cm².

of (1R,5S,8S,9S,10S)-**1** to the experimental ECD spectrum of (+)-**1** and by comparison of a TDDFT calculation of the ORD of (1R,5S,8S,9S,10S)-**1** to the experimental ORD (589–436 nm) [6].

An isomer of plumericin, in which the C_{14} methyl group is on the other side of the C_{11}=C_{13} double bond, named isoplumericin, was also obtained by Albers-Schönberg and Schmid from *Plumeria rubra var. alba* [2]. The VCD of isoplumericin also confirmed the AC assigned by Albers-Schönberg and Schmid, which is the same as the AC of plumericin [6].

REFERENCES

1. J.E. Little, D.B. Johnstone, Plumericin—An Antimicrobial Agent from *Plumeria Multiflora*, *Arch. Biochem.*, 30, 445–452, 1951.
2. (a) G. Albers-Schönberg, H. Schmid, The Structure of Plumericin and Related Compounds, *Chimia*, 14, 127–128, 1960; (b) G. Albers-Schönberg, H. Schmid, Uber Die Struktur Von Plumericin, Isoplumericin, Beta-Dihydroplumericin Und Der Beta-Dihydroplumericinsaure, *Helv. Chim. Acta*, 44, 1447–1473, 1961.
3. (a) L.B.S. Kardono, S. Tsauri, K. Padmawinata, J.M. Pezzuto, A.D. Kinghorn, Studies cn Indonesian Medicinal-Plants 2. Cytotoxic Constituents of the Bark of Plumeria-Rubra Collected in Indonesia, *J. Nat. Prod.*, 53, 1447–1455, 1990; (b) M.O. Hamburger, G.A. Cordell, N. Ruangrungsi, Traditional Medicinal-Plants of Thailand 17. Biologically-Active Constituents of Plumeria-Rubra, *J. Ethnopharm.*, 33, 289–292, 1991; (c) M.P. Dobhal, A.M. Hasan, M.C. Sharma, B.C. Joshi, Ferulic Acid Esters from Plumeria Bicolor, *Phytochemistry*, 51, 319–321, 1999; (d) B.R. Pai, P.S. Subramaniam, U.R Rao, Isolation of Plumericin and Isoplumericin from Allamanda-Cathartica Linn, *Indian J. Chem.*, 8, 851, 1970; (e) S.M. Kupchan, A.L. Dessertine, B.T. Blaylock, R.F. Bryan, Isolation and Structural Elucidation of Allamandin, and Antileukemic Iridoid Lactone from *Allamanda Cathartica*, *J. Org. Chem.*, 39, 2477–2482, 1974; (f) T. Yamauchi, F. Abe, M. Taki, Protoplumericin, An Iridoid Bis-Glucoside in Allamanda-Neriifolia, *Chem. Pharm. Bull.*, 29, 3051–3055, 1981; (g) J. Bhattacharyya, M.S.Q. Morais, 5,6-Dimethoxy-7-hydroxycoumarin (Umckalin) from Allamanda blanchetti: Isolation and ^{13}C-nmr Characteristics, *J. Nat. Prod.*, 49, 354–371, 1986; (h) J.E. Anderson, C.J. Chang, J.L. McLaughlin, Bioactive Components of *Allamanda schottii*, *J. Nat. Prod.*, 51, 307–308, 1988; (i) J.H.Y. Vilegas, E.M. Hachich, M. Garcia, A. Brasileiro, M.A.G. Carneiro, V.L.B. Campos, *Rev. Latinoam. Quim.*, 22, 44–45, 1992; (j) J.H.Y. Vilegas, E.M. Hachich, M. Garcia, A. Basileiro, M.A.G. Carneiro, V.L.B. Campos, *Rev. Latinoam. Quim.*, 23, 73–75, 1994; (k) M.S. Abdel-Kader, J. Wisse, R. Evans, H. Werff, D.G.I. Kingston, Bioactive Iridoids and a New Lignan from *Allamanda cathartica and Himatanthus fallax* from the Suriname Rainforest, *J. Nat. Prod.*, 60, 1294–1297, 1997; (l) D. Basu, A. Chatterjee, Occurrence of Plumericin in Nerium-Indicum, *Indian J. Chem.*, 11, 297, 1973; (m) M.F. Vanderlei, M.S. Silva, H.E. Gottlieb, R. Braz-Filho, Iridoids and Triterpenes from *Himatanthus phagedaenica*: The Complete Assignment of the ^{1}H and ^{13}C NMR spectra of Two Iridoid Glycosides, *J. Braz. Chem. Soc.*, 2, 51–55, 1991; (n) C.A. Wood, K. Lee, A.J. Vaisberg, D.G.I. Kingston, C.C. Neto, G.B. Hammond, A Bioactive Spirolactone Iridoid and Triterpenoids from *Himatanthus sucuuba*, *Chem. Pharm. Bull.*, 49, 1477–1478, 2001; (o) J.E. Page, S. Madrinan, G.H.N. Towers, Identification of a Plant-Growth Inhibiting Iridoid Lactone from *Duroia hirsuta*, The Allelopathic Tree of the Devil's Garden, *Experientia*, 50, 840–842, 1994.

4. B. Elsasser, K. Krohn, M.N. Akhtar, U. Florke, S.F. Kouam, M.G. Kuigoua, B.T. Ngadjui, B.M. Abegaz, S. Antus, T. Kurtan, Revision of the Absolute Configuration of Plumericin and Isoplumericin from *Plumeria rubra, Chem. Biodiversity*, 2, 799–808, 2005.
5. J.W. Downing, *DZDO.* Department of Chemistry and Biochemistry, University of Colorado, Boulder, CO.
6. P.J. Stephens, J.J. Pan, F.J. Devlin, K. Krohn, and T. Kurtan, Determination of the Absolute Configurations of Natural Products via Density Functional Theory Calculations of Vibrational Circular Dichroism, Electronic Circular Dichroism, and Optical Rotation: The Iridoids Plumericin and Isoplumericin, *J. Org. Chem.*, 72, 3521–3536, 2007.

THE CHIRAL ALKANE, D₃-ANTI-TRANS-ANTI-TRANS-ANTI-TRANS-PERHYDROTRIPHENYLENE

The anti-trans-anti-trans-anti-trans isomer of the perhydro derivative of triphenylene, **1**, has D_3 symmetry and is therefore chiral. The two enantiomers are:

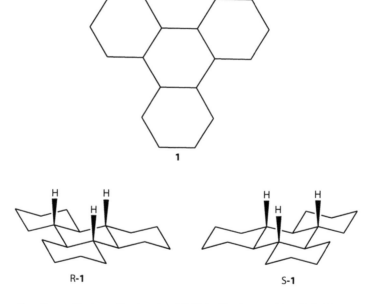

Optically active **1** was first synthesized in 1967 by Farina and Audisio [1]. Racemic **1** was converted to the carboxylic acid derivative, **2**, which was then resolved by fractional crystallization of its dehydroabietylamine salt. Decarboxylation of the two enantiomers of **2** led to (+)-**1** and (–)-**1**. Using the approach for predicting the $[\alpha]_D$ values of chiral molecules proposed by Brewster in 1959 [2], Farina and Audisio predicted that the AC of *R*-**1** is negative and concluded that the AC of **1** is *R*(–)/*S*(+). In 1970, Farina and Audisio provided further support for this AC by converting the carboxyl group of (+)-**2** to a carbonyl group and determining the AC of the resulting ketone to be *S* from the ORD and ECD of its n → π* carbonyl electronic transition, using the octant rule [3]. It followed thence that the AC of (+)-**2** is *S* and, since decarboxylation of (+)-**2** gives (+)-**1**, that the AC of (+)-**1** is *S*, consistent with the previously determined AC.

COOH

2

One of the interesting properties of racemic **1** is that it forms crystalline inclusion complexes [4]. It can be expected that the enantiomers of **1** will exhibit similar behavior. In 2001, in order to facilitate the study of the inclusion complexes of (+)-**1** and (−)-**1**, Schürch et al. applied chiral gas chromatography, using a chiral cyclodextrin stationary phase, to the resolution of racemic **1** [5]. The enantiomers of **1** were shown to be obtainable with ee's greater than 99%.

In 2004, Professors Schürch and Hulliger provided us with a sample of (+)-**1**, obtained by gas chromatography, of ee 93.5%, in order to confirm the Farina-Audisio AC of **1**, by using VCD [6]. The IR and VCD spectra of (+)-**1** were measured using a 0.08 M solution in CCl_4 and a 597 μ pathlength cell. Given the unavailability of (−)-**1** and (±)-**1**, the VCD baseline was the VCD spectrum of the CCl_4 solvent. The solvent-baseline-subtracted IR and VCD spectra of the solution are shown in Figure 7.58; the VCD spectrum is normalized to 100% ee. The signal-to-noise ratio of the VCD spectrum is excellent, showing that many of the transitions of (+)-**1** possess quite large anisotropy ratios.

As discussed in Chapter 5, of the four conformations of **1** predicted using MMFF94 and DFT (B3LYP/6-31G*), only the lowest-energy conformation, **a**, in which all four cyclohexane rings possess chair conformations, is significantly populated at room temperature. **1** is therefore effectively a conformationally rigid molecule. The analysis of its IR and VCD spectra is therefore based on DFT calculations for conformation **a**. The B3LYP/6-31G* structure of conformation **a** of *S*-**1** is:

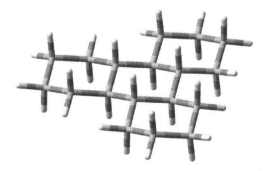

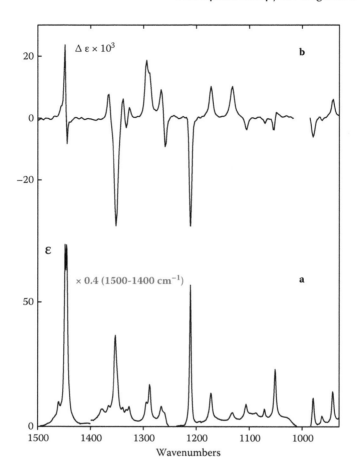

FIGURE 7.58 (a) The IR absorption spectrum of (+)-**1** in CCl$_4$; 0.08 M, 1,500–930 cm^{-1}; pathlength 597 µ. (b) The VCD spectrum of (+)-**1** in CCl$_4$; 0.08 M, 1,500–930 cm^{-1}; pathlength 597 µ. The VCD spectrum is normalized to 100% ee.

In Table 7.15, the B3LYP/cc-pVTZ and B3PW91/cc-pVTZ harmonic frequencies, dipole strengths, and rotational strengths of modes 42–108 of **1a** are given. Due to its D$_3$ symmetry, the vibrational modes have either a$_1$, a$_2$, or e symmetry; transitions of a$_1$ modes are symmetry forbidden and therefore have zero dipole and rotational strengths. In Figure 7.59, the B3LYP/cc-pVTZ and B3PW91/cc-pVTZ IR spectra are compared to the experimental IR spectrum. The B3LYP and B3PW91 spectra are very similar, but not identical. For example, B3PW91 predicts that the modes 66/67 and 68/69 are strong and weak, respectively, while B3LYP predicts the modes 68/69 to be stronger than 66/67. Comparison to the experimental band at ~1,210 cm^{-1}, which is quite sharp and shows no sign of multiple transitions, leads to the conclusion that the B3PW91 prediction is in better agreement with experiment. As a result, we assign the experimental IR spectrum using the B3PW91/cc-pVTZ spectrum, as shown in Figure 7.59. Lorentzian fitting of the experimental IR spectrum,

TABLE 7.15

Calculated and Experimental Frequencies, Dipole Strengths, and Rotational Strengths for 1[a]

Mode	Calculation[b]						Experiment[c]					
	B3LYP			B3PW91			IR			VCD		
	ν	D	R	ν	D	R	ν	D	γ	ν	R	γ
108	1,507	1.2	−0.2	1,496	2.2	0.0						
							1,460	12.8	3.7			
107	1,507	1.2	−0.2	1,496	2.2	0.0						
106	1,506	0.0	0.0	1,495	0.0	0.0						
105	1,498	37.6	48.0	1,485	47.4	57.4	1,448	43.5	1.6	1,449	29.8	2.2
104	1,495	5.2	−15.2	1,482	4.5	−15.9						
103	1,495	5.2	−15.2	1,482	4.5	−15.9						
102	1,492	15.9	−20.9	1,480	18.0	−24.9	1,445	65.8	2.1	1,445	−17.2	2.2
101	1,492	13.2	12.4	1,480	17.8	12.6						
100	1,492	13.2	12.4	1,480	17.8	12.6						
99	1,489	0.4	−1.6	1,476	0.8	−2.8						
										1,437	−1.3	2.3
98	1,489	0.4	−1.6	1,476	0.8	−2.8						
97	1,488	0.0	0.0	1,475	0.0	0.0						
96	1,407	0.0	0.0	1,402	0.0	0.0						
95	1,400	0.1	1.6	1,394	0.2	3.7						
							1,366	4.0	4.6	1,366	19.3	3.4
94	1,400	0.1	1.6	1,394	0.2	3.7						
93	1,388	23.1	−53.1	1,383	24.7	−57.9						
92	1,385	1.6	6.9	1,381	0.6	10.1						
91	1,385	1.6	6.9	1,381	0.6	10.1	1,353	26.7	3.7	1,352	−78.3	4.0
90	1,382	0.7	−17.5	1,380	1.6	−24.4						
89	1,382	0.7	−17.5	1,380	1.6	−24.4						
88	1,379	1.4	3.0	1,371	0.0	0.0						
87	1,378	0.1	−2.8	1,371	0.1	1.7						
86	1,378	0.1	−2.8	1,367	0.3	3.4	1,340	3.9	6.3	1,339	16.1	2.9
85	1,373	0.0	0.0	1,367	0.3	3.4						
84	1,365	0.1	2.2	1,361	0.7	−6.2						
							1,332	2.9	5.9	1,333	−6.4	2.6
83	1,365	0.1	2.2	1,361	0.7	−6.2						
82	1,364	0.0	0.0	1,359	0.0	0.0						
81	1,363	1.2	−2.8	1,354	1.7	5.1						
							1,326	4.1	4.2	1,327	5.6	2.4
80	1,363	1.2	−2.8	1,354	1.7	5.1						
79	1,357	0.0	0.0	1,349	0.0	0.0						
78	1,331	2.0	23.1	1,319	3.1	30.8						
77	1,326	2.0	15.8	1,317	0.0	2.0	1,295	5.7	3.8	1,295	27.7	3.0
76	1,326	2.0	15.8	1,317	0.0	2.0						
75	1,322	3.5	−0.9	1,313	6.1	10.3						
							1,288	8.5	2.6	1,289	19.7	3.0
74	1,322	3.5	−0.9	1,313	6.1	10.3						
73	1,300	3.6	21.7	1,294	5.0	26.5	1,267	6.4	4.3	1,267	20.6	3.5
72	1,293	0.0	0.0	1,286	0.0	0.0						

(*Continued*)

TABLE 7.15 (CONTINUED)
Calculated and Experimental Frequencies, Dipole Strengths, and Rotational Strengths for 1[a]

| | Calculation[b] | | | | | | Experiment[c] | | | | | |
| | B3LYP | | | B3PW91 | | | IR | | | VCD | | |
Mode	ν	D	R	ν	D	R	ν	D	γ	ν	R	γ
71	1,291	0.1	−3.3	1,285	0.1	−3.5 ⎫						
						⎬ 1,261	1.9	2.5	1,259	−16.5	2.4	
70	1,291	0.1	−3.3	1,285	0.1	−3.5 ⎭						
69	1,247	14.0	3.8	1,241	0.1	3.5						
68	1,247	14.0	3.8	1,241	0.1	3.5						
67	1,238	7.6	−29.6	1,235	18.2	−34.3 ⎫						
						⎬ 1,211	21.9	1.6	1,211	−50.0	2.2	
66	1,238	7.6	−29.6	1,235	18.2	−34.3 ⎭						
65	1,218	0.0	0.0	1,216	0.0	0.0						
64	1,195	2.2	8.2	1,193	8.5	18.8	1,172	12.2	4.1	1,173	22.8	3.5
63	1,189	0.0	0.0	1,184	0.0	0.0						
62	1,188	4.6	14.2	1,180	0.0	−0.1						
61	1,144	2.3	13.5	1,151	2.4	15.8 ⎫						
						⎬ 1,133	8.4	7.3	1,132	27.3	4.0	
60	1,144	2.3	13.5	1,151	2.4	15.8 ⎭						
59	1,125	2.8	−3.8	1,126	2.5	−3.3 ⎫						
						⎬ 1,106	9.6	5.4	1,106	−7.7	2.9	
58	1,125	2.8	−3.8	1,126	2.5	−3.3 ⎭						
57	1,104	0.0	0.0	1,118	0.0	0.0						
56	1,094	0.6	0.0	1,106	1.5	−0.1 ⎫						
						⎬ 1,089	9.0	7.8	1,089	−0.2	2.2	
55	1,094	0.6	0.0	1,106	1.5	−0.1 ⎭						
54	1,086	1.9	−1.8	1,088	1.2	−1.3 ⎫						
						⎬ 1,071	5.9	4.8	1,070	−2.7	2.3	
53	1,086	1.9	−1.8	1,088	1.2	−1.3 ⎭						
52	1,069	6.9	6.7	1,083	0.0	0.0						
51	1,069	6.9	6.7	1,067	9.2	4.8 ⎫						
50	1,068	1.2	−12.3	1,067	9.2	4.8					⎫ 1,054	−9.0	2.1
49	1,066	0.0	0.0	1,066	1.0	−13.0						
48	1,054	0.1	0.3	1,063	0.8	0.9 ⎬ 1,051	17.4	3.1			
									⎬ 1,050	9.3	4.9	
47	1,054	0.1	0.3	1,063	0.8	0.9 ⎭						
46	1,037	0.0	0.0	1,053	0.0	0.0						
45	995	9.4	−14.1	991	8.5	−13.2	979	5.5	1.6	979	−12.0	2.4
44	973	0.4	−0.2	980	2.7	−3.2	962	2.0	2.1	962	−1.1	1.9
43	949	5.0	5.5	959	4.9	6.3 ⎫						
						⎬ 942	7.2	1.8	942	16.6	3.5	
42	949	5.0	5.5	959	4.9	6.3 ⎭						

[a] Frequencies ν and bandwidths γ in cm^{-1}; dipole strengths D in 10^{-40} esu^2 cm^2; rotational strengths R in 10^{-44} esu^2 cm^2. Experimental rotational strengths (normalized to 100% ee) are for (+)-**1**; calculated rotational strengths are for (S)-**1**.

[b] Using the cc-pVTZ basis set.

[c] From Lorentzian fitting of the IR and the VCD spectra.

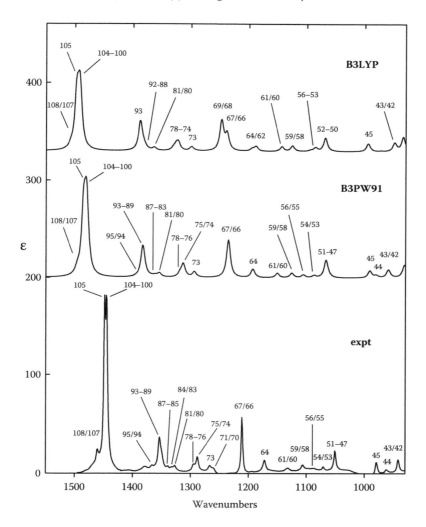

FIGURE 7.59 Comparison of the B3LYP/cc-pVTZ and B3PW91/cc-pVTZ IR spectra of **1**, calculated using Lorentzian bandshapes ($\gamma = 4.0$ cm^{-1}), to the experimental IR spectrum of (+)-**1**. The assignment of the experimental spectrum is based on the B3PW91/cc-pVTZ spectrum.

based on this assignment, shown in Figure 7.60, leads to the frequencies, dipole strengths, and bandwidths given in Table 7.15. Comparisons of the B3LYP/cc-pVTZ and B3PW91/cc-pVTZ frequencies and dipole strengths to the experimental frequencies and dipole strengths are shown in Figures 7.61 and 7.62. As usual, the calculated frequencies are 0–4% greater than the experimental frequencies, due to the absence of anharmonicity. Both B3PW91 and B3LYP dipole strengths are in good agreement with the experimental dipole strengths, supporting the reliability of the assignment of the experimental IR spectrum.

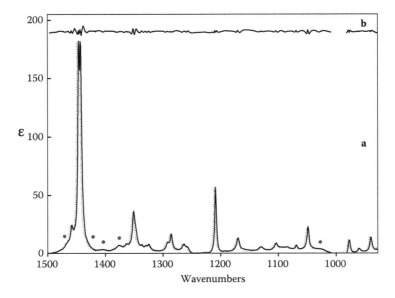

FIGURE 7.60 (a) Lorentzian fit of the experimental IR spectrum of (+)-**1** over the range 1,500–930 cm^{-1}. The solid line is the experimental spectrum; the dotted line is the Lorentzian fit. (b) The difference spectrum: the experimental spectrum minus the Lorentzian fit. The asterisks indicate bands not assigned to fundamental modes.

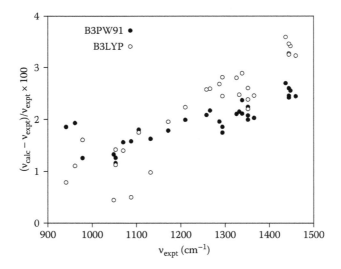

FIGURE 7.61 Comparison of the B3LYP/cc-pVTZ and B3PW91/cc-pVTZ frequencies to the experimental IR frequencies of modes 42–108. For bands assigned to multiple vibrational modes, all calculated frequencies are compared to the corresponding experimental frequencies.

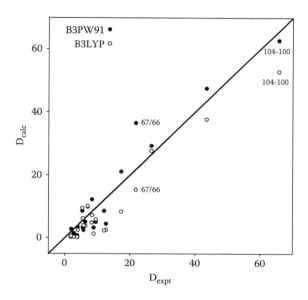

FIGURE 7.62 Comparison of the B3LYP/cc-pVTZ and B3PW91/cc-pVTZ dipole strengths to the experimental dipole strengths of modes 42–108. Dipole strengths D are in 10^{-40} esu^2 cm^2. For bands assigned to multiple vibrational modes, calculated dipole strengths are the sums of the dipole strengths of contributing modes.

In Figure 7.63, the B3LYP/cc-pVTZ and B3PW91/cc-pVTZ VCD spectra of S-**1** are compared to the experimental VCD spectrum of (+)-**1**. As with the IR spectra, the B3LYP and B3PW91 VCD spectra are very similar overall, but not identical. For example, the predicted VCD spectra of modes 80–87 differ substantially, the B3PW91 spectrum being in much better agreement with the experimental VCD spectrum. The B3PW91/cc-pVTZ assignment of the IR spectrum, together with the B3PW91/cc-pVTZ VCD spectrum of S-**1**, leads to the assignment of the experimental VCD spectrum of (+)-**1** detailed in Figure 7.63. Lorentzian fitting of the experimental VCD spectrum, shown in Figure 7.64, gives the experimental frequencies, rotational strengths, and bandwidths given in Table 7.15. The VCD frequencies are essentially identical to the IR frequencies, confirming the consistency of the assignments of the IR and VCD spectra. Comparisons of the B3PW91/cc-pVTZ and B3LYP/cc-pVTZ rotational strengths of S-**1** to the experimental rotational strengths of (+)-**1** are shown in Figure 7.65. The excellent agreement of the B3PW91/cc-pVTZ VCD spectrum of S-**1** with the experimental VCD spectrum of (+)-**1**, and of the B3PW91/cc-pVTZ rotational strengths of S-**1** with the experimental rotational strengths of (+)-**1**, unambiguously proves the AC of **1** to be $S(+)/R(-)$, the AC assigned by Farina and Audisio [1,3]. The excellent agreements of the B3PW91/cc-pVTZ IR and VCD spectra with the experimental IR and VCD spectra also confirm the reliability of the conformational analysis of **1**.

In order to extend the studies of the functional dependence of the accuracy of DFT rotational strengths, discussed in Chapter 6, we have compared the mean absolute deviations of rotational strengths, calculated for S-**1** using the 19 functionals in

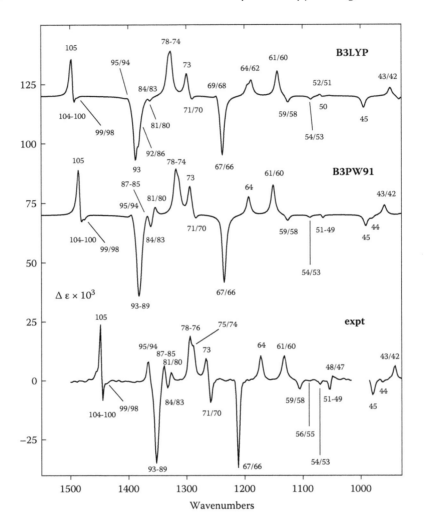

FIGURE 7.63 Comparison of the B3LYP/cc-pVTZ and B3PW91/cc-pVTZ VCD spectra of (S)-**1**, calculated using Lorentzian bandshapes ($\gamma = 4.0$ cm^{-1}), to the experimental VCD spectrum of (+)-**1**. The assignment of the experimental spectrum is based on the B3PW91/cc-pVTZ spectrum.

Chapter 6 and the cc-pVTZ basis set, and the experimental rotational strengths of (+)-**1**. The results are given in Table 7.16 and Figure 7.66. For this molecule only one functional, MPW1PW91, is slightly more accurate than B3PW91. Thirteen functionals are more accurate than B3LYP. Except for BB95 and τHCTH, all are hybrid functionals. The five functionals less accurate than B3LYP are all pure functionals. The number of incorrect signs varies from zero to seven. For B3PW91 and B3LYP the numbers of incorrect signs are zero and four, respectively

To evaluate the accuracies of other basis sets, relative to cc-pVTZ and cc-pVQZ, calculations have been repeated using the 10 additional basis sets listed in Table 7.17,

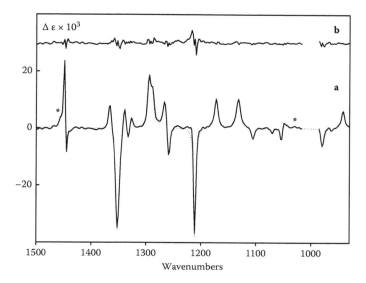

FIGURE 7.64   (a) Lorentzian fit of the experimental VCD spectrum of (+)-**1** over the range 1,500–930 cm⁻¹. The solid line is the experimental spectrum; the dotted line is the Lorentzian fit. (b) The difference spectrum: the experimental spectrum minus the Lorentzian fit. The asterisks indicate bands not assigned to fundamental modes.

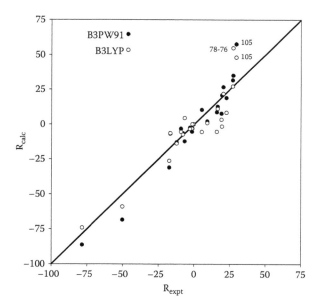

FIGURE 7.65   Comparison of the B3LYP/cc-pVTZ and B3PW91/cc-pVTZ rotational strengths of *S*-**1** to the experimental rotational strengths of modes 42–105 of (+)-**1**. For bands assigned to multiple vibrational modes, calculated rotational strengths are the sums of the rotational strengths of contributing modes. Rotational strengths *R* are in 10⁻⁴⁴ esu² cm².

TABLE 7.16

Functional Dependence of the Rotational Strengths of 1 Calculated Using the cc-pVTZ Basis Set

Functional	Type[a]	MAD[b]	No. Wrong Signs
BLYP	GGA	14.80	7
BP86	GGA	10.47	2
OLYP	GGA	10.30	5
BPW91	GGA	9.98	2
HCTH	GGA	8.86	4
B3LYP	H-GGA	8.62	4
CAMB3LYP	H-X	8.39	2
τHCTH	M-GGA	8.38	4
B1B95	HM-GGA	8.01	2
PBE1PBE	H-GGA	8.00	2
HSE1PBE	H-S	7.88	2
O3LYP	H-GGA	7.75	2
BB95	M-GGA	7.08	0
BMK	HM-GGA	7.05	3
M05-2X	HM-GGA	6.91	2
B3P86	H-GGA	6.82	0
τHCTHHYB	HM-GGA	6.80	2
B3PW91	H-GGA	6.80	0
MPW1PW91	H-GGA	6.46	0

[a] GGA, generalized gradient approximation; M-GGA, meta-generalized gradient approximation; H-GGA, hybrid generalized gradient approximation; HM-GGA, hybrid meta-generalized gradient approximation; H-X, range-separated hybrid (with short-range DFT exchange and long-range HF exchange); H-S, range-separated hybrid (short-range HF exchange and long-range DFT exchange).

[b] Mean absolute deviation of calculated rotational strengths of S-**1** relative to experimental rotational strengths of modes 42–105 of (+)-**1**.

and the B3PW91 functional. The values of the mean absolute deviations of the calculated rotational strengths from the cc-pVQZ rotational strengths of modes 42–105 are given in Table 7.17. In Figure 7.67 the dependence of the mean absolute deviation on the number of basis functions is plotted. As the number of basis functions diminishes, the mean absolute deviation increases substantially.

In reference 6, the B3LYP and B3PW91 calculations of the IR and VCD spectra of **1** used the basis set TZ2P. Comparison of the B3PW91/TZ2P and B3PW91/cc-pVTZ rotational strengths of S-**1**, shown in Figure 7.68 for modes 42–105, demonstrates that the TZ2P and cc-pVTZ rotational strengths are very similar, except for modes 49, 51, 76, and 77/78.

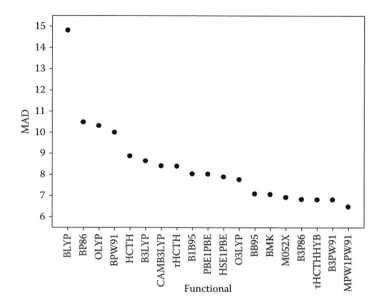

FIGURE 7.66 Mean absolute deviation (MAD) functional dependence of the rotational strengths of **1** calculated using the cc-pVTZ basis set.

TABLE 7.17
The Basis Set Dependence of the Mean Absolute Deviations of B3PW91-Calculated and B3PW91/cc-pVQZ Rotational Strengths of Modes 42–105 of (*S*)-1

Basis Set	Contracted Set[a]	No. Fns.	MAD[b]
cc-pVQZ	[5s4p3d2f1g/4s3p2d1f] (5d,7f,9g)	1,890	0.00
aug-cc-pVTZ	[5s4p3d2f/4s3p2d] (5d,7f)	1,518	0.19
cc-pVTZ	[4s3p2d1f/3s2p1d] (5d,7f)	960	0.19
6-311++G(2d,2p)	[5s4p2d/4s2p] (5d)	786	0.41
TZ2P	[5s4p2d/3s2p] (6d)	792	1.77
6-31G**	[3s2p1d/2s1p] (6d)	420	3.28
6-311++G**	[5s4p1d/4s1p] (5d)	606	3.45
aug-cc-pVDZ	[4s3p2d/3s2p] (5d)	684	3.51
6-31G*	[3s2p1d/2s] (6d)	330	3.64
6-311G**	[4s3p1d/3s1p] (5d)	504	3.95
cc-pVDZ	[3s2p1d/2s1p] (5d)	402	5.21
6-31G	[3s2p/2s]	222	6.04

[a] Of the form [1/2] where the first term refers to the contracted set on carbon and the second refers to the contracted set on hydrogen. The numbers in parentheses refer to the number of functions used for each type of polarization function.

[b] Mean absolute deviation relative to cc-pVQZ for modes 42–105.

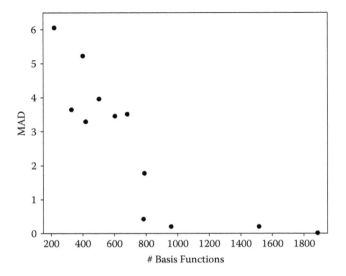

FIGURE 7.67   B3PW91 basis set dependence of (*S*)-**1** rotational strengths.

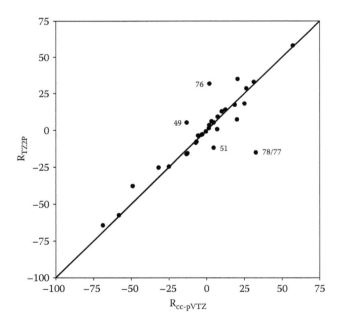

FIGURE 7.68   Comparison of the B3PW91/TZ2P and B3PW91/cc-pVTZ rotational strengths of modes 42–105 for *S*-**1**. The line has a slope of +1.

The basis sets TZ2P and 6-31G* are frequently used in calculating VCD spectra. The B3PW91/TZ2P and B3PW91/6-31G* VCD spectra of (S)-**1** are compared to the B3PW91/cc-pVQZ VCD spectrum in Figure 7.69. The TZ2P VCD spectrum is quantitatively very nearly identical to the cc-pVQZ VCD spectrum, proving that the basis set error of TZ2P is not significantly greater than the basis set error of cc-pVTZ. The 6-31G* VCD spectrum is somewhat different from the cc-pVQZ and TZ2P VCD spectra. However, since the qualitative differences are not great, the 6-31G* basis set error is not much greater than the basis set errors of TZ2P and cc-pVTZ.

To confirm that DFT is more accurate than HF theory, the IR and VCD spectra have been calculated using HF theory and the cc-pVTZ basis set. The spectra obtained are compared to the experimental spectra in Figures 7.70 and 7.71. The experimental rotational strengths are compared to the HF/cc-pVTZ and B3PW91/cc-pVTZ rotational strengths in Figure 7.72. The B3PW91/cc-pVTZ rotational strengths are clearly in better agreement with the experimental rotational strengths than are the HF/cc-pVTZ rotational strengths, especially for modes 78–76, 93–89, and 95/94. The mean absolute deviations between calculated and experimental rotational strengths of modes 42–105 are 6.8 and 10.5×10^{-44} esu^2 cm^2 for B3PW91/cc-pVTZ and HF/cc-pVTZ, respectively, confirming this conclusion.

The redetermination of the AC of the chiral alkane, perhydrotriphenylene, using VCD, illustrates the utility of VCD for evaluating the reliability of previously assigned ACs.

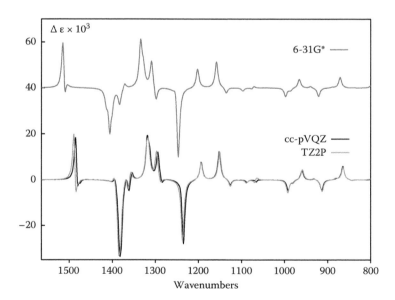

FIGURE 7.69 Comparison of the B3PW91/TZ2P and B3PW91/6-31G* VCD spectra of (S)-**1** to the B3PW91/cc-pVQZ VCD spectrum.

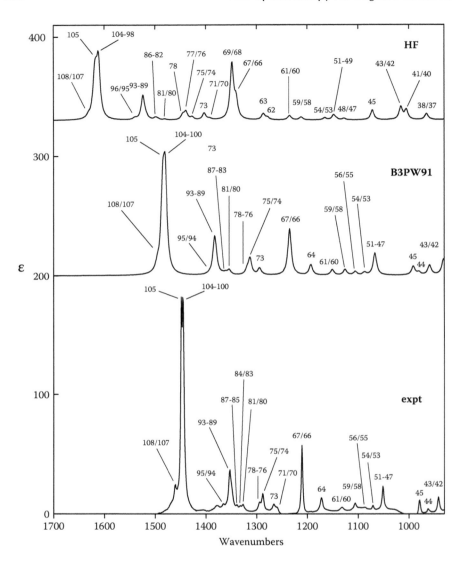

FIGURE 7.70 Comparison of the HF/cc-pVTZ and B3PW91/cc-pVTZ IR spectra of (*S*)-**1** to the experimental IR spectrum of (+)-**1**.

References

1. M. Farina and G. Audisio, Optically Active Perhydrotriphenylene: The First Resolution of a D$_3$ Organic Molecule, *Tet. Lett.*, 1285–1288, 1967.
2. J.H. Brewster, A Useful Model of Optical Activity. I. Open Chain Compounds, *J. Am. Chem. Soc.*, 81, 5475–5483, 1959.
3. M. Farina and G. Audisio, Stereochemistry of Perhydrotriphenylene. II. Absolute Rotation and Configuration of Optically Active Anti-Trans-Anti-Trans-Anti-Trans-Perhydrotriphenylene, *Tetrahedron*, 26, 1839–1844, 1970.

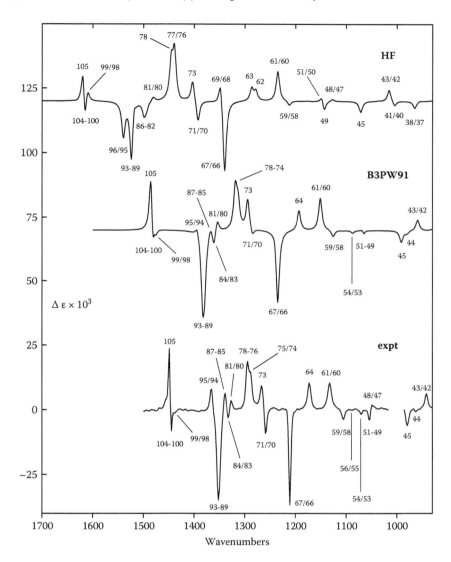

FIGURE 7.71 Comparison of the HF/cc-pVTZ and B3PW91/cc-pVTZ VCD spectra of (*S*)-**1** to the experimental VCD spectrum of (+)-**1**.

4. (a) R. Hoss, O. König, V. Kramer-Hoss, U. Berger, P. Rogin, and J. Hulliger, Crystalization of Supramolecular Materials: Perhydrotriphenylene (PHTP) Inclusion Compounds with Nonlinear Optical Properties, *Angew. Chem. Int. Ed. Engl.*, 35, 1664–1666, 1996; (b) O. König, H.B. Bürgi, T. Armbruster, J. Hulliger, and T. Weber, A study in Crystal Engineering: Structure, Crystal Growth, and Physical Properties of a Polar Perhydrotriphenylene Inclusion Compound. *J. Am. Chem. Soc.*, 119, 10632–10640, 1997.

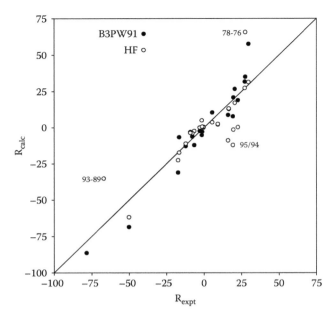

FIGURE 7.72   Comparison of the HF/cc-pVTZ and B3PW91/cc-pVTZ rotational strengths to the experimental rotational strengths of modes 42–105. Rotational strengths R are in 10^{-44} esu^2 cm^2.

5. S. Schürch, A. Saxer, S. Claude, R. Tabacchi, B. Trusch, and J. Hulliger, Semi-Preparative Gas Chromatographic Separation of All-Trans-Perhydrotriphenylene Enantiomers on a Chiral Cyclodextrin Stationary Phase, *J. Chromatogr. A*, 905, 175–182, 2001.
6. P.J. Stephens, F.J. Devlin, S. Schürch, and J. Hulliger, Determination of the Absolute Configurations of Chiral Molecules via Density Functional Theory Calculations of Vibrational Circular Dichroism and Optical Rotation. The Chiral Alkane, D$_3$-Anti-Trans-Anti-Trans-Anti-Trans-Perhydrotriphenylene, *Theor. Chem. Acc.*, 119, 19–28, 2008.

THE ISOTOPICALLY CHIRAL SULFOXIDE, PERDEUTERIOPHENYL-PHENYL-SULFOXIDE

The determination of the ACs of chiral molecules whose chirality is due to isotopic substitution is extremely challenging. X-ray crystallography is ineffective, as is ECD, since isotopic substitution does not change the equilibrium geometry of a molecule or the electronic wavefunctions. The largest effect of isotopic substitution is on the vibrational states, since the normal modes and vibrational frequencies are a function of the nuclear masses. Consequently, VCD provides the optimum methodology for determining the ACs of isotopically chiral molecules. Here, we discuss the determination of the AC of the isotopically chiral sulfoxide, $(C_6D_5)(C_6H_5)SO$, perdeuteriophenyl-phenyl-sulfoxide, **1**, using VCD [1]. The enantiomers of **1** were synthesized by Professor Drabowicz (Lodz, Poland) by reaction of the enantiomers of O-menthyl-benzenesulfinate with the Grignard reagent $(C_6D_5)MgBr$ [1]. The ee's of (+)-**1** and (–)-**1** were 98.4 and 77.7%, respectively. Racemic **1** was synthesized by reaction of racemic (OEt)(C$_6$H$_5$)SO with (C$_6$D$_5$)MgBr [1].

1

EXPERIMENTAL IR AND VCD SPECTRA

The IR and VCD spectra of CCl$_4$ solutions of (+)-**1**, (–)-**1**, and (±)-**1**, of concentrations 0.1 M, were measured using a 597 μ pathlength cell. The IR spectra of the two enantiomers and racemate of **1** were perfectly superposable, demonstrating that their chemical purities were identical, and supporting the conclusion that they were ~100% pure. The IR spectrum of (+)-**1**, obtained over the frequency range 1,150–1,550 cm^{-1}, is shown in Figure 7.73. The VCD spectra of (+)-**1** and (–)-**1** were obtained using the VCD spectrum of (±)-**1** as the baseline. The VCD spectrum of (–)-**1** was normalized to the ee of (+)-**1**. The half-difference VCD spectrum of (+)-**1** resulting, over the range 1,150–1,550 cm^{-1}, the mid-IR frequency range with the best signal-to-noise ratios, is also shown in Figure 7.73.

CONFORMATIONAL ANALYSIS

The PES of **1** is identical to that of the undeuterated biphenyl sulfoxide, (C$_6$H$_5$)$_2$SO, **2**. Since the phenyl rings of **1** and **2** can rotate about the C-S bonds connecting them to the sulfoxide group, the possibility of conformational flexibility exists. In order to determine the number of stable conformations, a 2D PES scan was carried out at the B3LYP/6 31G* level, varying the dihedral angles OSC$_1$C$_2$ and OSC$_3$C$_4$ from –60° to +60°, with the results shown in Figure 7.74. A single valley is observed in the PES, leading to the conclusion that **1** and **2** are in fact conformationally rigid. A 2D PES scan varying the dihedral angles OSC$_1$C$_2$ and OSC$_3$C$_4$ from –180° to +180° found no additional stable conformations, confirming the conformational rigidity of **1** and **2**.

Optimization of the minimum energy structure in the B3LYP/6-31G* PES scan gives a structure with dihedral angles OSC$_1$C$_2$ = –8.8° and OSC$_3$C$_4$ = +8.8°. The structure of **2** has been determined using x-ray crystallography [2]. The angles OSC$_1$C$_2$ and OSC$_3$C$_4$ of the x-ray structure are –11.7° and +11.4°, respectively. The reliability of the calculated structure is thus supported by comparison to the

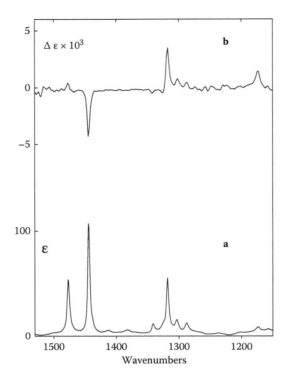

FIGURE 7.73   (a) The IR spectrum of a 0.10 M CCl$_4$ solution of (+)-**1**. Pathlength 597 μ. (b) The VCD half-difference spectrum of (+)-**1**, ½ [Δε(+) − Δε(−)], obtained from the VCD spectra of 0.10 M CCl$_4$ solutions of (+)-**1** and (−)-**1**. Given the 27% higher ee of (+)-**1**, compared to (−)-**1**, the VCD of (−)-**1** was multiplied by 1.27.

x-ray structure. Further optimizations of the predicted conformation at the B3LYP/ TZ2P and B3PW91/TZ2P levels change the OSC$_1$C$_2$/OSC$_3$C$_4$ dihedral angles to −10.8°/+10.8° and −11.0°/+11.0°, respectively, in better agreement with the x-ray structure values.

ANALYSIS OF THE IR AND VCD SPECTRA

The IR spectrum of **1** and the VCD spectra of *R*-**1** and *S*-**1** have been predicted at the B3LYP/TZ2P and B3PW91/TZ2P levels, with the results for modes 44–52 shown in Figures 7.75 and 7.76, together with the experimental spectra. The VCD spectra of *S*-**1** are in much better agreement with the experimental VCD

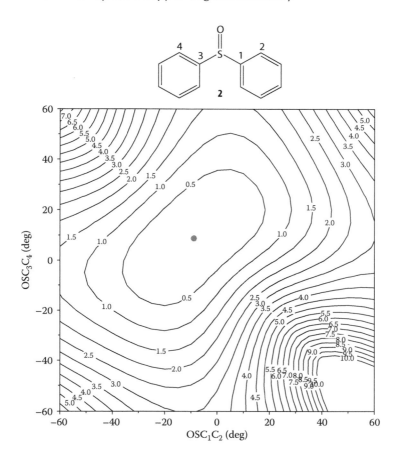

FIGURE 7.74 The B3LYP/6-31G* PES of **2**, as a function of the dihedral angles OSC_1C_2 and OSC_3C_4. The two dihedral angles were varied in 5° steps over a range from −60° to +60°. Contours are shown at 0.5 kcal/mol intervals. ● is the B3LYP/6-31G* optimized geometry.

spectrum of (+)-**1** than are the VCD spectra of *R*-**1**. As a result, the AC of **1** is assigned as *S*(+). The B3LYP/TZ2P VCD of *S*-**1** is in slightly better agreement with the experimental VCD spectrum of (+)-**1** than the B3PW91/TZ2P VCD of *S*-**1**. The assignments of the experimental IR and VCD spectra of (+)-**1** in Figures 7.75 and 7.76 are therefore based on the B3LYP/TZ2P IR and VCD spectra.

The VCD study of **1** was the first application of VCD to the determination of the AC of an isotopically chiral molecule, using the DFT methodology, and demonstrates the utility of VCD in determining the ACs of isotopically chiral molecules.

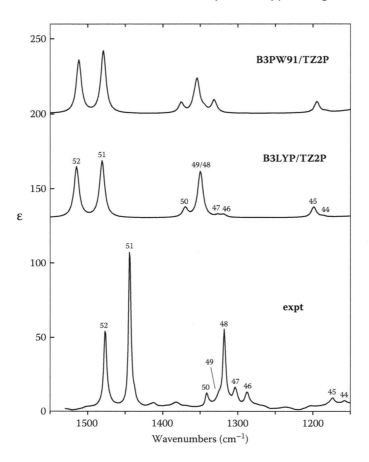

FIGURE 7.75 Comparison of the B3LYP/TZ2P and B3PW91/TZ2P IR spectra of **1** to the experimental IR spectrum. The assignment of the experimental spectrum is based on the B3LYP/TZ2P spectrum. The calculated IR spectra have Lorentzian bandshapes (γ = 4.0 cm^{-1}).

REFERENCES

1. J. Drabowicz, A. Zajac, P. Lyzwa, P.J. Stephens, J.J. Pan, and F.J. Devlin, Determination of the Absolute Configurations of Isotopically Chiral Molecules Using Vibrational Circular Dichroism (VCD) Spectroscopy: The Isotopically Chiral Sulfoxide, Perdeuteriophenyl-Phenyl-Sulfoxide, *Tet. Asymm.*, 19, 288–294, 2008.
2. D. Casarini, L. Lunazzi, and A. Mazzanti, Unprecedented Detection of Distinct Barriers Involving Formally Enantiotopic Substituents: Phenyl Rotation in Solid Diphenyl Sulfoxide, *Angew. Chem., Int. Ed.*, 40, 2536, 2001.

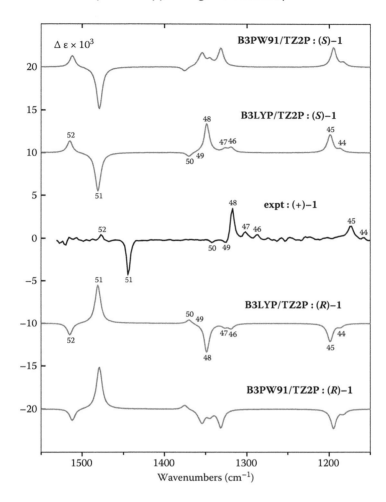

FIGURE 7.76   Comparison of the B3LYP/TZ2P and B3PW91/TZ2P VCD spectra of (*R*)-**1** and (*S*)-**1** to the experimental VCD spectrum of (+)-**1**. The assignment of the experimental spectrum is based on the B3LYP/TZ2P spectrum of (*S*)-**1**. The calculated VCD spectra have Lorentzian bandshapes ($\gamma = 4.0$ cm^{-1}).

THE SEX PHEROMONE OF THE OBSCURE MEALYBUG

In 2005, Professor Jocelyn Millar, Department of Entomology, University of California, Riverside, identified the composition of the sex pheromone of the obscure mealybug insect, *Pseudococcus viburni*, which is a worldwide pest of grapevines, and numerous other crops and ornamental plants, to be 1-acetoxymethyl-2,3,4,4-tetramethylcyclopentane, **1** [1] :

1

The C atoms 1, 2, and 3 are stereogenic. It was determined that C_1 and C_2 have the same AC, and that the AC of C_3 is opposite. The two enantiomers of **1** therefore have 1*R*,2*R*,3*S* and 1*S*,2*S*,3*R* ACs.

In order to determine the AC of the biologically active sex pheromone, in 2007 Professor Millar and Professor Figadère (University of Paris) provided us with synthetic samples of (+)-**1** and (±)-**1**. The VCD spectrum of (+)-**1** was measured using 0.9 M solutions of (+)-**1** and (±)-**1** in $CDCl_3$ [2], the VCD spectrum of (±)-**1** being the baseline for the VCD spectrum of (+)-**1**, in order to minimize artefacts. The resulting VCD spectrum of (+)-**1** over the frequency range 1,300–1,550 cm^{-1} is shown in Figure 7.77.

1 is conformationally flexible, due to the flexibilities of the cyclopentane ring and the acetoxymethyl group. Conformational analysis was carried out using DFT, at the B3PW91/TZ2P level, with the results given in Table 7.18. Fourteen conformations have free energies within 3 kcal/mol of the lowest free energy conformation. The five lowest free energy conformations have free energies within 0.5 kcal/mol and are predicted to be 80% of the room temperature populations, and therefore to dominate the VCD spectrum.

In Figure 7.77, the conformationally averaged B3PW91/TZ2P VCD spectra of 1*S*,2*S*,3*R*-**1** and 1*R*,2*R*,3*S*-**1** are compared to the experimental VCD spectrum of (+)-**1**. The most intense VCD bands of the experimental spectrum can be successfully assigned, using the 1*S*,2*S*,3*R* VCD spectrum, but not using the 1*R*,2*R*,3*S* VCD spectrum. As a result, the AC of **1** is determined to be 1*S*,2*S*,3*R* (+) [2].

According to Professor Millar, the determination of the AC of the mealybug pheromone provides the basis for the preparation of an efficient pesticide for eliminating mealybugs [3].

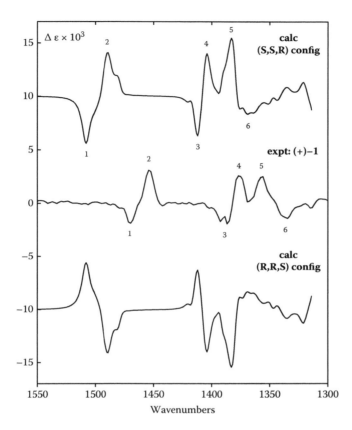

FIGURE 7.77   Comparison of the conformationally averaged B3PW91/TZ2P VCD spectra of (1*S*,2*S*,3*R*)-**1** and (1*R*,2*R*,3*S*)-**1** to the experimental VCD spectrum of (+)-**1**. The calculated VCD spectra have Lorentzian bandshapes ($\gamma = 4.0$ cm^{-1}). The calculated VCD spectra include the contributions of the 15 conformations (Table 7.18). Bands 1–6 of the experimental spectrum are assigned as bands 1–6 of the 1*S*,2*S*,3*R* spectrum.

REFERENCES

1. (a) J.G. Millar, S.L. Midland, J.S. McElfresh, K.M. Daane, (2,3,4,4-Tetramethylc yclopentyl) Methyl Acetate, a Sex Pheromone from the Obscure Mealybug: First Example of a New Structural Class of Monoterpenes, *J. Chem. Ecol.*, 31, 2999–3005, 2005. (b) J.G. Millar, S.L. Midland, Synthesis of the Sex Pheromone of the Obscure Mealybug, the First Example of a New Class of Monoterpenoids, *Tet. Lett.*, 48, 6377–6379, 2007.
2. B. Figadère, F.J. Devlin, J.G. Millar, P.J. Stephens, Determination of the Absolute Configuration of the Sex Pheromone of the Obscure Mealybug by Vibrational Circular Dichroism Analysis, *J. Chem. Soc. Chem. Comm.*, 1106–1108, 2008.
3. RSC Publishing, Mealybugs Look in the Mirror, *Chemical Science*, January 22, 2008.

TABLE 7.18
B3PW91/TZ2P Relative Energies,
Relative Free Energies, and
Populations of the Conformations of 1

Conformer	ΔE^a	ΔG^a	$P(\%)^b$
a	0.00	0.00	23.6
b	0.24	0.19	17.2
c	0.46	0.24	15.6
d	0.39	0.31	14.0
e	0.67	0.50	10.2
f	0.61	0.93	4.9
g	1.34	0.95	4.8
h	0.99	1.09	3.7
i	0.97	1.43	2.1
j	1.07	1.49	1.9
k	1.58	2.02	0.8
l	2.32	2.26	0.5
m	2.64	2.62	0.3
n	2.38	2.70	0.3
o	2.59	3.00	0.2

[a] ΔE and ΔG in kcal/mol.
[b] Populations based on ΔG values, T = 298 K.

A THIA-BRIDGED TRIARYLAMINE HETEROHELICENE

Helicenes are chiral, due to their helicity. In 2007, Professor Menichetti (University of Florence, Italy), synthesized the heterohelicene **1**, which was subsequently resolved by Professors Gasparrini and Villani (University of Rome, Italy), using chiral chromatography. A new study was initiated in order to determine its AC using VCD spectroscopy [1]. The heterohelicene is:

1

Chiral chromatography resolved the two enantiomers of **1**, *P*-**1**, and *M*-**1**:

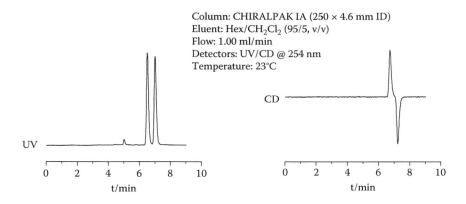

P-**1**   *M*-**1**

as shown in Figure 7.78. The two enantiomers were chromatographed to determine their ee's. The ee's of the first-eluted and second-eluted enantiomers were determined to be >99.9 and 97.6%, respectively. The $[\alpha]_D$ signs of the first-eluted and second-eluted enantiomers were (+) and (–), respectively.

The IR and VCD spectra of (+), (–), and (±)-**1** were measured using $CDCl_3$ solutions of concentrations 0.05 M, and cell pathlengths 546 and 236 μ. The baselines of the VCD spectra of (+) and (–)-**1** were the VCD spectrum of (±)-**1**. The VCD spectrum of (–)-**1** was normalized to 100% ee. The half-difference VCD spectrum, ½ [Δε(+) – Δε(–)], is shown in Figure 7.79, together with the IR spectrum of (–)-**1**.

Conformational analysis of *P*-**1** using MMFF94 found only one conformation within 20 kcal/mol. **1** is therefore predicted to be conformationally rigid. Reoptimizations of the MMFF94 conformation and harmonic frequency calculations using B3LYP/TZ2P and B3PW91/TZ2P verified the stability of the conformation. The B3LYP/TZ2P and B3PW91/TZ2P harmonic frequencies, dipole strengths, and rotational strengths of modes 99–44 are given in Table 7.19. The B3LYP/TZ2P and B3PW91/TZ2P IR and VCD spectra of *P*-**1** are compared to the experimental IR spectrum of (–)-**1** and the VCD spectrum of (+)-**1** in

Column: CHIRALPAK IA (250 × 4.6 mm ID)
Eluent: Hex/CH_2Cl_2 (95/5, v/v)
Flow: 1.00 ml/min
Detectors: UV/CD @ 254 nm
Temperature: 23°C

FIGURE 7.78   Enantioresolution of heterohelicene **1**.

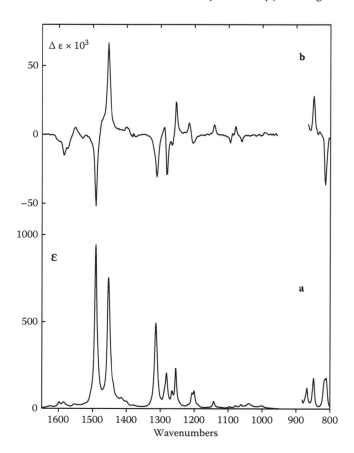

FIGURE 7.79   (a) IR spectrum of (–)-**1**; (b) VCD spectrum of (+)-**1**. The gap at ~900 cm⁻¹ is due to strong absorption of the CDCl₃ solvent.

Figures 7.80 and 7.81. The most intense bands in the experimental IR and VCD spectra are in the frequency range 1,400–1,500 cm⁻¹. The B3PW91/TZ2P IR and VCD spectra are in good agreement with these bands of the experimental spectra. The B3LYP/TZ2P spectra are in poor agreement. As a result, the assignments of the experimental IR and VCD spectra, shown in Figures 7.80 and 7.81, are based on the B3PW91/TZ2P spectra. The experimental frequencies, dipole strengths, and rotational strengths, obtained by Lorentzian fitting of the experimental IR and VCD spectra, are given in Table 7.19. The excellent agreement of the experimental IR and VCD frequencies confirms that the assignments of the experimental IR and VCD spectra are consistent. In Figure 7.82, the calculated and experimental rotational strengths are compared.

TABLE 7.19
Calculated and Experimental Frequencies, Dipole Strengths, and Rotational Strengths of 1[a]

	Calculation[b]						Experiment[c]					
	B3LYP			B3PW91			IR			VCD		
Mode	ν	D	R	ν	D	R	ν	D	γ	ν	R	γ
99	1,636	16.3	−6.6	1,649	22.3	−15.4	1,625	13.0	6.5			
98	1,620	47.4	−43.2	1,632	39.8	−44.4	1,598	27.3	5.3	1,599	−14.8	8.7
97	1,607	8.9	−34.1	1,621	8.2	−32.5						
96	1,605	1.2	−6.0	1,620	1.8	−10.1	1,585	29.4	5.5	1,585	−27.6	4.5
95	1,581	21.2	33.6	1,594	29.0	52.5	1,552	24.3	7.1	1,552	16.4	5.4
94	1,529	48.2	13.9	1,534	41.0	14.4	1,499	39.2	6.0	1,498	−15.4	9.3
93	1,524	679.3	−144.3	1,532	776.7	−171.1	1,491	734.6	4.2	1,491	−97.4	3.8
92	1,500	26.5	−43.5	1,493	8.2	28.9						
91	1,499	113.3	47.5	1,491	123.3	44.2	1,455	111.7	5.2			
90	1,496	20.9	23.4	1,489	621.8	75.8	1,453	597.2	4.6	1,454	154.4	4.8
89	1,493	8.4	−31.7	1,488	61.5	34.5						
88	1,493	8.6	24.1	1,484	21.7	−6.7						
87	1,493	51.9	41.7	1,484	11.0	14.5	1,440	79.4	5.6	1,446	11.6	10.7
86	1,477	549.0	115.5	1,484	36.1	7.3						
85	1,434	15.7	21.5	1,438	12.8	17.0	1,416	68.4	7.9	1,420	10.3	8.6
84	1,430	22.0	4.3	1,432	17.3	3.0						
83	1,420	0.8	0.9	1,419	5.7	−12.8	1,400	27.1	5.6	1,402	8.7	3.7
82	1,419	0.1	0.0	1,408	0.0	0.0						
81	1,418	1.5	−1.7	1,408	0.3	−0.7						
80	1,418	3.4	−8.0	1,407	1.5	0.3						
79	1,339	129.9	14.9	1,352	196.2	−45.8	1,313	510.4	4.7	1,312	−90.5	5.3
78	1,334	309.9	−100.6	1,352	68.5	−7.6						
77	1,316	83.6	26.1	1,334	230.3	49.9				1,289	32.3	4.4
							1,283	191.6	4.6			
76	1,307	181.8	−101.4	1,330	277.1	−123.6				1,282	−62.6	3.3
75	1,300	44.1	−14.7	1,323	7.2	−3.0						
74	1,294	49.9	−14.2	1,290	38.6	−15.0	1,266	56.2	4.2	1,266	−19.8	4.1
73	1,275	266.7	110.1	1,282	138.8	69.0	1,255	166.5	3.4	1,255	50.9	3.4
72	1,245	0.9	1.0	1,256	3.0	4.0				1,232	4.9	8.0
71	1,235	2.8	17.5	1,241	8.2	20.1	1,218	10.4	4.5	1,216	26.4	4.2
70	1,234	16.9	1.0	1,237	0.5	3.8						
69	1,233	156.5	−39.5	1,232	112.1	−41.7	1,206	61.5	4.0	1,204	−30.6	6.6
68	1,221	10.4	−2.0	1,227	10.3	1.6	1,200	65.2	3.6			

(Continued)

TABLE 7.19 (CONTINUED)
Calculated and Experimental Frequencies, Dipole Strengths, and Rotational Strengths of 1[a]

| | Calculation[b] | | | | | | Experiment[c] | | | | | |
| | B3LYP | | | B3PW91 | | | IR | | | VCD | | |
Mode	ν	D	R	ν	D	R	ν	D	γ	ν	R	γ
67	1,172	23.5	4.9	1,168	23.1	4.9 ⎤						
						⎬1,143	1,143	40.8	4.3	1,142	16.9	3.4
66	1,170	10.8	12.0	1,166	4.8	7.3 ⎦						
65	1,114	4.5	–11.0	1,117	4.5	–11.2	1,097	11.7	5.7	1,096	–12.2	3.1
64	1,096	7.1	9.5	1,100	7.9	8.4	1,079	17.3	5.2	1,080	13.3	3.0
63	1,082	15.0	–19.1	1,083	14.7	–18.2	1,062	27.5	5.4	1,062	–14.3	4.2
62	1,065	36.3	84.0	1,057	37.7	90.6 ⎤						
61	1,065	42.4	–92.7	1,057	44.8	–100.0 ⎬1,039	1,039	61.7	7.9			
60	1,064	38.4	3.4	1,055	42.1	4.0 ⎦						
59	1,024	9.6	5.1	1,018	10.2	4.9 ⎤						
58	1,020	17.7	–46.3	1,014	18.9	–50.2 ⎬1,002	1,002	42.5	9.4			
57	1,018	8.1	49.5	1,013	8.7	55.2 ⎦						
56	977	0.3	0.8	974	0.4	1.0						
55	974	2.1	–2.0	971	2.6	–1.7						
54	956	12.3	–2.8	958	10.1	–4.3						
53	950	0.8	0.6	955	0.1	0.1						
52	905	0.4	1.3	909	5.6	–1.8						
51	902	3.8	–5.5	902	0.2	–1.9						
50	899	15.5	11.4	896	5.6	8.5 ⎤						
						⎬876	876	44.0	4.7			
49	898	33.2	–22.2	895	31.1	–23.7 ⎦						
48	878	113.7	–10.8	884	124.8	–7.3	868	117.2	3.4			
47	874	113.7	–12.2	870	90.7	–21.4 ⎤						
						⎬849	849	196.8	3.3	849	86.0	3.5
46	858	48.5	92.0	864	55.8	96.7 ⎦						
45	835	23.8	25.4	833	28.5	26.8	816	193.2	3.9	820	32.5	3.5
44	833	156.1	–104.5	831	168.2	–111.1	811	150.3	3.1	814	–151.0	4.1

[a] Frequencies ν and bandwidths γ in cm⁻¹; dipole strengths D in 10^{-40} esu² cm²; rotational strengths R in 10^{-44} esu² cm². Experimental rotational strengths are for (+)-**1**; calculated rotational strengths are for P-**1**.

[b] Using the TZ2P basis set.

[c] From Lorentzian fitting of the IR and VCD spectra.

The excellent agreement of the B3PW91/TZ2P and experimental VCD spectra and of the B3PW91/TZ2P and experimental rotational strengths proves that the AC of (+)-**1** is P.

The x-ray structure of (±)-**1** was determined [1]. In Figure 7.83, the B3PW91/TZ2P ring dihedral angles are compared to the ring dihedral angles of the x-ray

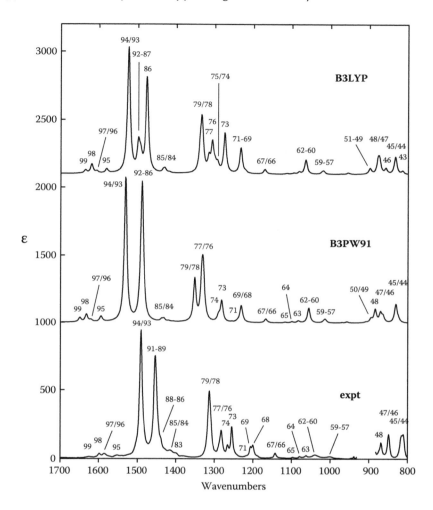

FIGURE 7.80   Comparison of the B3LYP/TZ2P and B3PW91/TZ2P IR spectra of **1** to the experimental IR spectrum of (–)-**1**. Assignment of the latter is based on the B3PW91/TZ2P spectrum. The calculated IR spectra have Lorentzian bandshapes ($\gamma = 4.0$ cm^{-1}).

structure. The excellent agreement confirms the reliability of the B3PW91/TZ2P structure of **1**.

The AC determination of heterohelicene **1**, using VCD, strongly supports the utility of VCD in determining the ACs of helicenes.

REFERENCE

1. G. Lamanna, C. Faggi, F. Gasparrini, A. Ciogli, C. Villani, P.J. Stephens, F.J. Devlin, and S. Menichetti, Efficient Thia-Bridged Triarylamine Heterohelicenes: Synthesis, Resolution and Absolute Configuration Determination, *Chem. Eur. J.*, 14, 5747–5750, 2008.

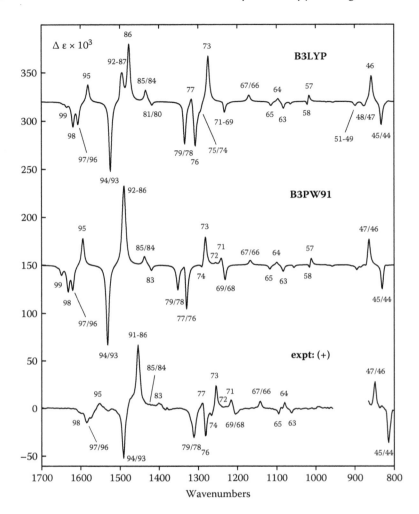

FIGURE 7.81 Comparison of the B3LYP/TZ2P and B3PW91/TZ2P VCD spectra of *P*-**1** to the experimental VCD spectrum of (+)-**1**. Assignment of the latter is based on the assignment of the IR spectrum (Figure 7.80) and on the B3PW91/TZ2P VCD spectrum. The calculated VCD spectra have Lorentzian bandshapes ($\gamma = 4.0$ cm^{-1}).

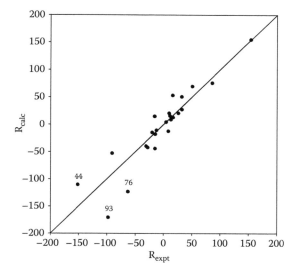

FIGURE 7.82 Comparison of the B3PW91/TZ2P rotational strengths of *P*-**1** to the experimental rotational strengths of modes 98–44 of (+)-**1**. Rotational strengths R are in 10^{-44} esu² cm².

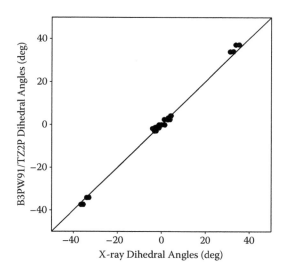

FIGURE 7.83 Comparison of ring dihedral angles of the x-ray structure of **1** and B3PW91/TZ2P-calculated dihedral angles for *P*-**1**. The RMS deviation between the calculated and x-ray dihedral angles is 1.24.

ENONE PRECURSOR TO NOR-SUBEROSENONE
AND NOR-SUBEROSANONE

The sesquiterpenes suberosenone, **1**, and suberosanone, **2**, were isolated from the marine gorgonians *Subergorgia suberosa*, and *Isis hippuris*, and shown to be cytotoxic [1,2]. Their structures are related to the cytotoxic sesquiterpene, quadrone, **3**, whose AC was determined using VCD, as discussed earlier in this chapter.

(1*R*,2*R*,8*R*,11*R*)-**1**   (1*R*,2*R*,5*S*,8*R*,11*R*)-**2**   (1*R*,2*R*,5*S*,8*R*,11*R*)-**3**

In 2004, syntheses of **1** and **2** were reported by Professor Dumas (Paris) and her student Jean-Charles, using asymmetric Michael reactions [3]. Subsequently, syntheses of nor-suberosenone, **4**, and nor-suberosanone, **5**, were attempted, using the reaction scheme:

4   (1*R*,11*S*)-**6**   5

The enone **6** was synthesized by Jean-Charles from *R*-α-phenylethylamine as a precursor to both **4** and **5**. Based upon the postulated mechanisms of asymmetric Michael reactions [4], the AC of the enone **6** was predicted to be 1*R*,11*S*. In order to evaluate this prediction, Dumas provided us with a sample of (–)-**6**, so as to determine the AC using VCD. According to Dumas, the ee of (–)-**6** was >95%.

The IR and VCD spectra of (–)-**6** were measured using 0.23 M solutions in CDCl₃ and CHCl₃ and 597 and 239 μ pathlength cells. Given the unavailability of (+)-**6** and (±)-**6**, the baselines of the VCD spectra of (–)-**6** were the VCD spectra of the solvents in the same cells. The solvent-baseline-subtracted IR and VCD spectra of (–)-**6** are

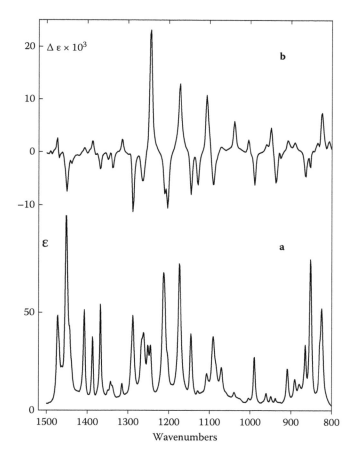

FIGURE 7.84 (a) The IR absorption spectrum of (–)-**6** in CDCl₃; 0.23 M, pathlength 597 μ, 1,500–800 cm⁻¹; except over the range 965–868 cm⁻¹, where a 0.23 M CHCl₃ solution was used, pathlength 597 μ. (b) The VCD spectrum of (–)-**6** in CDCl₃; 0.23 M, pathlength 597 μ, 800–850 cm⁻¹ and 984–1,377 cm⁻¹; pathlength 239 μ, 850–858 cm⁻¹ and 1,377–1,500 cm⁻¹; over the range 860–984 cm⁻¹ a 0.23 M CHCl₃ solution was used, pathlength 597 μ.

shown in Figure 7.84. The specific rotations of (–)-**6** were measured using the 0.23 M CDCl₃ solution; [α]$_D$ was –37.9° (c 4.4, CDCl₃).

CONFORMATIONAL ANALYSIS

Conformational analysis of **6** using the MMFF94 force field predicts only two conformations within 20 kcal/mol. The relative energies of these conformations, **a** and **b**, are given in Table 7.20. In **a** and **b** the cyclohexane ring has chair and twist-boat conformations, respectively. Optimizations of the MMFF94 conformations **a** and **b** at the B3LYP/6-31G* and B3PW91/cc-pVTZ levels lead to the relative energies given in Table 7.20. Harmonic frequency calculations for

TABLE 7.20
Conformational Analysis of 6

Conf.	$\Delta E^{a,b}$	$\Delta E^{a,c}$	$\Delta G^{a,c}$	$\Delta E^{a,d}$	$\Delta G^{a,d}$	$P(\%)^e$
a	0.00	0.00	0.00	0.00	0.00	99.98
b	7.71	5.69	5.10	5.61	4.95	0.02

^a kcal/mol.
^b MMFF94.
^c B3LYP/6-31G*.
^d B3PW91/cc-pVTZ.
^e Populations calculated using B3PW91/cc-pVTZ ΔG values at 298 K.

[a] kcal/mol.
[b] MMFF94.
[c] B3LYP/6-31G*.
[d] B3PW91/cc-pVTZ.
[e] Populations calculated using B3PW91/cc-pVTZ ΔG values at 298 K.

a and **b** confirm their stabilities and lead to the relative free energies given in Table 7.20.

The B3PW91/cc-pVTZ population of **a** is predicted to be >99.9%. **6** is thus predicted to be a conformationally rigid molecule. The B3PW91/cc-pVTZ structure of conformation **a** is:

ANALYSIS OF THE IR AND VCD SPECTRA

In Table 7.21, the B3LYP/cc-pVTZ and B3PW91/cc-pVTZ harmonic frequencies, dipole strengths, and rotational strengths of modes 25–70 of (1*R*,11*S*)-**6** are given. In Figure 7.85, the B3LYP/cc-pVTZ and B3PW91/cc-pVTZ IR spectra are compared to the experimental IR spectrum. The B3LYP and B3PW91 spectra are very similar, but not identical. The largest difference is for modes 44–49. The B3PW91 spectrum of modes 44–49 is in better agreement with the experimental spectrum than the B3LYP spectrum of modes 44–49. As a result, we assign the experimental IR spectrum using the B3PW91/cc-pVTZ spectrum, as shown in Figure 7.85. Lorentzian fitting of the experimental IR spectrum, based on this assignment, shown in Figure 7.86, leads to the frequencies, dipole strengths, and bandwidths given in Table 7.21. The B3PW91/cc-pVTZ and B3LYP/cc-pVTZ frequencies and dipole strengths are compared to the experimental frequencies and dipole strengths in Figures 7.87 and 7.88. The deviations of the calculated frequencies are standard. The deviations of the calculated dipole strengths are greater than normally observed.

TABLE 7.21
Calculated and Experimental Frequencies, Dipole Strengths, and Rotational Strengths of 6[a]

| | Calculation[b] | | | | | | Experiment[c] | | | | | |
| | B3LYP | | | B3PW91 | | | IR | | | VCD | | |
Mode	ν	D	R	ν	D	R	ν	D	γ	ν	R	γ
70	1,518	11.6	8.0	1,507	7.6	9.1				1,474	3.6	2.4
							1,472	29.5	3.6			
69	1,516	11.6	−4.9	1,505	19.9	−6.1				1,470	−1.6	2.1
68	1,508	1.6	0.7	1,498	2.3	1.8	1,464	2.3	2.4			
67	1,500	36.9	−3.6	1,488	36.6	−2.7	1,451	62.3	3.4			
66	1,497	10.3	−6.8	1,484	15.5	−8.1				1,450	−12.4	3.4
65	1,494	10.1	−0.8	1,482	14.1	−2.6	1,443	23.0	5.0			
64	1,490	3.6	−1.3	1,478	4.6	−2.3						
63	1,486	4.3	−1.6	1,473	4.9	−1.8				1,438	−3.0	3.3
62	1,459	21.9	1.5	1,446	24.5	1.9	1,408	24.9	2.6	1,408	1.0	2.3
61	1,427	18.3	2.8	1,415	22.8	4.3	1,387	13.6	2.0	1,388	3.2	2.5
60	1,406	17.8	−2.9	1,394	23.2	−6.5	1,368	23.0	2.2	1,369	−4.0	2.1
59	1,386	2.9	−10.2	1,380	1.2	−3.5	1,351	4.4	7.7	1,350	−2.2	3.0
58	1,377	6.9	1.3	1,371	6.6	0.9	1,343	5.7	4.0			
57	1,373	0.1	0.4	1,366	1.4	−3.8	1,337	4.5	4.7	1,338	−3.5	1.8
56	1,345	4.0	3.9	1,342	4.7	3.8	1,315	6.0	3.7	1,315	5.3	3.7
55	1,321	7.1	−1.1	1,319	9.3	−0.6				1,294	5.0	3.2
							1,288	40.2	4.2			
54	1,307	15.2	−12.8	1,311	21.2	−13.1				1,288	−25.7	4.0
53	1,292	20.5	5.1	1,290	17.4	3.7	1,267	19.0	3.8			
										1,264	−16.3	4.4
52	1,287	14.9	−20.9	1,285	16.0	−21.2	1,261	19.9	3.5			
51	1,279	3.7	−1.1	1,274	10.8	−6.7	1,251	15.2	3.3			
50	1,267	15.1	41.3	1,267	21.1	52.4	1,245	16.0	3.1	1,245	45.9	3.1
49	1,233	8.2	9.6	1,236	9.9	4.7				1,211	−13.3	3.0
							1,212	69.2	4.5			
48	1,228	8.2	−3.1	1,231	42.6	−17.7						
47	1,221	39.1	−40.8	1,223	11.7	−27.0	1,204	13.3	5.4	1,203	−18.1	2.9
46	1,199	27.7	13.9	1,196	27.3	12.5				1,173	35.7	4.5
							1,174	76.0	4.5			
45	1,182	54.6	29.5	1,187	71.0	26.1						
44	1,167	66.5	−13.1	1,163	42.8	−17.6	1,145	26.4	3.1	1,146	−14.4	2.7
43	1,143	11.0	−30.0	1,148	1.4	−12.1	1,127	5.0	5.5	1,129	−12.8	3.0
42	1,125	9.6	30.8	1,124	13.1	35.9	1,107	12.3	4.2	1,107	29.1	4.0
41	1,108	46.0	14.6	1,110	43.7	−25.4	1,091	44.5	5.4	1,091	−19.4	4.0
40	1,102	19.7	−21.2	1,102	6.4	2.2	1,083	5.0	3.1			
39	1,079	5.2	−0.5	1,090	7.7	2.4	1,071	18.0	4.4	1,071	3.3	4.6
38	1,054	2.4	9.7	1,056	3.1	12.4	1,040	7.1	5.7	1,039	13.1	3.1

(Continued)

TABLE 7.21 (CONTINUED)

Calculated and Experimental Frequencies, Dipole Strengths, and Rotational Strengths of 6[a]

	Calculation[b]						Experiment[c]					
	B3LYP			B3PW91			IR			VCD		
Mode	ν	D	R	ν	D	R	ν	D	γ	ν	R	γ
37	1,024	1.9	11.5	1,019	1.3	8.1	1,004	2.0	3.2	1,005	4.6	2.6
36	1,002	23.7	−16.5	1,005	22.5	−21.8	990	17.7	2.5	990	−13.0	2.7
35	969	3.0	2.1	975	4.5	2.3	961	4.8	3.1	960	3.3	3.2
34	959	5.1	14.5	962	11.8	8.8	949	2.7	2.5	949	10.0	2.6
33	948	7.0	−16.3	948	2.9	−13.3	938	2.4	3.2	937	−15.6	3.1
32	923	9.7	−12.2	923	12.8	−3.3	909	16.3	2.9	908	7.4	4.2
31	905	9.9	8.0	907	21.5	8.9	891	13.0	3.8	891	6.2	4.3
30	893	31.9	2.5	895	11.4	1.9	879	11.6	4.7			
29	880	54.0	15.4	883	57.3	16.0 ⎤						
						⎬ 865	24.0	2.9	865	−11.1	2.6
28	873	6.6	−18.2	878	5.6	−16.6 ⎦						
27	870	16.5	−12.3	870	26.2	−13.0	852	71.7	2.9	853	−6.3	2.5
26	835	2.2	2.5	844	4.4	−1.9	828	23.1	3.5			
25	819	95.8	10.1	831	88.5	13.1	824	47.0	3.3	824	22.1	3.3

[a] Frequencies ν and bandwidths γ in cm^{-1}; dipole strengths D in 10^{-40} esu^2 cm^2; rotational strengths R in 10^{-44} esu^2 cm^2. Experimental rotational strengths are for (−)-**6**; calculated rotational strengths are for (1R,11S)-**6**.

[b] Using the cc-pVTZ basis set.

[c] From Lorentzian fitting of the IR and the VCD spectra.

In Figure 7.89, the B3LYP/cc-pVTZ and B3PW91/cc-pVTZ VCD spectra of (1R,11S)-**6** are compared to the experimental VCD spectrum of (−)-**6**. As with the IR spectra, the B3LYP and B3PW91 VCD spectra are very similar, but not identical. The largest difference is for modes 40–49. The B3PW91 spectrum of modes 40–49 is in better agreement with the experimental spectrum than the B3LYP spectrum of modes 40–49. The assignment of the experimental VCD spectrum based on the B3PW91/cc-pVTZ VCD spectrum is shown in Figure 7.89. Lorentzian fitting of the experimental VCD spectrum based on this assignment, shown in Figure 7.90, leads to the experimental frequencies, rotational strengths, and bandwidths given in Table 7.21. The VCD frequencies are essentially identical to the IR frequencies, confirming the consistency of the assignments of the IR and VCD spectra. Comparisons of the B3PW91/cc-pVTZ and B3LYP/cc-pVTZ rotational strengths of (1R,11S)-**6** to the experimental rotational strengths of (−)-**6** are shown in Figure 7.91. The B3PW91 rotational strengths are substantially more accurate than the B3LYP rotational strengths. The excellent agreement of the B3PW91/cc-pVTZ VCD spectrum of (1R,11S)-**6** with the experimental VCD spectrum of (−)-**6**, and of the B3PW91/cc-pVTZ rotational strengths of (1R,11S)-**6** with the experimental

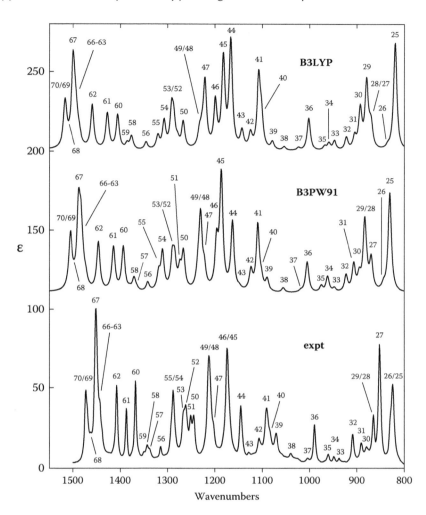

FIGURE 7.85 Comparison of the B3LYP/cc-pVTZ and B3PW91/cc-pVTZ IR spectra of **6**, calculated using Lorentzian bandshapes ($\gamma = 4.0$ cm^{-1}), to the experimental IR spectrum of (–)-**6**. The assignment of the experimental spectrum is based on the B3PW91/cc-pVTZ spectrum.

rotational strengths of (–)-**6**, unambiguously proves the AC of **6** to be (1R,11S)-(–), the AC predicted by Dumas on the basis of the mechanisms of the reactions used to synthesize **6**. The excellent agreement of the B3PW91/cc-pVTZ VCD spectrum of **6** with the experimental VCD spectrum of **6** also confirms the reliability of the conformational analysis of **6**.

In order to extend the studies of the functional dependence of the accuracy of DFT rotational strengths, discussed in Chapter 6, we have compared the mean absolute deviations of rotational strengths of (1R,11S)-**6**, calculated using the 19 functionals in Chapter 6 and the cc-pVTZ basis set, and the experimental rotational strengths

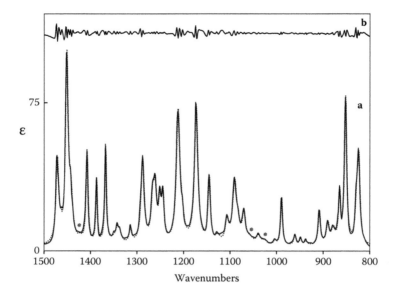

FIGURE 7.86 (a) Lorentzian fit of the experimental IR spectrum of (–)-**6** over the range 1,500–800 cm⁻¹. The solid line is the experimental spectrum; the dotted line is the Lorentzian fit. (b) The difference spectrum: the experimental spectrum minus the Lorentzian fit. The asterisks indicate bands that are not assigned to fundamental modes.

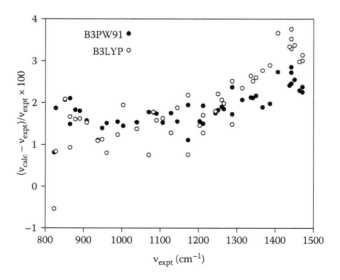

FIGURE 7.87 Comparison of the B3LYP/cc-pVTZ and B3PW91/cc-pVTZ frequencies to the experimental IR frequencies of modes 25–70.

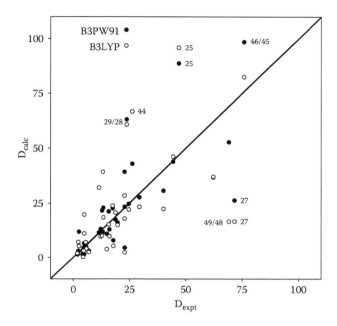

FIGURE 7.88 Comparison of the B3LYP/cc-pVTZ and B3PW91/cc-pVTZ dipole strengths to the experimental dipole strengths of modes 25–70. For bands assigned to multiple vibrational modes, calculated dipole strengths are the sums of the dipole strengths of contributing modes. Dipole strengths D are in 10^{-40} esu^2 cm^2.

of modes 25–70 of (–)-**6**. The results are given in Table 7.22 and Figure 7.92. Only one functional, BMK, is slightly more accurate than B3PW91. Seventeen functionals are more accurate than B3LYP for this molecule. Except for BB95, HCTH, BP86, BPW91, τHCTH, and OLYP, all are hybrid functionals. The functional less accurate than B3LYP is the nonhybrid functional BLYP. The number of incorrect signs varies from 1 to 7. B3PW91 and B3LYP have two and six incorrect signs, respectively.

To evaluate the accuracies of other basis sets, relative to cc-pVTZ and cc-pVQZ, calculations have been repeated using the 10 additional basis sets listed in Table 7.23, and the B3PW91 functional. The values of the mean absolute deviations of the calculated rotational strengths from the cc-pVQZ rotational strengths of modes 25–70 are given in Table 7.23. In Figure 7.93 the dependence of the mean absolute deviation on the number of basis functions is plotted. As the number of basis functions diminishes, the mean absolute deviation increases substantially.

The basis sets TZ2P and 6-31G* are frequently used in calculating VCD spectra. The B3PW91/TZ2P and B3PW91/6-31G* VCD spectra of (1R,11S)-**6** are compared to the B3PW91/cc-pVQZ VCD spectrum in Figure 7.94. The TZ2P VCD spectrum is quantitatively very nearly identical to the cc-pVQZ VCD spectrum, proving that the basis set error of TZ2P is not significantly greater than the basis set error of cc-pVTZ. The 6-31G* VCD spectrum is somewhat different from the cc-pVQZ and TZ2P VCD spectra. However, since the qualitative differences are not great, the 6-31G* basis set error is not much greater than the basis set errors of TZ2P and cc-pVTZ.

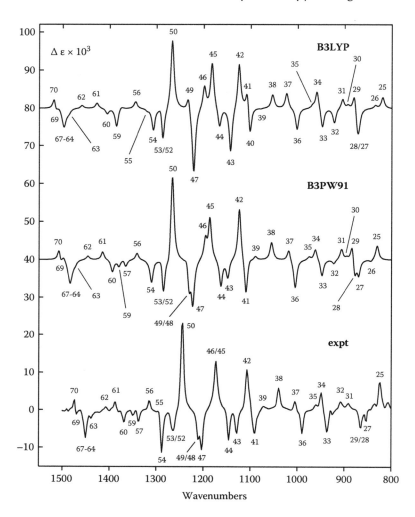

FIGURE 7.89   Comparison of the B3LYP/cc-pVTZ and B3PW91/cc-pVTZ VCD spectra of (1*R*,11*S*)-**6**, calculated using Lorentzian bandshapes ($\gamma = 4.0$ cm^{-1}), to the experimental VCD spectrum of (−)-**6**. The assignment of the experimental spectrum is based on the B3PW91/cc-pVTZ spectrum.

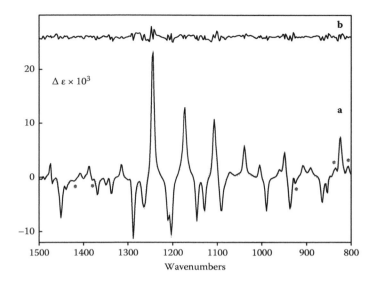

FIGURE 7.90 (a) Lorentzian fit of the experimental VCD spectrum of (–)-**6** over the range 1,500–800 cm^{-1}. The solid line is the experimental spectrum; the dotted line is the Lorentzian fit. (b) The difference spectrum: the experimental spectrum minus the Lorentzian fit. The asterisks indicate bands that are not assigned to fundamental modes.

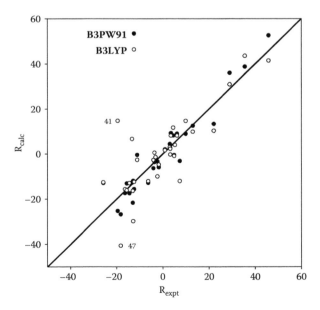

FIGURE 7.91 Comparison of the B3LYP/cc-pVTZ and B3PW91/cc-pVTZ rotational strengths to the experimental rotational strengths of modes 25–70. For bands assigned to multiple vibrational modes, calculated rotational strengths are the sums of the rotational strengths of contributing modes. Rotational strengths R are in 10^{-44} esu^2 cm^2.

TABLE 7.22
Functional Dependence of the Rotational Strengths of (1R,11S)-6 Calculated Using the cc-pVTZ Basis Set

Functional	Type[a]	MAD[b]	No. Wrong Signs
BLYP	GGA	9.86	7
B3LYP	H-GGA	6.81	6
OLYP	GGA	6.44	6
τHCTH	M-GGA	5.76	4
BPW91	GGA	5.42	4
τHCTHHYB	HM-GGA	5.40	3
O3LYP	H-GGA	5.32	4
CAMB3LYP	H-X	5.23	2
BP86	GGA	5.02	2
HCTH	GGA	4.99	2
PBE1PBE	H-GGA	4.90	1
M05–2X	HM-GGA	4.80	2
HSE1PBE	H-S	4.79	2
B1B95	HM-GGA	4.46	1
MPW1PW91	H-GGA	4.40	1
B3P86	H-GGA	4.33	2
BB95	M-GGA	4.32	3
B3PW91	H-GGA	4.09	2
BMK	HM-GGA	3.86	2

[a] GGA, generalized gradient approximation; M-GGA, meta-generalized gradient approximation; H-GGA, hybrid generalized gradient approximation; HM-GGA, hybrid meta-generalized gradient approximation; H-X, range-separated hybrid (with short-range DFT exchange and long-range HF exchange); H-S, range-separated hybrid (short-range HF exchange and long-range DFT exchange).

[b] Mean absolute deviation of calculated rotational strengths of (1R,11S)-**6** and experimental rotational strengths of modes 25–70.

To confirm that DFT is more accurate than HF theory, the IR and VCD spectra have been calculated using HF theory and the cc-pVTZ basis set. The spectra obtained are compared to the experimental spectra in Figures 7.95 and 7.96. The experimental rotational strengths are compared to the HF/cc-pVTZ and B3PW91/cc-pVTZ rotational strengths in Figure 7.97. The B3PW91/cc-pVTZ rotational strengths are clearly in better agreement with the experimental rotational strengths than are the HF/cc-pVTZ rotational strengths, especially for modes 25, 32, 41, 48/49, and 50. The mean absolute deviations between calculated and experimental rotational strengths of modes 25–70 are 4.1 and 7.4 × 10^{-44} esu^2 cm^2 for B3PW91/cc-pVTZ and HF/cc-pVTZ, respectively, confirming this conclusion.

The determination of the AC of the enone **6**, using VCD, provides further evidence of the utility of VCD in assessing the understanding of the stereochemistries of organic reactions used to synthesize chiral molecules.

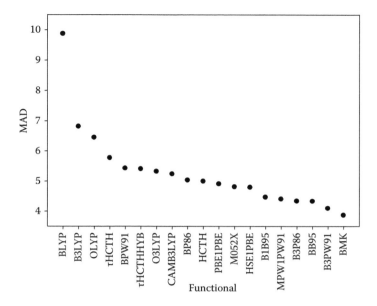

FIGURE 7.92   Functional dependence of the mean absolute deviation (MAD) of the rotational strengths of (1R,11S)-**6** calculated using the cc-pVTZ basis set.

TABLE 7.23
The Basis Set Dependence of the Mean Absolute Deviations of B3PW91-Calculated and B3PW91/cc-pVQZ Rotational Strengths of Modes 25–70 of (1R,11S)-6

Basis Set	Contracted Set[a]	No. Fns.[b]	MAD[c]
cc-pVQZ	[5s4p3d2f1g/4s3p2d1f] (5d,7f,9g)	1,310	0.00
cc-pVTZ	[4s3p2d1f/3s2p1d] (5d,7f)	672	0.34
aug-cc-pVTZ	[5s4p3d2f/4s3p2d] (5d,7f)	1,058	0.36
6-311++G(2d,2p)	[5s4p2d/4s2p] (5d)	558	0.96
6-311++G**	[5s4p1d/4s1p] (5d)	434	1.18
TZ2P	[5s4p2d/3s2p] (6d)	568	1.23
6-311G**	[4s3p1d/3s1p] (5d)	360	1.40
6-31G**	[3s2p1d/2s1p] (6d)	300	3.05
6—31G*	[3s2p1d/2s] (6d)	246	3.45
cc-pVDZ	[3s2p1d/2s1p] (5d)	286	4.73
aug-cc-pVDZ	[4s3p2d/3s2p] (5d)	484	4.81
6-31G	[3s2p/2s]	162	5.73

[a]   Of the form [1/2] where the first term refers to the contracted set on carbon and oxygen and the second refers to the contracted set on hydrogen. The numbers in parentheses refer to the number of functions used for each type of polarization function.

[b]   Number of basis functions.

[c]   Mean absolute deviation relative to cc-pVQZ for modes 25–70.

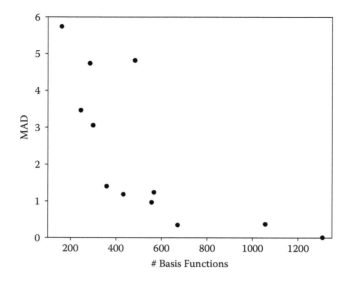

FIGURE 7.93 B3PW91 basis set dependence of (1*R*,11*S*)-**6** rotational strengths.

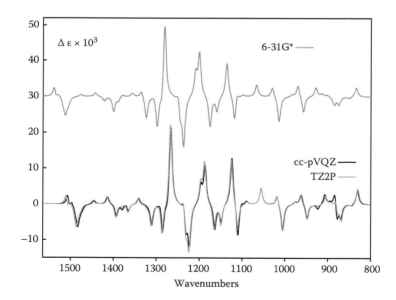

FIGURE 7.94 Comparison of the B3PW91/TZ2P and B3PW91/6-31G* VCD spectra of (1*R*,11*S*)-**6** to the B3PW91/cc-pVQZ VCD spectrum.

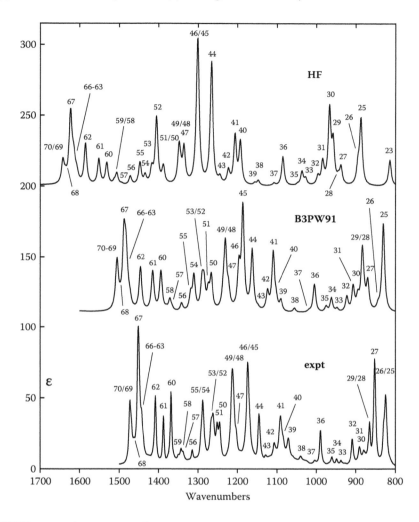

FIGURE 7.95 Comparison of the HF/cc-pVTZ and B3PW91/cc-pVTZ IR spectra of (1*R*,11*S*)-**6** to the experimental IR spectrum of (−)-**6**.

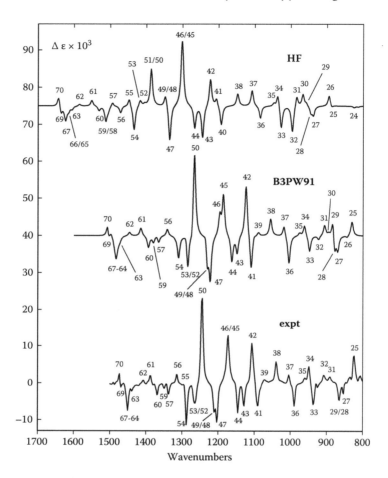

FIGURE 7.96 Comparison of the HF/cc-pVTZ and B3PW91/cc-pVTZ VCD spectra of (1*R*,11*S*)-**6** to the experimental VCD spectrum of (–)-**6**.

REFERENCES

1. H.R. Bokesch, T.C. McKee, J.H. Cardellina, M.R. Boyd, Suberosenone, a New Cytotoxin from *Subergorgia suberosa, Tet. Lett.*, 37, 3259–3262, 1996.
2. J.H. Sheu, K.C. Hung, G.H. Wang, C.Y. Duh, New Cytotoxic Sesquiterpenes from the *Gorgonian Isis hippuris, J. Nat. Prod.*, 63, 1603–1607, 2000.
3. K.L. Jean-Charles, C. Camara, F. Dumas, First Asymmetric Synthesis of Suberosenone and Suberosanone: Absolute Stereochemistry Assignment of Natural Suberosanes. poster at Chirality 2004, the 16th International Symposium on Chirality, ISCD 16, New York, July 2004.
4. (a) C. Camara, D. Joseph, F. Dumas, J. d'Angelo, A. Chiaroni, High Pressure Activation in the Asymmetric Michael Addition of Chiral Imines to Alkyl and Aryl Crotonates, *Tet. Lett.*, 43, 1445–1448, 2002; (b) S. Delarue-Cochin, J.J. Pan, A. Dauteloup, F. Hendra, R.G. Angoh, D. Joseph, P.J. Stephens, C. Cavé, Asymmetric Michael Reaction: Novel Efficient Access to Chiral β-ketophosphonates, *Tet. Asymm.*, 18, 685–691, 2007.

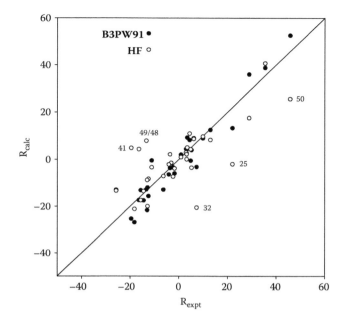

FIGURE 7.97 Comparison of the HF/cc-pVTZ and B3PW91/cc-pVTZ rotational strengths to the experimental rotational strengths of modes 25–70. Rotational strengths R are in 10^{-44} esu^2 cm^2.

CONCLUSIONS

The determination of the AC of a chiral molecule from its experimental VCD spectrum requires a methodology which reliably predicts the VCD spectra of its enantiomers. The prediction of the VCD spectra of the enantiomers of a chiral molecule using the Stephens equation for vibrational rotational strengths and the DFT methodology, together with optimally chosen basis sets and density functionals, is of sufficient reliability to permit the unambiguous determination of the ACs of organic molecules.

In the cases of conformationally-rigid molecules, predicted rotational strengths are in excellent quantitative agreement with experimental rotational strengths—the minor differences can be attributed to both experimental and computational errors. The accuracies of experimental VCD spectra depend on the magnitudes of artefact signals due to imperfections in the optics of the spectrometer. The magnitudes of these artefacts can be assessed by comparison of the VCD spectra of the two enantiomers, measured using the VCD spectrum of the racemate as baseline. Deviations from zero of the half-sum spectrum define the magnitudes of artefacts.

The accuracies of the predicted VCD spectra depend on the quality of the basis set and density functional used and also on the magnitudes of contributions to the vibrational rotational strengths due to vibrational anharmonicity and solvent effects, which are currently not included in the predicted spectra. The presence of anharmonicity is responsible for the overall shift of calculated frequencies from experimental frequencies and typically causes very small contributions to the rotational strengths of vibrational transitions in the mid-IR, the region universally used in determining ACs. Solvent effects typically cause small contributions to vibrational rotational strengths, as long as the solvent chosen for the measurement of the experimental VCD spectrum interacts weakly with the solute molecule.

In the case of conformationally flexible molecules VCD spectra are also dependent on the fractional populations of the populated conformers, which are determined by their relative free energies. For some conformationally flexible molecules, calculated and experimental rotational strengths agree less well than is the case for conformationally rigid molecules, presumably due to errors in predicted conformational populations. It is anticipated that solvent effects on conformer free energies and populations will give rise to greater solvent effects on VCD spectra than the solvent effects on rotational strengths. If the molecule is enormously flexible, and the number of populated conformations is very large, the prediction of its VCD spectrum becomes very time-consuming and less reliable. For very large bioorganic molecules, such as proteins and nucleic acids, DFT calculations are currently impractical. Nevertheless, for the majority of organic molecules, VCD spectroscopy is an extremely useful and practical technique for determining ACs.

The prediction of vibrational rotational strengths and VCD spectra will therefore be improved in accuracy in the future by the development of more accurate density functionals, the incorporation of anharmonicity and solvent effects and also by improvements in the prediction of the equilibrium populations of conformationally-flexible molecules. Such advances will further increase the reliability of the ACs of organic molecules determined using VCD spectroscopy.

Bibliography

A. Aamouche, F.J. Devlin, P.J. Stephens, J. Drabowicz, B. Bujnicki, and M. Mikolajczyk, Vibrational Circular Dichroism and Absolute Configuration of Chiral Sulfoxides: Tert-Butyl Methyl Sulfoxide, *Chem. Eur. J.* 6 (24), 4479 (2000).

A. Aamouche, F.J. Devlin, and P.J. Stephens, Structure, Vibrational Absorption and Circular Dichroism Spectra and Absolute Configuration of Tröger's Base, *J. Am. Chem. Soc.* 122 (10), 2346 (2000).

A. Aamouche, F.J. Devlin, and P.J. Stephens, Conformations of Chiral Molecules in Solution: *Ab Initio* Vibrational Absorption and Circular Dichroism Studies of 4,4a,5,6,7,8-Hexahydro-4a-Methyl-2(3H)-Naphthalenone and 3,4,8,8a-Tetrahydro-8a-Methyl-1,6(2H,7H)- Naphthalenedione, *J. Am. Chem. Soc.* 122 (30), 7358 (2000).

A. Aamouche, F.J. Devlin, and P.J. Stephens, Determination of Absolute Configuration Using Circular Dichroism: Tröger's Base Revisited Using Vibrational Circular Dichroism, *Chem. Comm.* (4), 361 (1999).

S. Abbate, F. Lebon, S. Lepri, G. Longhi, R. Gangemi, S. Spizzichino, G. Bellachioma, and R. Ruzziconi, Vibrational Circular Dichroism: A Valuable Tool for Conformational Analysis and Absolute Configuration Assignment of Chiral 1-Aryl-2,2,2-Trifluoroethanols, *Chem. Phys. Chem.* 12 (18), 3519 (2011).

S. Abbate, G. Longhi, F. Gangemi, R. Gangemi, S. Superchi, A.M. Caporusso, and R. Ruzziconi, Electrical and Mechanical Anharmonicities from NIR-VCD Spectra of Compounds Exhibiting Axial and Planar Chirality: The Cases of (S)-2,3-Pentadiene and Methyl-d$_3$ (R)- and (S)-[2.2]Paracyclophane-4-Carboxylate, *Chirality* 23 (9), 841 (2011).

S. Abbate, A. Ciogli, S. Fioravanti, F. Gasparrini, G. Longhi, L. Pellacani, E. Rizzato, D. Spinelli, and P.A. Tardella, Solving the Puzzling Absolute Configuration Determination of a Flexible Molecule by Vibrational and Electronic Circular Dichroism Spectroscopies and DFT Calculations: The Case Study of a Chiral 2,2´-Dinitro-2,2´-biaziridine, *Eur. J. Org. Chem.* (32), 6193 (2010).

S. Abbate, L.F. Burgi, F. Gangemi, R. Gangemi, F. Lebon, G. Longhi, V. Pultz, and D.A. Lightner, Comparative Analysis of IR and Vibrational Circular Dichroism Spectra for a Series of Camphor Related Molecules, *J. Phys. Chem. A*, 113 (42), 11390 (2009).

S. Abbate, F. Lebon, G. Longhi, F. Fontana, T. Caronna, and D.A. Lightner, Experimental and Calculated Vibrational and Electronic Circular Dichroism Spectra of 2-Br-Hexahelicene, *Phys. Chem. Chem. Phys.* 11 (40), 9039 (2009).

S. Abbate, F. Lebon, R. Gangemi, G. Longhi, S. Spizzichino, and R. Ruzziconi, Electronic and Vibrational Circular Dichroism Spectra of Chiral 4-X-[2.2]Paracyclophanes with X Containing Fluorine Atoms, *J. Phys. Chem. A*, 113 (52), 14851 (2009).

S. Abbate, G. Longhi, E. Castiglioni, F. Lebon, P.M. Wood, L.W.L. Woo, and B.V.L. Potter, Determination of the Absolute Configuration of Aromatase and Dual Aromatase-Sulfatase Inhibitors by Vibrational and Electronic Circular Dichroism Spectra Analysis, *Chirality* 21 (9), 802 (2009).

S. Abbate, F. Castiglione, F. Lebon, G. Longhi, A. Longo, A. Mele, W. Panzeri, A. Ruggirello, and V. Turco Liveri, Spectroscopic and Structural Investigation of the Confinement of D and L Dimethyl Tartrate in Lecithin Reverse Micelles, *J. Phys. Chem. B* 113 (10), 3024 (2009).

S. Abbate, L.F. Burgi, E. Castiglioni, F. Lebon, G. Longhi, E. Toscano, and S. Caccamese, Assessment of Configurational and Conformational Properties of Naringenin by Vibrational Circular Dichroism, *Chirality* 21 (4), 436 (2009).

S. Abbate, G. Longhi, A. Ruggirello, and V.T. Liveri, Confinement of Chiral Molecules in Reverse Micelles: FT-IR, Polarimetric and VCD Investigation on the State of Dimethyl Tartrate in Sodium Bis(2-Ethylhexyl) Sulfosuccinate Reverse Micelles Dispersed in Carbon Tetrachloride, *Colloids Surf. A Physicochem. Eng. Aspects* 327 (1–3), 44 (2008).

S. Abbate, E. Castiglioni, F. Gangemi, R. Gangemi, G. Longhi, R. Ruzziconi, and S. Spizzichino, Harmonic and Anharmonic Features of IR and NIR Absorption and VCD Spectra of Chiral 4-X-[2.2]-Paracyclophanes, *J. Phys. Chem. A* 111 (30), 7031 (2007).

S. Abbate, R. Gangemi, and G. Longhi, Dipole and Rotational Strengths for Overtone Transitions of a C_2-Symmetry HCCH Molecular Fragment Using Van Vleck Perturbation Theory, *J. Chem. Phys.* 117 (16), 7575 (2002).

S. Abbate, G. Longhi, and C. Santina, Theoretical and Experimental Studies for the Interpretation of Vibrational Circular Dichroism Spectra in the CH-Stretching Overtone Region, *Chirality* 12 (4), 180 (2000).

S. Abbate, G. Longhi, K. Kwon, and A. Moscowitz, The Use of Cross-Correlation Functions in the Analysis of Circular Dichroism Spectra, *J. Chem. Phys.* 108 (1), 50 (1998).

S. Abbate, G. Longhi, S. Boiadjiev, D.A. Lightner, C. Bertucci, and P. Salvadori, Analysis of Vibrational Circular Dichroism Data in the Near Infrared and Visible Range, *Enantiomer* 3, 337 (1998).

S. Abbate, G. Longhi, J.W. Givens III, S.E. Boiadjiev, D.A. Lightner, and A. Moscowitz, Observation of Vibrational Circular Dichroism for Overtone Transitions with Commercially Available CD Spectrometers, *Appl. Spectrosc.* 50 (5), 642 (1996).

S. Abbate, G. Longhi, L. Ricard, C. Bertucci, C. Rosini, P. Salvadori, and A. Moscowitz, Vibrational Circular Dichroism as a Criterion for Local-Mode versus Normal-Mode Behavior. Near-Infrared Circular Dichroism Spectra of Some Monoterpenes, *J. Am. Chem. Soc.* 111 (3), 836 (1989).

S. Abbate, H.A. Havel, L. Laux, V. Pultz, and A. Moscowitz, Vibrational Optical Activity in Deuteriated Phenylethanes, *J. Phys. Chem.* 92 (11), 3302 (1988).

S. Abbate, L. Laux, V. Pultz, H.A. Havel, J. Overend, and A. Moscowitz, The Charge Flow Model Applied to the Vibrational Circular Dichroism of Oriented Species, *Chem. Phys. Lett.* 113 (2), 202 (1985).

S. Abbate, L. Laux, J. Overend, and A. Moscowitz, A Charge Flow Model for Vibrational Rotational Strengths, *J. Chem. Phys.* 75 (7), 3161 (1981).

S. Abdali, K.J. Jalkanen, X. Cao, L.A. Nafie, and H. Bohr, Conformational Determination of [Leu]Enkephalin Based on Theoretical and Experimental VA and VCD Spectral Analyses, *Phys. Chem. Chem. Phys.* 6 (9), 2434 (2004).

S. Abdali, K.J. Jalkanen, H. Bohr, S. Suhai, and R.M. Nieminen, The VA and VCD Spectra of Various Isotopomers of L-Alanine in Aqueous Solution, *Chem. Phys.* 282, 219 (2002).

D. Abramavicius and S. Mukamel, Chirality-Induced Signals in Coherent Multidimensional Spectroscopy of Excitons, *J. Chem. Phys.* 124, 034113 (2006).

I. Ahmadi, H. Rahemi, and S.F. Tayyari, Structural, Potential Surface and Vibrational Spectroscopy Studies of Hypophosphorous Acid in the Gas Phase and Chain Conformation. A Theoretical Study, *J. Kor. Chem. Soc.* 49 (2), 129 (2005).

L. Alagna, S. DiFonzo, T. Prosperi, S. Turchini, P. Lazzeretti, M. Malagoli, R. Zanasi, C.R. Natoli, and P.J. Stephens, Random Phase Approximation Calculations of K-Edge Rotational Strengths of Chiral Molecules: Propylene Oxide, *Chem. Phys. Lett.* 223, 402 (1994).

I. Alkorta, K. Zborowski, and J. Elguero, Self-Aggregation as a Source of Chiral Discrimination, *Chem. Phys. Lett.* 427 (4–6), 289 (2006).

S. Allenmark and J. Gawronski, Determination of Absolute Configuration—An Overview Related to This Special Issue, *Chirality* 20 (5), 606 (2008).

R.D. Amos, N.C. Handy, and P. Palmieri, Vibrational Properties of (R)-Methylthiirane from Møller-Plesset Perturbation Theory, *J. Chem. Phys.* 93 (8), 5796 (1990).

R.D. Amos, K.J. Jalkanen, and P.J. Stephens, Alternative Formalism for the Calculation of Atomic Polar Tensors and Atomic Axial Tensors, *J. Phys. Chem.* 92 (20), 5571 (1988).

R.D. Amos, N.C. Handy, A.F. Drake, and P. Palmieri, The Vibrational Circular Dichroism of Dimethylcyclopropane in the C-H Stretching Region, *J. Chem. Phys.* 89 (12), 7287 (1988).

R.D. Amos, N.C. Handy, K.J. Jalkanen, and P.J. Stephens, Efficient Calculation of Vibrational Magnetic Dipole Transition Moments and Rotational Strengths, *Chem. Phys. Lett* 133, 21 (1987).

D.L. An, Q. Chen, J. Fang, H. Yan, A. Orita, N. Miura, A. Nakahashi, K. Monde, and J. Otera, Vibrational CD Spectroscopy as a Powerful Tool for Stereochemical Study of Cyclophynes in Solution, *Tetrahedron Lett.* 50 (15), 1689 (2009).

N.H. Andersen, N.J. Christensen, P.R. Lassen, T.B.N. Freedman, L.A. Nafie, K. Strømgaard, and L. Hemmingsen, Structure and Absolute Configuration of Ginkgolide B Characterized by IR- and VCD Spectroscopy, *Chirality* 22 (2), 217 (2010).

V. Andrushchenko, D. Tsankov, M. Krasteva, H. Wieser, and P. Bouř, Spectroscopic Detection of DNA Quadruplexes by Vibrational Circular Dichroism, *J. Am. Chem. Soc.* 133 (38), 15055 (2011).

V. Andrushchenko and P. Bouř, Applications of the Cartesian Coordinate Tensor Transfer Technique in the Simulations of Vibrational Circular Dichroism Spectra of Oligonucleotides, *Chirality* 22 (1E), E96 (2010).

V. Andrushchenko and P. Bour, Circular Dichroism Enhancement in Large DNA Aggregates Simulated by a Generalized Oscillator Model, *J. Comput. Chem.* 29 (16), 2693 (2008).

V. Andrushchenko, H. Wieser, and P. Bour, DNA Oligonucleotide-Cis-Platin Binding: *Ab Initio* Interpretation of the Vibrational Spectra, *J. Phys. Chem. A* 111 (39), 9714 (2007).

V. Andrushchenko, H. Wieser, and P. Bour, RNA Structural Forms Studied by Vibrational Circular Dichroism: *Ab Initio* Interpretation of the Spectra, *J. Phys. Chem. B* 108 (12), 3899 (2004).

V. Andrushchenko, D. Tsankov, and H. Wieser, Vibrational Circular Dichroism Spectroscopy and the Effects of Metal Ions on DNA Structure, *J. Mol. Struct.* 661–662, 541 (2003).

V. Andrushchenko, J.H.V.D. Sande, and H. Wieser, Vibrational Circular Dichroism and IR Absorption of DNA Complexes with Cu^{2+} Ions, *Biopolymers (Biospectroscopy)* 72 (5), 374 (2003).

V. Andrushchenko, H.V.D. Sande, and H. Wieser, DNA Interaction with Mn^{2+} Ions at Elevated Temperatures: VCD Evidence of DNA Aggregation, *Biopolymers* 69 (4), 529 (2003).

V. Andrushchenko, H. Wieser, and P. Bour, B-Z Conformational Transition of DNA Monitored by Vibrational Circular Dichroism. *Ab Initio* Interpretation of the Experiment, *J. Phys. Chem. B* 106 (48), 12623 (2002).

V. Andrushchenko, Y. Blagoi, J.H.V.D. Sande, and H. Wieser, Poly(Ra)·Poly(Ru) with Ni^{2+}Ions at Different Temperatures: Infrared Absorption and Vibrational Circular Dichroism Spectroscopy, *J. Biomol. Struct. Dyn.* 19 (5), 889 (2002).

V. Andrushchenko, J.L. McCann, J.H.V.D. Sande, and H. Wieser, Determining Structures of Polymeric Molecules by Vibrational Circular Dichroism (VCD) Spectroscopy, *Vibrational Spectrosc.* 22 (1–2), 101 (2000).

V.V. Andrushchenko, J.H.V.D. Sande, H. Wieser, S.V. Kornilova, and Y.P. Blagoi, Complexes of (dG-dC)$_{20}$ with Mn^{2+} Ions: A Study by Vibrational Circular Dichroism and Infrared Absorption Spectroscopy, *J. Biomol. Struct. Dyn.* 17 (3), 545 (1999).

V.V. Andrushchenko, J.H.V.D. Sande, and H. Wieser, Interaction of Deoxyribo-Oligonucleotides with Divalent Manganese Ions: Comparison of Vibrational Circular Dichroism and Absorption Spectroscopy, *Vibrational Spectrosc.* 19 (2), 341 (1999).

A. Annamalai, K.J. Jalkanen, U. Narayanan, M.C. Tissot, T.A. Keiderling, and P.J. Stephens, Theoretical Study of the Vibrational Circular Dichroism of 1,3-Dideuterioallene: Comparison of Methods, *J. Phys. Chem.* 94 (1), 194 (1990).

A. Annamalai and T.A. Keiderling, Vibrational Circular Dichroism of Poly(Ribonucleic Acids). A Comparative Study in Aqueous Solution, *J. Am. Chem. Soc.* 109 (10), 3125 (1987).

A. Annamalai, T.A. Keiderling, and J.S. Chickos, Vibrational Circular Dichroism of Trans-1,2-Dideuteriocyclobutane. Experimental and Calculational Results in the Mid-Infrared, *J. Am. Chem. Soc.* 107 (8), 2285 (1985).

A. Annamalai, T.A. Keiderling, and J.S. Chickos, Vibrational Circular Dichroism of Trans-1,2-Dideuteriocyclobutane. A Comparison of Fixed Partial Charge and Localized Molecular Orbital Theories with Different Force Fields, *J. Am. Chem. Soc.* 106 (21), 6254 (1984).

Y. Aoyagi, A. Yamazaki, C. Nakatsugawa, H. Fukaya, K. Takeya, S. Kawauchi, and H. Izumi, Salvileucalin B, a Novel Diterpenoid with an Unprecedented Rearranged Neoclerodane Skeleton from Salvia Leucantha Cav, *Org. Lett.* 10 (20), 4429 (2008).

D.W. Armstrong, F.A. Cotton, A.G. Petrovic, P.L. Polavarapu, and M.M. Warnke, Resolution of Enantiomers in Solution and Determination of the Chirality of Extended Metal Atom Chains, *Inorg. Chem.* 46 (5), 1535 (2007).

S.A. Asher, R.W. Bormett, P.J. Larkin, W.G. Gustafson, N. Ragunathan, T.B. Freedman, L.A. Nafie, N.T. Yu, K. Gersonde, et al., Vibrational Circular Dichroism Is a Sensitive Probe Study of Heme Protein-Ligand Interactions, in *Spectroscopy of Biological Molecules*, R.E. Hester and R.B. Girling (Eds.). Royal Society of Chemistry Special Publications, 94, 139 (1991).

C.S. Ashvar, F.J. Devlin, and P.J. Stephens, Molecular Structure in Solution: An *Ab Initio* Vibrational Spectroscopy Study of Phenyloxirane, *J. Am. Chem. Soc.* 121 (12), 2836 (1999).

C.S. Ashvar, P.J. Stephens, T. Eggimann, and H. Wieser, Vibrational Circular Dichroism Spectroscopy of Chiral Pheromones: Frontalin (1,5-Dimethyl-6,8-Dioxabicyclo [3.2.1] Octane), *Tetrahedron Asymm.* 9 (7), 1107 (1998).

C.S. Ashvar, F.J. Devlin, P.J. Stephens, K.L. Bak, T. Eggimann, and H. Wieser, Vibrational Absorption and Circular Dichroism of Mono- and Di-Methyl Derivatives of 6,8-Dioxabicyclo [3.2.1] Octane, *J. Phys. Chem. A*, 102 (34), 6842 (1998).

C.S. Ashvar, F.J. Devlin, K.L. Bak, P.R. Taylor, and P.J. Stephens, *Ab Initio* Calculation of Vibrational Absorption and Circular Dichroism Spectra: 6,8-Dioxabicyclo [3.2.1] Octane, *J. Phys. Chem.* 100 (22), 9262 (1996).

J. Autschbach and T. Ziegler, Double Perturbation Theory: A Powerful Tool in Computational Coordination Chemistry, *Coord. Chem. Rev.* 238, 83 (2003).

D.M. Back and P.L. Polavarapu, Fourier-Transform Infrared, Vibrational Circular Dichroism of Sugars. A Spectra-Structure Correlation, *Carbohyd. Res.* 133 (1), 163 (1984).

B.I. Baello, P. Pancoska, and T.A. Keiderling, Vibrational Circular Dichroism Spectra of Proteins in the Amide III Region: Measurement and Correlation of Bandshape to Secondary Structure, *Anal. Biochem.* 250 (2), 212 (1997).

K.L. Bak, A.E. Hansen, and P.J. Stephens, *Ab Initio* Calculations of Atomic Polar and Axial Tensors Using the Localized Orbital/Local Origin (Lorg) Approach, *J. Phys. Chem.* 99 (48), 17359 (1995).

K.L. Bak, F.J. Devlin, C.S. Ashvar, P.R. Taylor, M.J. Frisch, and P.J. Stephens, *Ab Initio* Calculation of Vibrational Circular Dichroism Spectra Using Gauge-Invariant Atomic Orbitals, *J. Phys. Chem.* 99 (41), 14918 (1995).

K.L. Bak, O. Bludsky, and P. Jørgensen, *Ab Initio* Calculations of Anharmonic Vibrational Circular Dichroism Intensities of Trans-2,3-Dideuteriooxirane, *J. Chem. Phys.* 103 (24), 10548 (1995).

K.L. Bak, P. Jørgensen, T. Helgaker, K. Ruud, and H.J.A. Jensen, Basis Set Convergence of Atomic Axial Tensors Obtained from Self-Consistent Field Calculations Using London Atomic Orbitals, *J. Chem. Phys.* 100 (9), 6620 (1994).

K.L. Bak, P. Jørgensen, T. Helgaker, K. Ruud, and H.J.A. Jensen, Gauge-Origin Independent Multiconfigurational Self-Consistent-Field Theory for Vibrational Circular Dichroism, *J. Chem. Phys.* 98 (11), 8873 (1993).

C.J. Barnett, A.F. Drake, and S.F. Mason, The Optical Activity of the H-Stretching Vibrational Modes of Calycanthine and IRCD Determinations of Absolute Configuration, *Tetrahedron Lett.* 21, 391 (1980).

C.J. Barnett, A.F. Drake, and S.F. Mason, Optical Activity of Characteristic Vibrational Modes and Absolute Configuration, *Chem. Comm.* 43 (1980).

C.J. Barnett, A.F. Drake, R. Kuroda, S.F. Mason, and S. Savage, Vibrational-Electronic Interaction in the Infrared Circular Dichroism Spectra of Transition-Metal Complexes, *Chem. Phys. Lett.* 70 (1), 8 (1980).

C.J. Barnett, A.F. Drake, R. Kuroda, and S.F. Mason, A Dynamic Polarization Model for Vibrational Optical Activity and the Infrared Circular Dichroism of a Dihydro[5] Helicene, *Mol. Phys.* 41 (2), 455 (1980).

C.J. Barnett, A.F. Drake, and S.F. Mason, The Polarized Luminescence and Vibrational Optical Activity of Calycanthine, *Bull. Soc. Chim. Belg.* 88 (11), 853 (1979).

L.D. Barron, *Molecular Light Scattering and Optical Activity,* Cambridge University Press, UK (2011).

D. Bas, T. Bürgi, J. Lacour, J. Vachon, and J. Weber, Vibrational and Electronic Circular Dichroism of Δ-TRISPHAT [Tris(Tetrachlorobenzenediolato)-Phosphate(V)] Anion, *Chirality* 17, S143 (2005).

J.M. Batista, Jr., A.N.L. Batista, J.S. Mota, Q.B. Cass, M.J. Kato, V.S. Bolzani, T.B. Freedman, S.N. López, M. Furlan, and L.A. Nafie, Structure Elucidation and Absolute Stereochemistry of Isomeric Monoterpene Chromane Esters, *J. Org. Chem.* 76 (8), 2603 (2011).

J.M. Batista, Jr., A.N.L. Batista, D. Rinaldo, W. Vilegas, D. L. Ambrósio, R.M.B. Cicarelli, V.S. Bolzani, M.J. Kato, L.A. Nafie, S.N. López, and M. Furlan, Absolute Configuration and Selective Trypanocidal Activity of Gaudichaudianic Acid Enantiomers, *J. Nat. Prod.* 74 (5), 1154 (2011).

J.M. Batista, Jr., A.N.L. Batista, D. Rinaldo, W. Vilegas, Q.B. Cass, V.S. Bolzani, M.J. Kato, S.N. López, M. Furlan, and L.A. Nafie, Absolute Configuration Reassignment of Two Chromanes from *Peperomia Obtusifolia* (Piperaceae) Using VCD and DFT Calculations, *Tet. Asymm.* 21 (19), 2402 (2010).

V. Baumruk, P. Pancoska, and T.A. Keiderling, Predictions of Secondary Structure Using Statistical Analyses of Electronic and Vibrational Circular Dichroism and Fourier Transform Infrared Spectra of Proteins in H_2O, *J. Mol. Biol.* 259 (4), 774 (1996).

V. Baumruk, D. Huo, R.K. Dukor, T.A. Keiderling, D. Lelievre, and A. Brack, Conformational Study of Sequential Lys and Leu Based Polymers and Oligomers Using Vibrational and Electronic CD Spectra, *Biopolymers* 34 (8), 1115 (1994).

V. Baumruk and T.A. Keiderling, Vibrational Circular Dichroism of Proteins in H_2O Solution, *J. Am. Chem. Soc.* 115 (15), 6939 (1993).

L. Bednárová, P. Maloň, and P. Bouř, Spectroscopic Properties of the Nonplanar Amide Group: A Computational Study, *Chirality* 19 (10), 775 (2007).

L. Bednárová, P. Bouř, and P. Maloň, Vibrational and Electronic Optical Activity of the Chiral Disulphide Group: Implications for Disulphide Bridge Conformation, *Chirality* 22 (5), 514 (2010).

S. Bercion, T. Buffeteau, L. Lespade, and M.A. Couppe Dek Martin, IR, VCD, ^{1}H and ^{13}C NMR Experimental and Theoretical Studies of a Natural Guaianolide: Unambiguous Determination of Its Absolute Configuration, *J. Mol. Struct.* 791 (1–3), 186 (2006).

P.U. Biedermann, J.R. Cheeseman, M.J. Frisch, V. Schurig, I. Gutman, and I. Agranat, Conformational Spaces and Absolute Configurations of Chiral Fluorinated Inhalation Anaesthetics. A Theoretical Study, *J. Org. Chem.* 64, 3878 (1999).

M. Bieri, C. Gautier, and T. Bürgi, Probing Chiral Interfaces by Infrared Spectroscopic Methods, *Phys. Chem. Chem. Phys.* 9 (6), 671 (2007).

S.S. Birke and M. Diem, Conformational Studies of the Smallest Structural Motifs of DNA Detectable via Vibrational Circular Dichroism: Cytidylyl-(3′-5′)-Guanosine and Guanylyl-(3′-5′)-Cytidine, *Biophys. J.* 68 (3), 1045 (1995).

S.S. Birke, M. Moses, B. Kagalovsky, D. Jano, M. Gulotta, and M. Diem, Infrared CD of Deoxy Oligonucleotides. Conformational Studies of 5′d(GCGC)3′, 5′d(CGCG)3′, 5′d(CCGG)3′, and 5′d(GGCC)3′ in Low and High Salt Aqueous Solution, *Biophys. J.* 65 (3), 1262 (1993).

S.S. Birke, I. Agbaje, and M. Diem, Experimental and Computational Infrared CD Studies of Prototypical Peptide Conformations, *Biochemistry* 31 (2), 450 (1992).

S. Birke, C. Farrell, O. Lee, L. Agbaje, G. Roberts, and M. Diem, Infrared (Vibrational) CD of Peptides in the Amide I and Amide III Spectral Regions, *Spec. Publ. R. Soc. Chem.* 94, 131 (1991).

T.M. Black, P.K. Bose, P.L. Polavarapu, L.D. Barron, and L. Hecht, Vibrational Optical Activity in Trans-2,3-Dimethyloxirane, *J. Am. Chem. Soc.* 112 (4), 1479 (1990).

Y. Blagoi, G. Gladchenko, L.A. Nafie, T.B. Freedman, V. Sorokin, V. Valeev, and Y. He, Phase Equilibrium in Poly(rA).Poly(rU) Complexes with Cd^{2+} and Mg^{2+} Ions, Studied by Ultraviolet, Infrared, and Vibrational Circular Dichroism Spectroscopy, *Biopolymers* 78 (5), 275 (2005).

J. Bloino and V. Barone, A Second-Order Perturbation Theory Route to Vibrational Averages and Transition Properties of Molecules: General Formulation and Application to Infrared and Vibrational Circular Dichroism Spectroscopies, *J. Chem. Phys.* 136 (12), 124108 (2012).

L.A. Bodack, T.B. Freedman, B.Z. Chowdhry, and L.A. Nafie, Solution Conformations of Cyclosporins and Magnesium-Cyclosporin Complexes Determined by Vibrational Circular Dichroism, *Biopolymers* 73 (2), 163 (2004).

H.G. Bohr, K. Frimand, K.J. Jalkanen, R.M. Nieminen, and S. Suhai, Neural-Network Analysis of the Vibrational Spectra of N-Acetyl L-Alanyl N′-Methyl Amide Conformational States, *Phys. Rev. E* 64, 1 (2001).

H.G. Bohr, K.J. Jalkanen, M. Elstner, K. Frimand, and S. Suhai, A Comparative Study of MP2, B3LYP, RHF and SCC-DFTB Force Fields in Predicting the Vibrational Spectra of N-Acetyl-L-Alanine-N′-Methyl Amide: VA and VCD Spectra, *Chem. Phys.* 246, 13 (1999).

A. Borics, R.F. Murphy, and S. Lovas, Conformational Analysis of Ac-Npgq-Nh2 and Ac-Vpah-Nh2 by Vibrational Circular Dichroism Spectroscopy Combined with Molecular Dynamics and Quantum Chemical Calculations, *Protein Peptide Lett.* 14 (4), 353 (2007).

A. Borics, R.F. Murphy, and S. Lovas, Optical Spectroscopic Elucidation of β-Turns in Disulfide Bridged Cyclic Tetrapeptides, *Biopolymers* 85 (1), 1 (2007).

A. Borics, R.F. Murphy, and S. Lovas, Fourier Transform Vibrational Circular Dichroism as a Decisive Tool for Conformational Studies of Peptides Containing Tyrosyl Residues, *Biopolymers (Biospectroscopy)* 72 (1), 21 (2003).

R.W. Bormett, S.A. Asher, P.J. Larkin, W.G. Gustafson, N. Ragunathan, T.B. Freedman, L.A. Nafie, S. Balasubramanian, S.G. Boxer, N.-T. Yu, K. Gersonde, R.W. Noble, B.A. Springer, and S.G. Sligar, Selective Examination of Heme Protein Azide Ligand-Distal Globin Interactions by Vibrational Circular Dichroism, *J. Am. Chem. Soc.* 114 (17), 6864 (1992).

P.K. Bose and P.L. Polavarapu, Evidence for Covalent Binding between Copper Ions and Cyclodextrin Cavity: A Vibrational Circular Dichroism Study, *Carbohyd. Res.* 323 (1–4), 63 (2000).

P.K. Bose and P.L. Polavarapu, Acetate Groups as Probes of the Stereochemistry of Carbohydrates: A Vibrational Circular Dichroism Study, *Carbohyd. Res.* 322 (1–2), 135 (1999).

P.K. Bose and P.L. Polavarapu, Vibrational Circular Dichroism of Monosaccharides, *Carbohyd. Res.* 319, 172 (1999).

P.K. Bose and P.L. Polavarapu, Vibrational Circular Dichroism Is a Sensitive Probe of the Glycosidic Linkage: Oligosaccharides of Glucose, *J. Am. Chem. Soc.* 121 (25), 6094 (1999).

P.K. Bose and P.L. Polavarapu, Circular Dichroism of the Benzene Group Vibrations and Molecular Stereochemistry, *Struct. Chem.* 1, 205 (1990).

P. Bouř, Cross-Polarization Detection Enables Fast Measurement of Vibrational Circular Dichroism, *Chem. Phys. Chem.* 10 (12), 1983 (2009).

P. Bouř, J. Kim, J. Kapitan, R.P. Hammer, R. Huang, L. Wu, and T.A. Keiderling, Vibrational Circular Dichroism and IR Spectral Analysis as a Test of Theoretical Conformational Modeling for a Cyclic Hexapeptide, *Chirality* 20 (10), 1104 (2008).

P. Bouř, D. Michalík, and J. Kapitán, Empirical Solvent Correction for Multiple Amide Group Vibrational Modes, *J. Chem. Phys.* 122 (14), 144501/1 (2005).

P. Bouř and T.A. Keiderling, Structure, Spectra and the Effects of Twisting of β-Sheet Peptides. A Density Functional Theory Study, *J. Mol. Struct. (Theochem)* 675 (1–3), 95 (2004).

P. Bouř, H. Navrátilová, V. Setnicka, M. Urbanová, and K. Volka, (3r,4s)-4-(4-Fluorophenyl)-3-Hydroxymethyl-1-Methylpiperidine: Conformation and Structure Monitoring by Vibrational Circular Dichroism, *J. Org. Chem.* 67 (1), 161 (2002).

P. Bouř, J. Kubelka, and T.A. Keiderling, *Ab Initio* Quantum Mechanical Models of Peptide Helices and Their Vibrational Spectra, *Biopolymers* 65, 45 (2002).

P. Bouř, K. Záruba, M. Urbanová, V. Setnicka, P. Matejka, Z. Fiedler, V. Král, and K. Volka, Vibrational Circular Dichroism of Tetraphenylporphyrin in Peptide Complexes? A Computational Study, *Chirality* 12 (4), 191 (2000).

P. Bouř, J. Kubelka, and T.A. Keiderling, Simulations of Oligopeptide Vibrational CD: Effects of Isotopic Labeling, *Biopolymers* 53 (5), 380 (2000).

P. Bouř, J. McCann, and H. Wieser, Measurement and Calculation of Absolute Rotational Strengths for Camphor, α-Pinene, and Borneol, *J. Phys. Chem. A* 102 (1), 102 (1998).

P. Bouř, J. McCann, and H. Wieser, The Excitation Scheme: A New Method for Calculation of Vibrational Circular Dichroism Spectra, *J. Chem. Phys.* 108 (21), 8782 (1998).

P. Bouř, J. Sopková, L. Bednárová, P. Malon, and T.A. Keiderling, Transfer of Molecular Property Tensors in Cartesian Coordinates: A New Algorithm for Simulation of Vibrational Spectra, *J. Comput. Chem.* 18 (5), 646 (1997).

P. Bouř, J. McCann, and H. Wieser, Vibrational Circular Dichroism Study of (–)-Sparteine, *J. Phys. Chem. A* 101 (50), 9783 (1997).

P. Bouř, C.N. Tam, M. Shaharuzzaman, J.S. Chickos, and T.A. Keiderling, Vibrational Optical Activity Study of Trans-Succinic-D_2 Anhydride, *J. Phys. Chem.* 100 (37), 15041 (1996).

P. Bouř, Anharmonic Corrections to Vibrational Energies of Molecules: Water and Dideuterioxirane, *J. Phys. Chem.* 98, 8862 (1994).

P. Bouř and T.A. Keiderling, *Ab Initio* Simulations of the Vibrational Circular Dichroism of Coupled Peptides, *J. Am. Chem. Soc.* 115 (21), 9602 (1993).

P. Bouř and T.A. Keiderling, Computational Evaluation of the Coupled Oscillator Model in the Vibrational Circular Dichroism of Selected Small Molecules, *J. Am. Chem. Soc.* 114 (23), 9100 (1992).

T. Brotin, D. Cavagnat, and T. Buffeteau, Conformational Changes in Cryptophane Having C1-Symmetry Studied by Vibrational Circular Dichroism, *J. Phys. Chem. A* 112 (36), 8464 (2008).

T. Brotin, D. Cavagnat, J.P. Dutasta, and T. Buffeteau, Vibrational Circular Dichroism Study of Optically Pure Cryptophane-A, *J. Am. Chem. Soc.* 128 (16), 5533 (2006).

A.D. Buckingham, P.W. Fowler, and P.A. Galwas, Velocity-Dependent Property Surfaces and the Theory of Vibrational Circular Dichroism, *Chem. Phys.* 112 (1), 1 (1987).

T. Buffeteau, D. Cavagnat, A. Bouchet, and T. Brotin, Vibrational Absorption and Circular Dichroism Studies of (–)-Camphanic Acid, *J. Phys. Chem. A* 111 (6), 1045 (2007).

T. Buffeteau, L. Ducasse, L. Poniman, N. Delsuc, and I. Huc, Vibrational Circular Dichroism and *Ab Initio* Structure Elucidation of an Aromatic Foldamer, *Chem. Comm.* (25), 2714 (2006).

T. Buffeteau, F. Lagugné-Labarthet, and C. Sourisseau, Vibrational Circular Dichroism in General Anisotropic Thin Solid Films: Measurement and Theoretical Approach, *Appl. Spectrosc.* 59 (6), 732 (2005).

T. Buffeteau, L. Ducasse, A. Brizard, I. Huc, and R. Oda, Density Functional Theory Calculations of Vibrational Absorption and Circular Dichroism Spectra of Dimethyl-L-Tartrate, *J. Phys. Chem. A* 108 (18), 4080 (2004).

T. Buffeteau and B. Desbat, Polarization Modulation FT-IR Spectroscopy: A Novel Optic Method for Studying Molecule Orientation and Conformation, *Actualité Chimique* 1, 18 (2003).

T. Bürgi, A. Urakawa, B. Behzadi, K.H. Ernst, and A. Baiker, The Absolute Configuration of Heptahelicene: A VCD Spectroscopy Study, *New J. Chem.* 28 (3), 332 (2004).

T. Bürgi, A. Vargas, and A. Baiker, VCD Spectroscopy of Chiral Cinchona Modifiers Used in Heterogeneous Enantioselective Hydrogenation: Conformation and Binding of Non-Chiral Acids, *J. Chem. Soc. Perkin Trans.* 2 (9), 1596 (2002).

E. Burgueño-Tapia, C.M. Cerda-García-Rojas, and P. Joseph-Nathan, Conformational Analysis of Perezone and Dihydroperezone Using Vibrational Circular Dichroism, *Phytochemistry* 74, 190 (2012).

E. Burgueño-Tapia and P. Joseph-Nathan, Absolute Configuration of Eremophilanoids by Vibrational Circular Dichroism, *Phytochemistry* 69 (11), 2251 (2008).

R. Bursi and P.J. Stephens, Ring Current Contributions to Vibrational Circular Dichroism? *Ab Initio* Calculations for Methyl Glycolate-D_1 and -D_4, *J. Phys. Chem.* 95 (17), 6447 (1991).

R. Bursi, F.J. Devlin, and P.J. Stephens, Vibrationally Induced Ring Currents? The Vibrational Circular Dichroism of Methyl Lactate, *J. Am. Chem. Soc.* 112 (25), 9430 (1990).

S. Bussieres, T. Buffeteau, B. Desbat, R. Breton, and C. Salesse, Secondary Structure of a Truncated Form of Lecithin Retinol Acyltransferase in Solution and Evidence for Its Binding and Hydrolytic Action in Monolayers, *Biochim. Biophys. Acta Biomembranes* 1778 (5), 1324 (2008).

E. Butkus, A. Zilinskas, S. Stoncius, R. Rozenbergas, M. Urbanová, V. Setnicka, P. Bour, and K. Volka, Synthesis and Chiroptical Properties of Enantiopure Tricyclo [4.3.0.0^{3,8}] Nonane-4.5-Dione (Twistbrendandione), *Tetrahedron Asymm.* 13, 633 (2002).

J. Calienni, J.B. Trager, M.A. Davies, U. Gunnia, and M. Diem, Vibrational Assignment, Normal-Coordinate Analysis, and Vibrational Optical Activity of Chlorofluoroacetic Acid and Chlorofluoroacetate, *J. Phys. Chem.* 93 (13), 5049 (1989).

X. Cao, R.K. Dukor, and L.A. Nafie, Reduction of Linear Birefringence in Vibrational Circular Dichroism Measurement: Use of a Rotating Half-Wave Plate, *Theor. Chem. Acc.* 119 (1–3), 69 (2008).

X. Cao, R.D. Shah, R.K. Dukor, C. Guo, T.B. Freedman, and L.A. Nafie, Extension of Fourier Transform Vibrational Circular Dichroism into the Near-Infrared Region: Continuous Spectral Coverage from 800 to 10,000 cm^{-1}, *Appl. Spectrosc.* 58 (9), 1057 (2004).

H. Cao, B. Teng, X.C. Liu, X.G. Zhao, N. Liu, X. Wang, W.J. Zhang, and Y. Wei, Studies on Vibrational Circular Dichroism Spectroscopy of Chiral Cavity in Rigid Cyclic Oligomers Containing 1,1'-Bi-2-Naphthyl Moiety, *Gaodeng Xuexiao Huaxue Xuebao* 24 (9), 1724 (2003).

H. Cao, C. Chen, B. Teng, X. Wang, X. Liu, and W. Zhang, Vibrational Absorption and Circular Dichroism of Chiral Rigid Cyclic Oligomers Containing 1,1'-Bi-2-Naphthyl Moiety, *Polym. Prepr.* 43 (2), 1254 (2002).

C. Cappelli and B. Mennucci, Modeling the Solvation of Peptides. The Case of (S)-N-Acetylproline Amide in Liquid Water, *J. Phys. Chem. B* 112 (11), 3441 (2008).

C. Cappelli, S. Monti, and A. Rizzo, Effect of the Environment on Vibrational Infrared and Circular Dichroism Spectra of (S)-Proline, *Int. J. Quantum Chem.* 104, 744 (2005).

C. Cappelli, S. Corni, B. Mennucci, R. Cammi, and J. Tomasi, Vibrational Circular Dichroism within the Polarizable Continuum Model: A Theoretical Evidence of Conformation Effects and Hydrogen Bonding for (S)-(−)-3-Butyn-2-ol in CCl_4 Solution, *J. Phys. Chem. A* 106 (51), 12331 (2002).

E. Carosati, R. Budriesi, P. Ioan, G. Cruciani, F. Fusi, M. Frosini, S. Saponara, F. Gasparrini, A. Ciogli, C. Villani, P.J. Stephens, F.J. Devlin, D. Spinelli, and A. Chiarini, Stereoselective Behavior of the Functional Diltiazem Analogue 1-[(4-Chlorophenyl) Sulfonyl]-2-(2-Thienyl)Pyrrolidine, a New L-Type Calcium Channel Blocker, *J. Med. Chem.* 52 (21), 6637 (2009).

E. Carosati, G. Cruciani, A. Chiarini, R. Budriesi, P. Ioan, R. Spisani, D. Spinelli, B. Cosimelli, F. Fusi, M. Frosini, R. Matucci, F. Gasparrini, A. Ciogli, P.J. Stephens, and F.J. Devlin, Calcium Channel Antagonists Discovered by a Multidisciplinary Approach, *J. Med. Chem.* 49 (17), 5206 (2006).

E. Castiglioni, F. Lebon, G. Longhi, and S. Abbate, Vibrational Circular Dichroism in the Near Infrared: Instrumental Developments and Applications, *Enantiomer* 7, 161 (2002).

D. Cavagnat, T. Buffeteau, and T. Brotin, Synthesis and Chiroptical Properties of Cryptophanes Having C1-Symmetry, *J. Org. Chem.* 73 (1), 66 (2008).

D. Cavagnat, L. Lespade, and T. Buffeteau, Vibrational Absorption and Circular Dichroism Studies of Trans-(3s,4s)-D_6-Cyclopentene in the Gas Phase, *J. Phys. Chem. A* 111 (30), 7014 (2007).

J.C. Cedron, A. Estevez-Braun, A.G. Ravelo, D. Gutierrez, N. Flores, M.A. Bucio, N. Perez-Hernandez, and P. Joseph-Nathan, Bioactive Montanine Derivatives from Halide-Induced Rearrangements of Haemanthamine-Type Alkaloids. Absolute Configuration by VCD, *Org. Lett.* 11 (7), 1491 (2009).

C.M. Cerda-Garcia-Rojas, C.A.N. Catalan, A.C. Muro, and P. Joseph-Nathan, Vibrational Circular Dichroism of Africanane and Lippifoliane Sesquiterpenes from Lippia Integrifolia, *J. Nat. Prod.* 71 (6), 967 (2008).

C.M. Cerda-García-Rojas, H.A. García-Gutiérrez, J.D. Hernández-Hernández, L.U. Román-Marín, and P. Joseph-Nathan, Absolute Configuration of Verticillane Diterpenoids by Vibrational Circular Dichroism, *J. Nat. Prod.* 70 (7), 1167 (2007).

V. Cerè, F. Peri, S. Pollicino, A. Ricci, F.J. Devlin, P.J. Stephens, F. Gasparrini, R. Rompietti, and C. Villani, Synthesis, Chromatographic Separation, VCD Spectroscopy and *Ab Initio* DFT Studies of Chiral Thiepane Tetraol Derivatives, *J. Org. Chem.* 70, 664 (2005).

I. Chabay and G. Holzwarth, Infrared Circular Dichroism and Linear Dichroism Spectrophotometer, *Appl. Opt.* 14 (2), 454 (1975).

I. Chabay, Absorptive and Scattering Circular Dichroism of Cholesteric Liquid Crystals in the Infrared, *Chem. Phys. Lett.* 17 (2), 283 (1972).

A.-C. Chamayou, S. Lüedeke, V. Brecht, T.B. Freedman, L.A. Nafie, and C. Janiak, Chirality and Diastereoselection of Δ/Λ-Configured Tetrahedral Zinc Complexes through Enantiopure Schiff Base Complexes: Combined Vibrational Circular Dichroism, Density Functional Theory, ^1H NMR, and X-ray Structural Studies, *Inorg. Chem.* 50 (22), 11363 (2011).

B. Chavali, K. Krishnamurthy, and J. Dage, Mid IR CD Spectroscopy for Medicinal Chemistry: A Pharmaceutical Perspective, *Am. Pharm. Rev.* 10 (3), 94 (2007).

J.R. Cheeseman, M.J. Frisch, F.J. Devlin, and P.J. Stephens, *Ab Initio* Calculation of Atomic Axial Tensors and Vibrational Rotational Strengths Using Density Functional Theory, *Chem. Phys. Lett.* 252 (3,4), 211 (1996).

G.-C. Chen, P.L. Polavarapu, and S. Weibel, New Design for Fourier Transform Infrared Vibrational Circular Dichroism Spectrometers, *Appl. Spectrosc.* 48 (10), 1218 (1994).

J.C. Cheng, L.A. Nafie, S.D. Allen, and A.I. Braunstein, Photoelastic Modulator for the 0.55–13 μm Range, *Appl. Optics* 15 (8), 1960 (1976).

J.C. Cheng, L.A. Nafie, and P.J. Stephens, Polarization Scrambling Using a Photoelastic Modulator: Application to Circular Dichroism Measurement, *J. Opt. Soc. Am.* 65, 1031 (1975).

F. Cherblanc, Y.-P. Lo, E. De Gussem, L. Alcazar-Fuoli, E. Bignell, Y. He, N. Chapman-Rothe, P. Bultinck, W.A. Herrebout, R. Brown, H.S. Rzepa, and M.J. Fuchter, On the Determination of the Stereochemistry of Semisynthetic Natural Product Analogues Using Chiroptical Spectroscopy: Desulfurization of Epidithiodioxopiperazine Fungal Metabolites, *Chem. Eur. J.* 17 (42), 11868 (2011).

A.C. Chernovitz, T.B. Freedman, and L.A. Nafie, Vibrational CD Studies of the Solution Conformation of N-Urethanyl L-Amino Acid Derivatives, *Biopolymers* 26 (11), 1879 (1987).

J.S. Chickos, A. Annamalai, and T.A. Keiderling, Thermolysis of (1R,2R)-1,2-Dideuteriocyclobutane. An Application of Vibrational Circular Dichroism to Kinetic Analysis, *J. Am. Chem. Soc.* 108 (15), 4398 (1986).

J.-H. Choi and M. Cho, Direct Calculations of Mid- and Near-IR Absorption and Circular Dichroism Spectra of Chiral Molecules Using QM/MM Molecular Dynamics Simulation Method, *J. Chem. Theory Comput.* 7 (12), 4097 (2011).

J.-H. Choi and M. Cho, Two-Dimensional Circularly Polarized IR Photon Echo Spectroscopy of Polypeptides: Four-Wave-Mixing Optical Activity Measurement, *J. Phys. Chem. A* 111 (24), 5176 (2007).

J.H. Choi, H. Lee, K.K. Lee, S. Hahn, and M. Cho, Computational Spectroscopy of Ubiquitin: Comparison between Theory and Experiments, *J. Chem. Phys.* 126 (4), 045102/1 (2007).

J.H. Choi, S. Hahn, and M. Cho, Vibrational Spectroscopic Characteristics of Secondary Structure Polypeptides in Liquid Water: Constrained MD Simulation Studies, *Biopolymers* 83 (5), 519 (2006).

J.H. Choi, J.S. Kim, and M. Cho, Amide I Vibrational Circular Dichroism of Polypeptides: Generalized Fragmentation Approximation Method, *J. Chem. Phys.* 122 (17), 174903/1 (2005).

J.H. Choi and M. Cho, Amide I Vibrational Circular Dichroism of Dipeptide: Conformation Dependence and Fragment Analysis, *J. Chem. Phys.* 120 (9), 4383 (2004).

S.-W. Choi, S. Kawauchi, S. Tanaka, J. Watanabe, and H. Takezoe, Vibrational Circular Dichroism Spectroscopic Study on Circularly Polarized Light-Induced Chiral Domains in the B4 Phase of a Bent Mesogen, *Chem. Lett.* 36 (8), 1018 (2007).

S.J. Cianciosi, N. Ragunathan, T.B. Freedman, L.A. Nafie, D.K. Lewis, D.A. Glenar, and J.E. Baldwin, Racemization and Geometrical Isomerization of (2S,3S)-Cyclopropane-1-^{13}C-1,2,3-D$_3$ at 407°C: Kinetically Competitive One-Center and Two-Center Thermal Epimerizations in an Isotopically Substituted Cyclopropane, *J. Am. Chem. Soc.* 113, 1864 (1991).

S.J. Cianciosi, N. Ragunathan, T.B. Freedman, L.A. Nafie, and J.E. Baldwin, Racemization and Geometrical Isomerization of (–)-(R,R)-Cyclopropane-1,2-^{2}H$_2$, *J. Am. Chem. Soc.* 112 (22), 8204 (1990).

S.J. Cianciosi, K.M. Spencer, T.B. Freedman, L.A. Nafie, and J.E. Baldwin, Synthesis and Gas-Phase Vibrational Circular Dichroism of (+)-(S,S)-Cyclopropane-1,2-^{2}H$_2$, *J. Am. Chem. Soc.* 111 (5), 1913 (1989).

R.H. Cichewicz, L.J. Clifford, P.R. Lassen, X. Cao, T.B. Freedman, L.A. Nafie, J.D. Deschamps, V.A. Kenyon, J.R. Flanary, T.R. Holman, and P. Crews, Stereochemical Determination and Bioactivity Assessment of (S)-(+)-Curcuphenol Dimers Isolated from the Marine Sponge *Didiscus aceratus* and Synthesized through Laccase Biocatalysis, *Bioorg. Med. Chem.* 13 (19), 5600 (2005).

M.J. Citra, M.G. Paterlini, T.B. Freedman, A. Fissi, and O. Pieroni, Vibrational Circular Dichroism Studies of 3_{10}-Helical Solution Conformers in Dehydro-Peptides, *Proc. SPIE* 2089, 478 (1993).

M.J. Citra, M.G. Paterlini, T.B. Freedman, L.A. Nafie, A. Shanzer, Y. Tor, C. Pratesi, and O. Pieroni, Fourier Transform Vibrational Circular Dichroism Studies of Model Peptide Molecules, *Proc. SPIE* 1575, 406 (1992).

R. Clark and P.J. Stephens, Vibrational Optical Activity, *Proc. SPIE* 112, 127 (1977).

J. Copps, R.F. Murphy, and S. Lovas, VCD Spectroscopic and Molecular Dynamics Analysis of the TRP-Cage Miniprotein TC5B, *Biopolymers* 88 (3), 427 (2007).

J. Copps, R.F. Murphy, and S. Lovas, Avian Pancreatic Polypeptide Fragments Refold to Native App Conformation When Combined in Solution: A CD and VCD Study, *Biopolymers* 83 (1), 32 (2006).

D.P. Craig and T. Thirunamachandran, The Adiabatic Approximation in the Ground-State Manifold, *Can. J. Chem.* 63, 1773 (1985).

D.P. Craig and T. Thirunamachandran, A Theory of Vibrational Circular Dichroism in Terms of Vibronic Interactions, *Mol. Phys.* 35 (3), 825 (1978).

T.D. Crawford, *Ab Initio* Calculation of Molecular Chiroptical Properties, *Theor. Chem. Acc.* 115 (4), 227 (2006).

E. Debie, E. De Gussem, R.K. Dukor, W. Herrebout, L.A. Nafie, and P. Bultinck, A Confidence Level Algorithm for the Determination of Absolute Configuration Using Vibrational Circular Dichroism or Raman Optical Activity, *Chem. Phys. Chem.* 12 (8), 1542 (2011).

E. Debie, L. Jaspers, P. Bultinck, W. Herrebout, and B. Van Der Veken, Induced Solvent Chirality: A VCD Study of Camphor in $CdCl_3$, *Chem. Phys. Lett.* 450 (4–6), 426 (2008).

E. Debie, P. Bultinck, W. Herrebout, and B. van der Veken, Solvent Effects on IR and VCD Spectra of Natural Products: An Experimental and Theoretical VCD Study of Pulegone, *Phys. Chem. Chem. Phys.* 10 (24), 3498 (2008).

E. Debie, T. Kuppens, K. Vandyck, J. Van der Eycken, B. Van Der Veken, W. Herrebout, and P. Bultinck, Vibrational Circular Dichroism DFT Study on Bicyclo [3.3.0] Octane Derivatives, *Tetrahedron Asymm.* 17 (23), 3203 (2006).

S. Delarue-Cochin, J.J. Pan, A. Dauteloup, F. Hendra, R.G. Angoh, D. Joseph, P.J. Stephens, and C. Cavé, Asymmetric Michael Reaction: Novel Efficient Access to Chiral β-Ketophosphonates, *Tetrahedron Asymm.* 18 (5), 685 (2007).

W.-Y. Deng and W.-Y. Qiu, Helical Chirality in Model Mirror-Imaged Carbyne Trefoil Knots, *J. Mol. Struct.* 875 (1–3), 515 (2008).

E. Deplazes, W. van Bronswijk, F. Zhu, L.D. Barron, S. Ma, L.A. Nafie, and K.J. Jalkanen, A Combined Theoretical and Experimental Study of the Structure and Vibrational Absorption, Vibrational Circular Dichroism, Raman and Raman Optical Activity Spectra of the L-Histidine Zwitterion, *Theor. Chem. Acc.* 119 (1–3), 155 (2008).

C.W. Deutsche and A. Moscowitz, Optical Activity of Vibrational Origin. II. Consequences of Polymer Conformation, *J. Chem. Phys.* 53 (7), 2630 (1970).

C.W. Deutsche and A. Moscowitz, Optical Activity of Vibrational Origin. I. A Model Helical Polymer, *J. Chem. Phys.* 49 (7), 3257 (1968).

F.J. Devlin, P.J. Stephens, and B. Figadère, Determination of the Absolute Configuration of the Natural Product Klaivanolide via Density Functional Calculations of Vibrational Circular Dichroism (VCD), *Chirality*, 21 (1E) E48 (2009).

F.J. Devlin, P.J. Stephens, and O. Bortolini, Determination of Absolute Configuration Using Vibrational Circular Dichroism Spectroscopy: Phenyl Glycidic Acid Derivatives Obtained via Asymmetric Epoxidation Using Oxone and a Keto Bile Acid, *Tetrahedron Asymm.* 16 (15), 2653 (2005).

F.J. Devlin, P.J. Stephens, and P. Besse, Are the Absolute Configurations of 2-(1-Hydroxyethyl)-Chromen-4-One and Its 6-Bromo Derivative Determined by X-Ray Crystallography Correct? A Vibrational Circular Dichroism (VCD) Study of Their Acetate Derivatives, *Tetrahedron Asymm.* 16, 1557 (2005).

F.J. Devlin, P.J. Stephens, and P. Besse, Conformational Rigidification via Derivatization Facilitates the Determination of Absolute Configuration Using Chiroptical Spectroscopy: A Case Study of the Chiral Alcohol Endo-Borneol, *J. Org. Chem.* 70, 2980 (2005).

F.J. Devlin, P.J. Stephens, P. Scafato, S. Superchi, and C. Rosini, Determination of Absolute Configuration Using Vibrational Circular Dichroism Spectroscopy: The Chiral Sulfoxide 1-Thiochromanone S-Oxide, *Chirality* 14 (5), 400 (2002).

F.J. Devlin, P.J. Stephens, P. Scafato, S. Superchi, and C. Rosini, Conformational Analysis Using Infrared and Vibrational Circular Dichroism Spectroscopies: The Chiral Cyclic Sulfoxides 1-Thiochroman-4-One S-Oxide, 1-Thiaindan S-Oxide and 1-Thiochroman S-Oxide, *J. Phys. Chem. A* 106 (44), 10510 (2002).

F.J. Devlin, P.J. Stephens, C. Österle, K.B. Wiberg, J.R. Cheeseman, and M.J. Frisch, Configurational and Conformational Analysis of Chiral Molecules Using IR and VCD Spectroscopies: Spiropentylcarboxylic Acid Methyl Ester and Spiropentyl Acetate, *J. Org. Chem.* 67 (23), 8090 (2002).

F.J. Devlin, P.J. Stephens, P. Scafato, S. Superchi, and C. Rosini, Determination of Absolute Configuration Using Vibrational Circular Dichroism Spectroscopy: The Chiral Sulfoxide 1-Thiochroman S-Oxide, *Tetrahedron Asymm.* 12, 1551 (2001).

F.J. Devlin and P.J. Stephens, *Ab Initio* Density Functional Theory Study of the Structure and Vibrational Spectra of Cyclohexanone and Its Isotopomers, *J. Phys. Chem. A* 103 (4), 527 (1999).

F.J. Devlin and P.J. Stephens, Conformational Analysis Using *Ab Initio* Vibrational Spectroscopy: 3-Methylcyclohexanone, *J. Am. Chem. Soc.* 121 (32), 7413 (1999).

F.J. Devlin, P.J. Stephens, J.R. Cheeseman, and M.J. Frisch, *Ab Initio* Prediction of Vibrational Absorption and Circular Dichroism Spectra of Chiral Natural Products Using Density Functional Theory: Camphor and Fenchone, *J. Phys. Chem. A* 101 (35), 6322 (1997).

F.J. Devlin, P.J. Stephens, J.R. Cheeseman, and M.J. Frisch, *Ab Initio* Prediction of Vibrational Absorption and Circular Dichroism Spectra of Chiral Natural Products Using Density Functional Theory: α-Pinene, *J. Phys. Chem. A* 101 (51), 9912 (1997).

F.J. Devlin, P.J. Stephens, J.R. Cheeseman, and M.J. Frisch, Prediction of Vibrational Circular Dichroism Spectra Using Density Functional Theory: Camphor and Fenchone, *J. Am. Chem. Soc.* 118 (26), 6327 (1996).

F.J. Devlin, J.W. Finley, P.J. Stephens, and M.J. Frisch, *Ab Initio* Calculation of Vibrational Absorption and Circular Dichroism Spectra Using Density Functional Force Fields: A Comparison of Local, Non-Local, and Hybrid Density Functionals, *J. Phys. Chem.* 99 (46), 16883 (1995).

F.J. Devlin and P.J. Stephens, *Ab Initio* Calculation of Vibrational Circular Dichroism Spectra of Chiral Natural Products Using MP2 Force Fields: Camphor, *J. Am. Chem. Soc.* 116 (11), 5003 (1994).

F.J. Devlin and P.J. Stephens, Vibrational Circular Dichroism Measurement in the Frequency Range of 800–650 cm^{-1}, *Appl. Spectrosc.* 41 (7), 1142 (1987).

M. Diem, Determining the Solution Conformation of Biological Molecules by Infrared Circular Dichroism, *Spectroscopy* 10 (4), 38 (1995).

M. Diem, Application of Infrared CD to the Analysis of the Solution Conformation of Biological Molecules, *Tech. Instrum. Anal. Chem.* 14, 91 (1994).

M. Diem, O. Lee, and G.M. Roberts, Vibrational Studies, Normal-Coordinate Analysis, and Infrared VCD of Alanylalanine in the Amide III Spectral Region, *J. Phys. Chem.* 96 (2), 548 (1992).

M. Diem, Instrumentation for the Observation of Circular Dichroism in IR Vibrational Transitions, *Proc. SPIE* 1681, 67 (1992).

M. Diem, Advances in Instrumentation for the Observation of Vibrational Optical Activity, *Vibrational Spectra Struct.* 19, 1 (1991).

M. Diem, Solution Conformation of Biomolecules from Infrared Vibrational Circular Dichroism (VCD) Spectroscopy, *Proc. SPIE* 1432 (2), 28 (1991).

M. Diem, G.M. Roberts, O. Lee, and A. Barlow, Design and Performance of an Optimized Dispersive Infrared Dichrograph, *Appl. Spectrosc.* 42 (1), 20 (1988).

M. Diem, Infrared Vibrational Circular Dichroism of Alanine in the Mid-Infrared Region: Isotopic Effects, *J. Am. Chem. Soc.* 110 (21), 6967 (1988).

M. Diem, P.L. Polavarapu, M. Oboodi, and L.A. Nafie, Vibrational Circular Dichroism in Amino Acids and Peptides. 4. Vibrational Analysis, Assignments, and Solution-Phase Raman Spectra of Deuterated Isotopomers of Alanine, *J. Am. Chem. Soc.* 104 (12), 3329 (1982).

M. Diem, E. Photos, H. Khouri, and L.A. Nafie, Vibrational Circular Dichroism in Amino Acids and Peptides. 3. Solution- and Solid-Phase Spectra of Alanine and Serine, *J. Am. Chem. Soc.* 101 (23), 6829 (1979).

M. Diem, P.J. Gotkin, J.M. Kupfer, and L.A. Nafie, Vibrational Circular Dichroism in Amino Acids and Peptides. 2. Simple Alanyl Peptides, *J. Am. Chem. Soc.* 100 (18), 5644 (1978).

M. Diem, P.J. Gotkin, J.M. Kupfer, A.G. Tindall, and L.A. Nafie, Vibrational Circular Dichroism in Amino Acids and Peptides. 1. Alanine, *J. Am. Chem. Soc.* 99 (24), 8103 (1977).

J. Döbler, N. Peters, C. Larsson, A. Bergman, E. Geidel, and H. Hühnerfuss, The Absolute Structures of Separated PCB-Methylsulfone Enantiomers Determined by Vibrational Circular Dichroism and Quantum Chemical Calculations, *J. Mol. Struct. (Theochem)* 586, 159 (2002).

J.C. Dobrowolski, M.H. Jamroz, R. Kolos, J.E. Rode, and J. Sadlej, Theoretical Prediction and the First IR Matrix Observation of Several L-Cysteine Molecule Conformers, *Chem. Phys. Chem.* 8 (7), 1085 (2007).

J.C. Dobrowolski, On Theoretical VCD, IR and Raman Spectra of (+)-Hexahelicene in the 3(C-H) Region, *J. Mol. Struct.* 651–653, 607 (2003).

J.C. Dobrowolski and A.P. Mazurek, On Theoretical VCD, IR and Raman Spectra of Model Chiral Right-Hand Trefoil Knots, *J. Mol. Struct.* 563–564, 309 (2001).

S.R. Domingos, M.R. Panman, B.H. Bakker, F. Hartl, W. J. Buma, and S. Woutersen, Amplifying Vibrational Circular Dichroism by Manipulation of the Electronic Manifold, *Chem. Comm.* 48 (3), 353 (2012).

M.I. Donnoli, S. Superchi, and C. Rosini, Recent Progress in Application of Spectroscopic Methods for Assigning Absolute Configuration of Optically Active Sulfoxides, *Mini Rev. Organic Chem.* 3 (1), 77 (2006).

H. Dothe, M.A. Lowe, and J.S. Alper, Calculations of the Infrared and Vibrational Circular Dichroism Spectra of Ethanol and Its Deuterated Isotopomers, *J. Phys. Chem.* 93 (18), 6632 (1989).

H. Dothe, M.A. Lowe, and J.S. Alper, Vibrational Circular Dichroism of Methylthiirane, *J. Phys. Chem.* 92 (22), 6246 (1988).

J. Drabowicz, A. Zajac, P. Lyzwa, P.J. Stephens, J.J. Pan, and F.J. Devlin, Determination of the Absolute Configurations of Isotopically Chiral Molecules Using Vibrational Circular Dichroism (VCD) Spectroscopy: The Isotopically Chiral Sulfoxide, Perdeuteriophenyl-Phenyl-Sulfoxide, *Tetrahedron Asymm.* 19 (3), 288 (2008).

J. Drabowicz, A. Zajac, D. Krasowska, B. Bujnicki, B. Dudzinski, M. Janicka, M. Mikolajczyk, M. Chmlelewski, Z. Czarnocki, J. Gawronski, P.L. Polavarapu, M.W. Wieczorek, B. Marciniak, and E. Sokolowska-Rozycka, Chiral Sulfur-Containing Structures: Selected Synthetic and Structural Aspects, *Heteroatom Chem.* 18 (5), 527 (2007).

J. Drabowicz, W. Kudelska, A. Lopusinski, and A. Zajac, The Chemistry of Phosphinic and Phosphinous Acid Derivatives Containing T-Butyl Group as a Single Bulky Substituent: Synthetic, Mechanistic and Stereochemical Aspects, *Curr. Org. Chem.* 11 (1), 3 (2007).

J. Drabowicz, B. Dudzinski, M. Mikolajczyk, F. Wang, A. Dehlavi, J. Goring, M. Park, C.J. Rizzo, P.L. Polavarapu, P. Biscarini, M.W. Wieczorek, and W.R. Majzner, Absolute Configuration, Predominant Conformations, and Vibrational Circular Dichroism Spectra of Enantiomers of N-Butyl Tert-Butyl Sulfoxide, *J. Org. Chem.* 66 (4), 1122 (2001).

A.F. Drake, G. Siligardi, D.H.G. Crout, and D.L. Rathbone, Applications of Vibrational Infrared Circular Dichroism to Biological Problems: Stereochemistry of Proton Exchange in Acetoin (3-Hydroxybutan-2-One) Catalysed by Acetolactate Decarboxylase, *Chem. Comm.* 1834 (1987).

H.H. Drews, Circular Dichroism. Chiral Spectroscopy in the IR, *Nachrichten Chemie* 51 (9), 999 (2003).

L. Ducasse, F. Castet, A. Fritsch, I. Huc, and T. Buffeteau, Density Functional Theory Calculations and Vibrational Circular Dichroism of Aromatic Foldamers, *J. Phys. Chem. A* 111 (23), 5092 (2007).

R.J. Dudley, S.F. Mason, and R.D. Peacock, Electronic and Vibrational Linear and Circular Dichroism of Nematic and Cholesteric *Faraday Trans.* 2 71 (5), 997 (1975).

R.J. Dudley, S.F. Mason, and R.D. Peacock, Infrared Vibrational Circular Dichroism, *Chem. Comm.* (19), 1084 (1972).

R.K. Dukor and T.A. Keiderling, Mutarotation Studies of Poly-L-Proline Using FTIR, Electronic and Vibrational Circular Dichroism, *Biospectroscopy* 2 (2), 83 (1996).

R.K. Dukor, P. Pancoska, T.A. Keiderling, S.J. Prestrelski, and T. Arakawa, Vibrational Circular Dichroism Studies of Epidermal Growth Factor and Basic Fibroblast Growth Factor, *Arch. Biochem. Biophys.* 298 (2), 678 (1992).

R.K. Dukor, T.A. Keiderling, and V. Gut, Vibrational Circular Dichroism Spectra of Unblocked Proline Oligomers, *Int. J. Peptide Protein Res.* 38 (3), 198 (1991).

R.K. Dukor and T.A. Keiderling, Reassessment of the Random Coil Conformation: Vibrational CD Study of Proline Oligopeptides and Related Polypeptides, *Biopolymers* 31 (14), 1747 (1991).

D. Dunmire, T.B. Freedman, L.A. Nafie, C. Aeschlimann, J.G. Gerber, and J. Gal, Determination of the Absolute Configuration and Solution Conformation of the Antifungal Agents Ketoconazole, Itraconazole, and Miconazole with Vibrational Circular Dichroism, *Chirality* 17, S101 (2005).

R. Dutler and A. Rauk, Calculated Infrared Absorption and Vibrational Circular Dichroism Intensities of Oxirane and Its Deuterated Analogues, *J. Am. Chem. Soc.* 111 (18), 6957 (1989).

A.B. Dyatkin, T.B. Freedman, X. Cao, R.K. Dukor, B.E. Maryanoff, C.A. Maryanoff, J.M. Matthews, R.D. Shah, and L.A. Nafie, Determination of the Absolute Configuration of a Key Tricyclic Component of a Novel Vasopressin Receptor Antagonist by Use of Vibrational Circular Dichroism, *Chirality* 14, 215 (2002).

T. Eggimann and H. Wieser, Vibrational Circular Dichroism Spectra of Five Methyl-Substituted 6,8-Dioxabicyclo [3.2.1] Octanes, *Proc. SPIE* 1575, 412 (1991).

T. Eggimann, R.A. Shaw, and H. Wieser, Measurement and Model Calculations of the Vibrational Circular Dichroism Spectrum of 6,8-Dioxabicyclo[3.2.1]Octane, *J. Phys. Chem.* 95 (2), 591 (1991).

T. Eggimann, N. Ibrahim, and H. Wieser, VCD Spectrum of 6,8-Dioxabicyclo[3.2.1]Octane: Comparison of Experimental and Model Spectra, *Proc. SPIE* 1145, 271 (1989).

F. Eker, K. Griebenow, X. Cao, L.A. Nafie, and R. Schweitzer-Stenner, Preferred Peptide Backbone Conformations in the Unfolded State Revealed by the Structure Analysis of Alanine-Based (Axa) Tripeptides in Aqueous Solution, *Proc. Natl. Acad. Sci. USA* 101 (27), 10054 (2004).

F. Eker, K. Griebenow, X. Cao, L.A. Nafie, and R. Schweitzer-Stenner, Tripeptides with Ionizable Side Chains Adopt a Perturbed Polyproline II Structure in Water, *Biochemistry* 43 (3), 613 (2004).

F. Eker, X. Cao, L. Nafie, Q. Huang, and R. Schweitzer-Stenner, The Structure of Alanine Based Tripeptides in Water and Dimethyl Sulfoxide Probed by Vibrational Spectroscopy, *J. Phys. Chem. B* 107 (1), 358 (2003).

F. Eker, X. Cao, L. Nafie, and R. Schweitzer-Stenner, Tripeptides Adopt Stable Structures in Water. A Combined Polarized Visible Raman, FTIR and VCD Spectroscopy Study, *J. Am. Chem. Soc.* 124 (48), 14330 (2002).

H. Elgavi, C. Krekeler, R. Berger, and D. Avnir, Chirality in Copper Nanoalloy Clusters, *J. Phys. Chem. C* 116 (1), 330 (2012).

M.W. Ellzy, J.O. Jensen, H.F. Hameka, and J.G. Kay, Correlation of Structure and Vibrational Spectra of the Zwitterion L-Alanine in the Presence of Water: An Experimental and Density Functional Analysis, *Spectrochim. Acta* 59A (11), 2619 (2003).

J.R. Escribano and L.D. Barron, Valence Optical Theory of Vibrational Circular Dichroism and Raman Optical Activity, *Mol. Phys.* 65 (2), 327 (1988).

J.R. Escribano, T.B. Freedman, and L.A. Nafie, A Bond-Origin-Independent Formulation of the Bond Dipole Model of Vibrational Circular Dichroism, *J.Phys. Chem.* 91 (1), 46 (1987).

F. Faglioni, P. Lazzeretti, M. Malagoli, R. Zanasi, and T. Prosperi, Calculation of Infrared and Vibrational Circular Dichroism Intensities via Nuclear Electromagnetic Shieldings, *J. Phys. Chem.* 97 (11), 2535 (1993).

T.R. Faulkner, C. Marcott, A. Moscowitz, and J. Overend, Anharmonic Effects in Vibrational Circular Dichroism, *J. Am. Chem. Soc.* 99 (25), 8160 (1977).

T.R. Faulkner, A. Moscowitz, G. Holzwarth, E.C. Hsu, and H.S. Mosher, Infrared Circular Dichroism of Carbon-Hydrogen and Carbon-Deuterium Stretching Modes. Calculations, *J. Am. Chem. Soc.* 96 (1), 252 (1974).

D.E.C. Ferreira, W.B. De Almeida, and H.F. Dos Santos, A Theoretical Investigation of Structural, Spectroscopic and Thermodynamic Properties of Cyclodecane, *J. Theor. Comput. Chem.* 6 (2), 281 (2007).

B. Figadère, F.J. Devlin, J.G. Millar, and P.J. Stephens, Determination of the Absolute Configuration of the Sex Pheromone of the Obscure Mealybug by Vibrational Circular Dichroism Analysis, *Chem. Comm.* (9), 1106 (2008).

L.P. Fontana, T. Chandramouly, H.E. Smith, and P.L. Polavarapu, Vibrational Circular Dichroism and Absolute Configuration of 1-Substituted Indans, *J. Org. Chem.* 53 (14), 3379 (1988).

T.B. Freedman, X. Cao, Z. Luz, H. Zimmermann, R. Poupko, and L.A. Nafie, Isotopic Difference Spectra as an Aide in Determining Absolute Configuration Using Vibrational Optical Activity: Vibrational Circular Dichroism of ^{13}C- and ^2H-Labelled Nonamethoxy Cyclotriveratrylene, *Chirality* 20 (5), 673 (2008).

T.B. Freedman, X. Cao, L.M. Phillips, P.T.W. Cheng, R. Dalterio, Y.Z. Shu, H. Zhang, N. Zhao, R.B. Shukla, A. Tymiak, S.K. Gozo, L.A. Nafie, and J.Z. Gougoutas, Determination of the Absolute Configuration and Solution Conformation of a Novel Disubstituted Pyrrolidine Acid A by Vibrational Circular Dichroism, *Chirality* 18 (9), 746 (2006).

T.B. Freedman, X. Cao, L.A. Nafie, M. Kalbermatter, A. Linden, and A.J. Rippert, Determination of the Atropisomeric Stability and Solution Conformation of Asymmetrically Substituted Biphenyls by Means of Vibrational Circular Dichroism (VCD), *Helv. Chim. Acta* 88 (8), 2302 (2005).

T.B. Freedman, X. Cao, L.A. Nafie, A. Solladié-Cavallo, L. Jierry, and L. Bouerat, VCD Configuration of Enantiopure/-Enriched Tetrasubstituted α-Fluoro Cyclohexanones and Their Use for Epoxidation of Trans-Olefins, *Chirality* 16 (7), 467 (2004).

T.B. Freedman, X. Cao, A. Rajca, H. Wang, and L.A. Nafie, Determination of Absolute Configuration in Molecules with Chiral Axes by Vibrational Circular Dichroism: A C_2-Symmetric Annelated Heptathiophene and a D_2-Symmetric Dimer of 1,1'-Binaphthyl, *J. Phys. Chem. A* 107 (39), 7692 (2003).

T.B. Freedman, X. Cao, R.V. Oliveira, Q.B. Cass, and L.A. Nafie, Determination of the Absolute Configuration and Solution Conformation of Gossypol by Vibrational Circular Dichroism, *Chirality* 15 (2), 196 (2003).

T.B. Freedman, X. Cao, L.A. Nafie, M. Kalbermatter, A. Linden, and A.J. Rippert, An Unexpected Atropisomerically Stable 1,1-Biphenyl at Ambient Temperature in Solution, Elucidated by Vibrational Circular Dichroism (VCD), *Helv. Chim. Acta* 86 (9), 3141 (2003).

T.B. Freedman, X. Cao, R.K. Dukor, and L.A. Nafie, Absolute Configuration Determination of Chiral Molecules in the Solution State Using Vibrational Circular Dichroism, *Chirality* 15 (9), 743 (2003).

T.B. Freedman, R.K. Dukor, van Hoof P.J.C.M, E.R. Kellenbach, and L.A. Nafie, Determination of the Absolute Configuration of (–)-Mirtazapine by Vibrational Circular Dichroism, *Helv. Chim. Acta* 85 (4), 1160 (2002).

T.B. Freedman, X. Cao, D.A. Young, and L.A. Nafie, Density Functional Theory Calculations of Vibrational Circular Dichroism in Transition Metal Complexes: Identification of Solution Conformations and Mode of Chloride Ion Association for (+)-Tris(Ethylenediaminato) Cobalt(III). *J. Phys. Chem. A* 106 (14), 3560 (2002).

T.B. Freedman, E. Lee, and L.A. Nafie, Vibrational Transition Current Density in (S)-Methyl Lactate: Visualizing the Origin of the Methine-Stretching Vibrational Circular Dichroism Intensity, *J. Phys. Chem. A* 104 (17), 3944 (2000).

T.B. Freedman, E. Lee, and L.A. Nafie, Vibrational Transition Current Density in (2S,3S)-Oxirane-d_2: Visualizing Electronic and Nuclear Contributions to IR Absorption and Vibrational Circular Dichroism Intensities, *J. Mol. Struct.* 550–551, 123 (2000).

T.B. Freedman, F. Long, M. Citra, and L.A. Nafie, Hydrogen-Stretching Vibrational Circular Dichroism Spectroscopy: Absolute Configuration and Solution Conformation of Selected Pharmaceutical Molecules, *Enantiomer* 4 (2), 103 (1999).

T.B. Freedman, D.L. Hausch, S.J. Cianciosi, and J.E. Baldwin, Kinetics of Thermal Racemization of (2S,3S)-1-^{13}C-1,2,3-D_3-Cyclopropane Followed by Vibrational Circular Dichroism Spectroscopy, *Can. J. Chem.* 76, 806 (1998).

T.B. Freedman, M.-L. Shih, E. Lee, and L.A. Nafie, Electron Transition Current Density in Molecules. 3. *Ab Initio* Calculations for Vibrational Transitions in Ethylene and Formaldehyde, *J. Am. Chem. Soc.* 119 (44), 10620 (1997).

T.B. Freedman, L.A. Nafie, and T.A. Keiderling, Vibrational Optical Activity of Oligopeptides, *Biopolymers (Peptide Sci.)* 37 (4), 265 (1995).

T.B. Freedman, N. Ragunathan, and S. Alexander, Vibrational Circular Dichroism in Ephedra Molecules. Experimental Measurement and *Ab Initio* Calculation, *Faraday Discuss* 99, 131 (1994).

T.B. Freedman, L.A. Nafie, and D. Yang, *Ab Initio* Locally Distributed Origin Gauge Calculations of Vibrational Circular Dichroism Intensity: Formulation and Application to (S,S)-Oxirane-2,3-$^2H_2$, *Chem. Phys. Lett.* 227 (4–5), 419 (1994).

T.B. Freedman and L.A. Nafie, Theoretical Formalism and Models for Vibrational Circular Dichroism Intensity, in *Modern Nonlinear Optics*, Part 3, Advances in Chemical Physics Series, ed. M. Evans and S. Kielich (Wiley, 1994), vol. 85, p. 207.

T.B. Freedman and L.A. Nafie, Infrared Circular Dichroism, *Methods Enzymol.* 226, 306 (1993).

T.B. Freedman, D.M. Gigante, M.J. Citra, and M.G. Paterlini, Vibrational Circular Dichroism in Molecules of Pharmaceutical Interest, *Proc. SPIE* 1890, 40 (1993).

T.B. Freedman, N. Ragunathan, S.J. Cianciosi, J.E. Baldwin, L.A. Nafie, J.A. Moore, and J.M. Schwab, Spectral and Kinetic Investigations of Chirally-Deuterated Three-Membered Ring Molecules Using Fourier Transform Vibrational Circular Dichroism Spectroscopy, *Proc. SPIE* 1575, 408 (1992).

T.B. Freedman, K.M. Spencer, N. Ragunathan, L.A. Nafie, J.A. Moore, and J.M. Schwab, Vibrational Circular Dichroism of (S,S)-[2,3-^{2}H$_2$]Oxirane in the Gas Phase and in Solution, *Can. J. Chem.* 69 (11), 1619 (1991).

T.B. Freedman, S.J. Cianciosi, N. Ragunathan, J.E. Baldwin, and L.A. Nafie, Optical Activity Arising from ^{13}C Substitution: Vibrational Circular Dichroism Study of (2S,3S)-Cyclopropane-1-^{13}C,^{2}H-2,3-^{2}H$_2$, *J. Am. Chem. Soc.* 113 (22), 8298 (1991).

T.B. Freedman, K.M. Spencer, C. McCarthy, S.J. Cianciosi, J.E. Baldwin, L.A. Nafie, J.A. Moore, and J.M. Schwab, Vibrational Circular Dichroism of Simple Chiral Molecules in the Gas Phase, *Proc. SPIE* 1145, 273 (1989).

T.B. Freedman and L.A. Nafie, in *Biomolecular Spectroscopy*, ed. R.R Birge and H.H. Mantsch (SPIE, 1989), vol. 1057, p. 15.

T.B. Freedman and L.A. Nafie, Vibronic Coupling Calculations of Vibrational Circular Dichroism Intensities Using Floating Basis Sets, *J. Chem. Phys.* 89 (1), 374 (1988).

T.B. Freedman, J. Kallmerten, E.D. Lipp, D.A. Young, and L.A. Nafie, Vibrational Circular Dichroism in the CH Stretching Region of (+)-3(R)-Methylcyclohexanone and Chiral Deuteriated Isotopomers, *J. Am. Chem. Soc.* 110 (3), 689 (1988).

T.B. Freedman, A.C. Chernovitz, W.M. Zuk, M.G. Paterlini, and L.A. Nafie, Vibrational Circular Dichroism in the Methine Bending Modes of Amino Acids and Dipeptides, *J. Am. Chem. Soc.* 110 (21), 6970 (1988).

T.B. Freedman, D.A. Young, M.R. Oboodi, and L.A. Nafie, Vibrational Circular Dichroism in Transition-Metal Complexes. 3. Ring Currents and Ring Conformations of Amino Acid Ligands, *J. Am. Chem. Soc.* 109 (5), 1551 (1987).

T.B. Freedman, M.G. Paterlini, N.-S. Lee, L.A. Nafie, J.M. Schwab, and T. Ray, Vibrational Circular Dichroism in the Carbon-Hydrogen and Carbon-Deuterium Stretching Modes of (S,S)-[2,3-^{2}h$_2$]Oxirane, *J. Am. Chem. Soc.* 109 (15), 4727 (1987).

T.B. Freedman and L.A. Nafie, in *Topics in Stereochemistry*, ed. E.L. Eliel and S.H. Wilen (Wiley, New York, 1987), vol. 17, p. 113.

T.B. Freedman and L.A. Nafie, Molecular Orbital Calculations of Infrared Intensities beyond the Born-Oppenheimer Approximation Using the Dipole Momentum Operator, *Chem. Phys. Lett.* 126 (5), 441 (1986).

T.B. Freedman, G.A. Balukjian, and L.A. Nafie, Enhanced Vibrational Circular Dichroism via Vibrationally Generated Electronic Ring Currents, *J. Am. Chem. Soc.* 107 (22), 6213 (1985).

T.B. Freedman and L.A. Nafie, Molecular Orbital Approaches to the Calculation of Vibrational Circular Dichroism, *J. Phys. Chem.* 88 (3), 496 (1984).

T.B. Freedman and L.A. Nafie, Vibrational Optical Activity Calculations Using Infrared and Raman Atomic Polar Tensors, *J. Chem. Phys.* 78 (1), 27 (1983).

T.B. Freedman, M. Diem, P.L. Polavarapu, and L.A. Nafie, Vibrational Circular Dichroism in Amino Acids and Peptides. 6. Localized Molecular Orbital Calculations of the Carbon-Hydrogen Stretching Vibrational Circular Dichroism in Deuterated Isotopomers of Alanine, *J. Am. Chem. Soc.* 104 (12), 3343 (1982).

K. Frimand and K.J. Jalkanen, SCC-TB, DFT/B3LYP, MP2, AM1, PM3 and RHF Study of Ethylene Oxide and Propylene Oxide Structures, VA and VCD Spectra, *Chem. Phys.* 279, 161 (2002).

K. Frimand, H. Bohr, K.J. Jalkanen, and S. Suhai, Structures, Vibrational Absorption and Vibrational Circular Dichroism Spectra of L-Alanine in Aqueous Solution: A Density Functional Theory and RHF Study, *Chem. Phys.* 255 (2–3), 165 (2000).

M.J. Frisch, G.W. Trucks, and J.R. Cheeseman, Systematic Model Chemistries Based on Density Functional Theory: Comparison with Traditional Models and with Experiment in Recent Developments and Appplications of Modern Density Functional Theory, *Theor. Comput. Chem.* 4, 679 (1996).

P. Fristrup, P.R. Lassen, D. Tanner, and K.J. Jalkanen, Direct Determination of Absolute Configuration: A Vibrational Circular Dichroism Study on Dimethyl-Substituted Phenyloxiranes Synthesized by Shi Epoxidation, *Theor. Chem. Acc.* 119 (1–3), 133 (2008).

P. Fristrup, P.R. Lassen, C. Johannessen, D. Tanner, P.O. Norrby, K.J. Jalkanen, and L. Hemmingsen, Direct Determination of Absolute Configuration of Methyl-Substituted Phenyloxiranes: Combined Experimental and Theoretical Approach, *J. Phys. Chem. A* 110 (29), 9123 (2006).

T. Fujita, K. Obata, S. Kuwahara, N. Miura, A. Nakahashi, K. Monde, J. Decatur, and N. Harada, (R)-(+)-[VCD(+)945]-4-Ethyl-4-Methyloctane, the Simplest Chiral Saturated Hydrocarbon with a Quaternary Stereogenic Center, *Tetrahedron Lett.* 48 (24), 4219 (2007).

A. Fulara, A. Lakhani, S. Wójcik, H. Nieznańska, T.A. Keiderling, and W. Dzwolak, Spiral Superstructures of Amyloid-Like Fibrils of Polyglutamic Acid: An Infrared Absorption and Vibrational Circular Dichroism Study, *J. Phys. Chem. B* 115 (37), 11010 (2011).

T. Furo, T. Mori, Y. Origane, T. Wada, H. Izumi, and Y. Inoue, Absolute Configuration Determination of Donor-Acceptor [2.2] Paracyclophanes by Comparison of Theoretical and Experimental Vibrational Circular Dichroism Spectra, *Chirality* 18 (3), 205 (2006).

T. Furo, T. Mori, T. Wada, and Y. Inoue, Absolute Configuration of Chiral [2.2]Paracyclophanes with Intramolecular Charge-Transfer Interaction. Failure of the Exciton Chirality Method and Use of the Sector Rule Applied to the Cotton Effect of the CT Transition, *J. Am. Chem. Soc.* 127 (23), 8242 (2005).

F. Gangemi, R. Gangemi, G. Longhi, and S. Abbate, Experimental and *Ab-Initio* Calculated VCD Spectra of the First OH-Stretching Overtone of (1R)-(−) and (1S)-(+)-Endo-Borneol, *Phys. Chem. Chem. Phys.* 11 (15), 2683 (2009).

F. Gangemi, R. Gangemi, G. Longhi, and S. Abbate, Calculations of Overtone NIR and NIR-VCD Spectra in the Local Mode Approximation: Camphor and Camphorquinone, *Vibrational Spectrosc.* 50 (2), 257 (2009).

R. Gangemi, G. Longhi, F. Lebon, S. Abbate, and L. Laux, Vibrational Excitons in CH-Stretching Fundamental and Overtone Vibrational Circular Dichroism Spectra, *Monatshefte Chemie Chemical Monthly* 136 (3), 325 (2005).

R. Gangemi, G. Longhi, and S. Abbate, Calculated Absorption and Vibrational Circular Dichroism Spectra of Fundamental and Overtone Transitions for a Chiral HCCH Molecular Fragment in the Hypothesis of Coupled Dipoles, *Chirality* 17 (9), 530 (2005).

C. Gautier and T. Bürgi, Chiral Gold Nanoparticles, *Chem. Phys. Chem.* 10 (3), 483 (2009).

C. Gautier and T. Bürgi, Chiral Metal Surfaces and Nanoparticles, *Chimia* 62 (6), 465 (2008).

C. Gautier and T. Bürgi, Chiral N-Isobutyryl-Cysteine Protected Gold Nanoparticles: Preparation, Size Selection, and Optical Activity in the UV-VIS and Infrared, *J. Am. Chem. Soc.* 128 (34), 11079 (2006).

C. Gautier, M. Bieri, I. Dolamic, S. Angeloni, J. Boudon, and T. Bürgi, Probing Chiral Nanoparticles and Surfaces by Infrared Spectroscopy, *Chimia* 60 (11), 777 (2006).

C. Gautier and T. Bürgi, Vibrational Circular Dichroism of N-Acetyl-L-Cysteine Protected Gold Nanoparticles, *Chem. Comm.* (43), 5393 (2005).

D.M.P. Gigante, F. Long, L.A. Bodack, J.M. Evans, J. Kallmerten, L.A. Nafie, and T.B. Freedman, Hydrogen Stretching Vibrational Circular Dichroism in Methyl Lactate and Related Molecules, *J. Phys. Chem. A* 103, 1523 (1999).

I. Goljer, A. Molinari, Y. He, L. Nogle, W. Sun, B. Campbell, and O. McConnell, Unexpected Rearrangement of Enantiomerically Pure 3-Aminoquinuclidine as a Simple Way of Preparing Diastereomeric Octahydropyrrolo[2,3-C]Pyridine Derivatives, *Chirality* 21 (7), 681 (2009).

I. Goncharova and M. Urbanova, Vibrational and Electronic Circular Dichroism Study of Bile Pigments: Complexes of Bilirubin and Biliverdin with Metals, *Anal. Biochem.* 392 (1), 28 (2009).

I. Goncharova and M. Urbanova, Stereoselective Bile Pigment Binding to Polypeptides and Albumins: A Circular Dichroism Study, *Anal. Bioanal.Chem.* 392 (7–8), 1355 (2008).

I. Goncharova and M. Urbanova, Bile Pigment Complexes with Cyclodextrins: Electronic and Vibrational Circular Dichroism Study, *TetrahedronAsymm.* 18 (17), 2061 (2007).

L. Gontrani, B. Mennucci, and J. Tomasi, Glycine and Alanine: A Theoretical Study of Solvent Effects upon Energetics and Molecular Response Properties, *J. Mol. Struct. (Theochem)* 500, 113 (2000).

B. Gordillo-Román, J. Camacho-Ruiz, M.A. Bucio, and P. Joseph-Nathan, Chiral Recognition of Diastereomeric 6-Cedrols by Vibrational Circular Dichroism, *Chirality* 24 (2), 147 (2012).

M. Gulotta, W. Zhong, D.J. Goss, H. Votavova, and M. Diem, Infrared (Vibrational) CD of DNA Models, *Spec. Publ. R. Soc. Chem.* 94, 135 (1991).

M. Gulotta, D.J. Goss, and M. Diem, IR Vibrational CD in Model Deoxyoligonucleotides: Observation of the B→Z Phase Transition and Extended Coupled Oscillator Intensity Calculations, *Biopolymers* 28 (12), 2047 (1989).

C. Guo, R.D. Shah, J. Mills, R.K. Dukor, X. Cao, T.B. Freedman, and L.A. Nafie, Fourier Transform Near-Infrared Vibrational Circular Dichroism Used for On-Line Monitoring the Epimerization of 2,2-Dimethyl-1,3-Dioxolane-4-Methanol: A Pseudo Racemization Reaction, *Chirality* 18 (10), 775 (2006).

C. Guo, R.D. Shah, R.K. Dukor, T.B. Freedman, X. Cao, and L.A. Nafie, Fourier Transform Vibrational Circular Dichroism from 800 to 10,000 cm^{-1}: Near-IR-VCD Spectral Standards for Terpenes and Related Molecules, *Vibrational Spectrosc.* 42 (2), 254 (2006).

C. Guo, R.D. Shah, R.K. Dukor, X. Cao, T.B. Freedman, and L.A. Nafie,Enantiomeric Excess Determination by Fourier Transform Near-Infrared Vibrational Circular Dichroism Spectroscopy: Simulation of Real-Time Process Monitoring, *Appl. Spectrosc.* 59 (9), 1114 (2005).

C. Guo, R.D. Shah, R.K. Dukor, X. Cao, T.B. Freedman, and L.A. Nafie, Determination of Enantiomeric Excess in Samples of Chiral Molecules Using Fourier Transform Vibrational Circular Dichroism Spectroscopy: Simulation of Real-Time Reaction Monitoring, *Anal. Chem.* 76 (23), 6956 (2004).

H. Guo, Q. Liu, L. Yang, W. Weng, Q. Wang, and C. Zheng, Two Novel 3-D Chiral Coordination Polymers with Helical Chains Assembled from L(D)-Tartrate: [Cd_2(L(D)-$C_4H_4O_6$) (H_2O)$_5$(SO_4)].$3H_2O$, *Inorg. Chem.Comm.* 11 (8), 859 (2008).

V.P. Gupta and T.A. Keiderling, Vibrational CD of the Amide II Band in Some Model Polypeptides and Proteins, *Biopolymers* 32 (3), 239 (1992).

S. Hahn, H. Lee, and M. Cho, Theoretical Calculations of Infrared Absorption, Vibrational Circular Dichroism, and Two-Dimensional Vibrational Spectra of Acetylproline in Liquids Water and Chloroform, *J. Chem. Phys.* 121 (4), 1849 (2004).

U. Hahn, A. Kaufmann, M. Nieger, O. Julínek, M. Urbanova, and F. Vögtle, Preparation and Chiroptical Studies of Dendritic Alkaloid Derivatives, *Eur. J. Org. Chem.* (5), 1237 (2006).

W.G. Han, K.J. Jalkanen, M. Elstner, and S. Suhai, Theoretical Study of Aqueous N-Acetyl-L-Alanine N′-Methylamide: Structures and Raman, VCD, and ROA Spectra, *J. Phys. Chem. B* 102 (14), 2587 (1998).

Z.-B. Han, Y.-K. He, M.-L. Tong, Y.-J. Song, X.-M. Song, and L.-G. Yang, Spontaneously Resolved 3D Homochiral In(III) Coordination Polymer with Extended In-OH-In Helical Chains, *Cryst. Eng. Comm.* 10 (8), 1070 (2008).

A.E. Hansen, P.J. Stephens, and T.D. Bouman, Theory of Vibrational Circular Dichroism: Formalisms for Atomic Polar and Axial Tensors Using Non-Canonical Orbitals, *J. Phys. Chem.* 95 (11), 4255 (1991).

T. Hayashi and S. Mukamel, Vibrational-Exciton Couplings for the Amide I, II, III, and a Modes of Peptides, *J. Phys. Chem. B* 111 (37), 11032 (2007).

J. He, A.G. Petrovic, S.V. Dzyuba, N. Berova, K. Nakanishi, and P.L. Polavarapu, Spectroscopic Investigation of Ginkgo Biloba Terpene Trilactones and Their Interaction with Amyloid Peptide Aβ (25–35), *Spectrochim. Acta A* 69A (4), 1213 (2008).

J. He, F. Wang, and P.L. Polavarapu, Absolute Configurations of Chiral Herbicides Determined from Vibrational Circular Dichroism, *Chirality* 17, S1 (2005).

J. He and P.L. Polavarapu, Determination of the Absolute Configuration of Chiral α-Aryloxypropanoic Acids Using Vibrational Circular Dichroism Studies: 2-(2-Chlorophenoxy) Propanoic Acid and 2-(3-Chlorophenoxy) Propanoic Acid, *Spectrochim. Acta A* 61 (7), 1327 (2005).

J. He and P.L. Polavarapu, Determination of Intermolecular Hydrogen Bonded Conformers of α-Aryloxypropanoic Acids Using Density Functional Theory Predictions of Vibrational Absorption and Vibrational Circular Dichroism Spectra, *J. Chem. Theory Comput.* 1 (3), 506 (2005).

J. He, A. Petrovich, and P.L. Polavarapu, Quantitative Determination of Conformer Populations: Assessment of Specific Rotation, Vibrational Absorption, and Vibrational Circular Dichroism in Substituted Butynes, *J. Phys. Chem. A* 108 (10), 1671 (2004).

J. He, A.G. Petrovic, and P.L. Polavarapu, Determining the Conformer Populations of (R)-(+)-3-Methylcyclopentanone Using Vibrational Absorption, Vibrational Circular Dichroism, and Specific Rotation, *J. Phys. Chem. B* 108 (52), 20451 (2004).

Y. He, X. Cao, L.A. Nafie, and T.B. Freedman, *Ab Initio* VCD Calculation of a Transition-Metal Containing Molecule and a New Intensity Enhancement Mechanism for VCD, *J. Am. Chem. Soc.* 123 (45), 11320 (2001).

V.J. Heintz, W.A. Freeman, and T.A. Keiderling, Molecular Structure and Vibrational Circular Dichroism of Tris [Trans-1,2-Cyclopropanediyldimethylene]Diborate, a New, Stable Bridged Borate Ester, *Inorg. Chem.* 22 (16), 2319 (1983).

V.J. Heintz and T.A. Keiderling, Vibrational Circular Dichroism of Optically Active Cyclopropanes. Experimental Results, *J. Am. Chem. Soc.* 103 (9), 2395 (1981).

J. Helbing and M. Bonmarin, Time-Resolved Chiral Vibrational Spectroscopy, *Chimia* 63 (3), 128 (2009).

D.O. Henderson and P.L. Polavarapu, Fourier Transform Infrared Vibrational Circular Dichroism of Matrix-Isolated Molecules, *J. Am. Chem. Soc.* 108 (22), 7110 (1986).

C. Herrmann and M. Reiher, in *Topics in Current Chemistry: Atomistic Approaches in Modern Biology: From Quantum Chemistry to Molecular Simulations* (2007), vol. 268, p. 85. Springer-Verlag, Berlin.

C. Herse, D. Bas, F.C. Krebs, T. Bürgi, J. Weber, T. Wesolowski, B.W. Laursen, and J. Lacour, A Highly Configurationally Stable [4]Heterohelicenium Cation, *Angew. Chem. Int. Ed.* 42 (27), 3162 (2003).

M. Heshmat, V.P. Nicu, and E.J. Baerends, On the Equivalence of Conformational and Enantiomeric Changes of Atomic Configuration for Vibrational Circular Dichroism Signs, *J. Phys. Chem. A* 116 (13), 3454 (2012).

J. Hilario, J. Kubelka, and T.A. Keiderling, Optical Spectroscopic Investigations of Model β-Sheet Hairpins in Aqueous Solution, *J. Am. Chem. Soc.* 125 (25), 7562 (2003).

J. Hilario, J. Kubelka, F.A. Syud, S.H. Gellman, and T.A. Keiderling, Spectroscopic Characterization of Selected β-Sheet Hairpin Models, *Biopolymers (Biospectroscopy)* 67 (3), 233 (2002).

J. Hilario, D. Drapcho, R. Curbelo, and T.A. Keiderling, Polarization Modulation Fourier Transform Infrared Spectroscopy with Digital Signal Processing: Comparison of Vibrational Circular Dichroism Methods, *Appl. Spectrosc.* 55 (11), 1435 (2001).

G.G. Hoffmann, Infrared, Raman and VCD Spectra of (S)-(+)-Carvone-Comparison of Experimental and *Ab Initio* Theoretical Results, *J. Mol. Struct.* 661–662, 525 (2003).

G.G. Hoffmann, M. Niemeyer, and B. Schrader, Vibrational Circular Dichroism of (+)-5,6,7,8-Tetrahydro-8-Methylindan-1,5-Dione, *J. Mol. Struct.* 349, 239 (1995).

G.G. Hoffmann and H.J. Hochkamp, Fourier-Transform Spectrometer for the Measurement of Vibrational Circular Dichroism (VCD) Reaches Down to the Beginning of the Far Infrared Spectral Region, *Proc. SPIE* 1575, 400 (1991).

G.G. Hoffmann, B. Schrader, and G. Snatzke, Photoelastic Modulator for the Mid-IR Range Down to 33 mm with Inexpensive and Simple Control Electronics, *Rev. Sci. Instrum.* 58 (9), 1675 (1987).

A. Holmén, J. Oxelbark, and S. Allenmark, Direct Determination of the Absolute Configuration of a Cyclic Thiolsulfinate by VCD Spectroscopy, *Tetrahedron Asymm.* 14 (15), 2267 (2003).

G. Holzwarth, E.C. Hsu, H.S. Mosher, T.R. Faulkner, and A. Moscowitz, Infrared Circular Dichroism of Carbon-Hydrogen and Carbon-Deuterium Stretching Modes. Observations, *J. Am. Chem. Soc.* 96 (1), 251 (1974).

G. Holzwarth and N.A.W. Holzwarth, Circular Dichroism and Rotatory Dispersion Near Absorption Bands of Cholesteric Liquid Crystals, *J. Opt. Soc. Am.* 63 (3), 324 (1973).

G. Holzwarth, I. Chabay, and N.A.W. Holzwarth, Infrared Circular Dichroism and Linear Dichroism of Liquid Crystals, *J. Chem. Phys.* 58 (11), 4816 (1973).

G. Holzwarth and I. Chabay, Optical Activity of Vibrational Transitions: A Coupled Oscillator Model, *J. Chem. Phys.* 57 (4), 1632 (1972).

K.H. Hopmann, J. Šebestík, J. Novotná, W. Stensen, M. Urbanová, J. Svenson, J.S. Svendsen, P. Bouř, and K. Ruud, Determining the Absolute Configuration of Two Marine Compounds Using Vibrational Chiroptical Spectroscopy, *J. Org. Chem.* 77 (2), 858 (2012).

J. Horníček, P. Kaprálová, and P. Bour, Simulations of Vibrational Spectra from Classical Trajectories: Calibration with *Ab Initio* Force Fields, *J. Chem. Phys.* 127 (8), 084502/1 (2007).

E.C. Hsu and G. Holzwarth, Vibrational Circular Dichroism Observed in Crystalline α-NiSO$_4$-6H$_2$O and α-ZnSeO$_4$-6H$_2$O between 1900 and 5000 cm^{-1}, *J. Chem. Phys.* 59 (9), 4678 (1973).

R. Huang, J. Kubelka, W. Barber-Armstrong, R.A.G.D. Silva, S.M. Decatur, and T.A. Keiderling, Nature of Vibrational Coupling in Helical Peptides: An Isotopic Labeling Study, *J. Am. Chem. Soc.* 126 (8), 2346 (2004).

K.L.C. Hunt and R.A. Harris, Vibrational Circular Dichroism and Electric-Field Shielding Tensors: A New Physical Interpretation Based on Nonlocal Susceptibility Densities, *J. Chem. Phys.* 94 (11), 6995 (1991).

H. Ito, Linear Response Polarizability Bandshape Calculations of Vibrational Circular Dichroism, Vibrational Absorption, and Electronic Circular Dichroism of Cyclo (Gly-Pro-Gly-D-Ala-Pro): A Small Cyclic Pentapeptide Having β- and γ-Turns, *Biospectroscopy* 2 (1), 17 (1996).

H. Ito and Y.J. I'Haya, Linear Response Polarizability Tensor Theory for Vibrational Circular Dichroism: VCD and IR Absorption Bandshape Calculations of α-Helical and β-Sheet Polypeptides, *Bull. Chem. Soc. Jpn.* 67 (5), 1238 (1994).

H. Izumi, A. Ogata, L.A. Nafie, and R.K. Dukor, Structural Determination of Molecular Stereochemistry Using VCD Spectroscopy and a Conformational Code: Absolute Configuration and Solution Conformation of a Chiral Liquid Pesticide, (R)-(+)-Malathion, *Chirality* 21 (1E), E172 (2009).

H. Izumi, A. Ogata, L.A. Nafie, and R.K. Dukor, A Revised Conformational Code for the Exhaustive Analysis of Conformers with One-to-One Correspondence between Conformation and Code: Application to the VCD Analysis of (S)-Ibuprofen, *J. Org. Chem.* 74 (3), 1231 (2009).

H. Izumi, A. Ogata, L.A. Nafie, and R.K. Dukor, Vibrational Circular Dichroism Analysis Reveals a Conformational Change of the Baccatin III Ring of Paclitaxel: Visualization of Conformations Using a New Code for Structure-Activity Relationships, *J. Org. Chem.* 73 (6), 2367 (2008).

H. Izumi, S. Futamura, N. Tokita, and Y. Hamada, Fliplike Motion in the Thalidomide Dimer: Conformational Analysis of (R)-Thalidomide Using Vibrational Circular Dichroism Spectroscopy, *J. Org. Chem.* 72 (1), 277 (2007).

H. Izumi, S. Yamagami, S. Futamura, L.A. Nafie, and R.K. Dukor, Direct Observation of Odd-Even Effect for Chiral Alkyl Alcohols in Solution Using Vibrational Circular Dichroism Spectroscopy, *J. Am. Chem. Soc.* 126 (1), 194 (2004).

H. Izumi, S. Yamagami, and S. Futamura, 1-Azaadamantanes: Pharmacological Applications and Synthetic Approaches, *Curr. Med. Chem. Cardiovasc. Hematol. Agents* 1 (2), 99 (2003).

H. Izumi, S. Futamura, L.A. Nafie, and R.K. Dukor, Determination of Molecular Stereochemistry Using Vibrational Circular Dichroism Spectroscopy: Absolute Configuration and Solution Conformation of 5-Formyl-Cis,Cis-1,3,5-Trimethyl-3-Hydroxymethylcyclohexane-1-Carboxylic Acid Lactone, *Chem. Rec.* 3 (2), 112 (2003).

K.J. Jalkanen, J.D. Gale, P.R. Lassen, L. Hemmingsen, A. Rodarte, I.M. Degtyarenko, R.M. Nieminen, S. Broegger Christensen, M. Knapp-Mohammady, and S. Suhai, A Configurational and Conformational Study of Aframodial and Its Diasteriomers via Experimental and Theoretical VA and VCD Spectroscopies, *Theor. Chem. Acc.* 119 (1–3), 177 (2008).

K.J. Jalkanen, J.D. Gale, G.J. Jalkanen, D.F. McIntosh, A.A. El-Azhary, and G.M. Jensen, Trans-1,2-Dicyano-Cyclopropane and Other Cyano-Cyclopropane Derivatives: A Theoretical and Experimental VA, VCD, Raman and ROA Spectroscopic Study, *Theor. Chem. Acc.* 119 (1–3), 211 (2008).

K.J. Jalkanen, I.M. Degtyarenko, R.M. Nieminen, X. Cao, L.A. Nafie, F. Zhu, and L.D. Barron, Role of Hydration in Determining the Structure and Vibrational Spectra of L-Alanine and N-Acetyl L-Alanine N'-Methylamide in Aqueous Solution: A Combined Theoretical and Experimental Approach, *Theor. Chem. Acc.* 119 (1–3), 191 (2008).

K.J. Jalkanen, V.W. Jürgensen, A. Claussen, A. Rahim, G.M. Jensen, R.C. Wade, F. Nardi, C. Jung, I.M. Degtyarenko, R.M. Nieminen, F. Herrmann, M. Knapp-Mohammady, T.A. Niehaus, K. Frimand, and S. Suhai, Use of Vibrational Spectroscopy to Study Protein and DNA Structure, Hydration, and Binding of Biomolecules: A Combined Theoretical and Experimental Approach, *Int. J. Quantum Chem.* 106 (5), 1160 (2006).

K.J. Jalkanen, M. Elstner, and S. Suhai, Amino Acids and Small Peptides as Building Blocks for Proteins: Comparative Theoretical and Spectroscopic Studies, *J. Mol. Struct. (Theochem)* 675 (1–3), 61 (2004).

K.J. Jalkanen, R.M. Nieminen, M. Knapp-Mohammady, and S. Suhai, Vibrational Analysis of Various Isotopomers of L-Alanyl-L-Alanine in Aqueous Solution: Vibrational Absorption, Vibrational Circular Dichroism, Raman, and Raman Optical Activity Spectra, *Int. J. Quantum Chem.* 92 (2), 239 (2003).

K.J. Jalkanen, Energetics, Structures, Vibrational Frequencies, Vibrational Absorption, Vibrational Circular Dichroism and Raman Intensities of Leu-Enkephalin, *J. Phys. Condensed Matter* 15 (18), S1823 (2003).

K.J. Jalkanen, S. Suhai, and H. Bohr, in *Theoretical and Computational Methods in Genome Research,* S. Suhai (Ed.). Plenum Press, NY. pp. 255–277 (1997).

K.J. Jalkanen and S. Suhai, N-Acetyl-L-Alanine N'-Methylamide: A Density Functional Analysis of the Vibrational Absorption and Vibrational Circular Dichroism Spectra, *Chem. Phys.* 208 (1), 81 (1996).

K.J. Jalkanen, R.W. Kawiecki, P.J. Stephens, and R.D. Amos, Basis Set and Gauge Dependence of *Ab Initio* Calculations of Vibrational Rotational Strengths, *J. Phys. Chem.* 94, 7040 (1990).

K.J. Jalkanen, P.J. Stephens, P. Lazzeretti, and R. Zanasi, Random Phase Approximation Calculations of Vibrational Circular Dichroism: Trans-2,3-Dideuteriooxirane, *J. Phys. Chem.* 93 (18), 6583 (1989).

K.J. Jalkanen, P.J. Stephens, P. Lazzeretti, and R. Zanasi, Nuclear Shielding Tensors, Atomic Polar and Axial Tensors and Vibrational Dipole and Rotational Strengths of NHDT, *J. Chem. Phys.* 90, 3204 (1989).

K.J. Jalkanen, P.J. Stephens, R.D. Amos, and N.C. Handy, Theory of Vibrational Circular Dichroism: Trans-2,3-Dideuteriooxirane, *J. Am. Chem. Soc.* 110 (6), 2012 (1988).

K.J. Jalkanen, P.J. Stephens, R.D. Amos, and N.C. Handy, Gauge Dependence of Vibrational Rotational Strengths: NHDT, *J. Phys. Chem.* 92, 1781 (1988).

K.J. Jalkanen, P.J. Stephens, R.D. Amos, and N.C. Handy, Theory of Vibrational Circular Dichroism: Trans-1s,2s-Dicyanocyclopropane, *J. Am. Chem. Soc.* 109 (23), 7193 (1987).

K.J. Jalkanen, P.J. Stephens, R. Amos, D, and N.C. Handy, Basis Set Dependence of *Ab Initio*Predictions of Vibrational Rotational Strengths: NHDT, *Chem. Phys. Lett* 142, 153 (1987).

C. Johannessen and P.W. Thulstrup, Vibrational Circular Dichroism Spectroscopy of a Spin-Triplet Bis-(Biuretato) Cobaltate(III) Coordination Compound with Low-Lying Electronic Transitions, *Dalton Trans.* (10), 1028 (2007).

H. Joo, E. Kraka, and D. Cremer, Environmental Effects on Molecular Conformation: Bicalutamide Analogs, *J. Mol. Struct. (Theochem)* 862 (1–3), 66 (2008).

B. Jordanov, E.H. Korte, and B. Schrader, FTIR Measurements of Optical Rotatory Dispersion and Circular Dichroism, *Mikrochim. Acta* II, 275 (1988).

O. Julínek, V. Setnička, N. Miklášová, M. Putala, K. Ruud, and M. Urbanová, Determination of Molecular Structure of Bisphenylene Homologues of BINOL-Based Phosphoramidites by Chiroptical Methods, *J. Phys. Chem.* A 113 (40), 10717 (2009).

O. Julínek, M. Urbanova, and W. Lindner, Enantioselective Complexation of Carbamoylated Quinine and Quinidine with N-Blocked Amino Acids: Vibrational and Electronic Circular Dichroism Study, *Anal. Bioanal. Chem.* 393 (1), 303 (2009).

O. Julínek, I. Goncharova, and M. Urbanova, Chiral Memory and Self-Replication Study of Porphyrin and Bilirubin Aggregates Formed on Polypeptide Matrices, *Supramol. Chem.* 20 (7), 643 (2008).

O. Julínek and M. Urbanova, The Study of Induced Circular Dichroism in the Porphyrin-Poly(L-Glutamic Acid) Complexes, *Chemicke Listy* 101 (3), 246 (2007).

V.W. Jürgensen and K. Jalkanen, The VA, VCD, Raman and ROA Spectra of Tri-L-Serine in Aqueous Solution, *Phys. Biol.* 3 (1), S63 (2006).

K. Kadode, T. Taniguchi, N. Miura, and S. Nishimura, Method for Analyzing Saccharide Using VCD, *Jpn. Kokai Tokkyo Koho* 27 (2005).

T. Kawauchi, A. Kitaura, J. Kumaki, H. Kusanagi, and E. Yashima, Helix-Sense-Controlled Synthesis of Optically Active Poly(Methyl Methacrylate) Stereocomplexes, *J. Am. Chem. Soc.* 130 (36), 11889 (2008).

R.W. Kawiecki, F.J. Devlin, P.J. Stephens, and R.D. Amos, Vibrational Circular Dichroism of Propylene Oxide, *J. Phys. Chem.* 95 (24), 9817 (1991).

R.W. Kawiecki, F. Devlin, P.J. Stephens, R.D. Amos, and N.C. Handy, Vibrational Circular Dichroism of Propylene Oxide, *Chem. Phys. Lett.* 145 (5), 411 (1988).

T.A. Keiderling, J. Kubelba, and J. Hilario, in *Vibrational Spectroscopy of Polymers and Biological Systems*, ed. M. Braiman and V. Gregoriou (Taylor & Francis/CRC Press, Boca Raton, FL, 2006), p. 253.

T.A. Keiderling and Q. Xu, Optical Spectra of Unfolded Proteins: A Partially Ordered Polymer Problem, *Macromol. Symp.* 220, 17 (2005).

T.A. Keiderling and Q. Xu, Unfolded Peptides and Proteins Studied with Infrared Absorption and Vibrational Circular Dichroism Spectra, *Adv. Protein Chem.* 62, 111 (2002).

T.A. Keiderling, Protein and Peptide Secondary Structure and Conformational Determination with Vibrational Circular Dichroism, *Curr. Opin. Chem. Biol.* 6 (5), 682 (2002).

T.A. Keiderling, in *Infrared and Raman Spectroscopy of Biological Materials*, ed. B. Yan and H.U. Gremlich (Marcel Dekker, New York, 2001), p. 55.

T.A. Keiderling, in *Circular Dichroism: Principles and Applications*, ed. K. Nakanishi, N. Berova, and R.A. Woody (Wiley, New York, 2000), 2nd ed., p. 621.

T.A. Keiderling, R.A.G.D. Silva, G. Yoder, and R.K. Dukor, Vibrational Circular Dichroism Spectroscopy of Selected Oligopeptide Conformations, *Bioorg. Med. Chem.* 7 (1), 133 (1999).

T.A. Keiderling, in *Biomolecular Structure and Dynamics*, ed. G. Vergoten and T. Theophanides (Kluwer, Dordrecht, Netherlands, 1997), p. 299.

T.A. Keiderling, in *Circular Dichroism and the Conformational Analysis of Biomolecules*, ed. G.D. Fasman (Plenum, New York, 1996), p. 555.

T.A. Keiderling, in *Spectroscopic Methods for Determining Protein Structure in Solution*, ed. H.A. Havel (Wiley-VCH, New York, 1996), p. 163.

T.A. Keiderling, B. Wang, M. Urbanova, P. Pancoska, and R.K. Dukor, Empirical Studies of Protein Secondary Structure by Vibrational Circular Dichroism and Related Techniques. α-Lactalbumin and Lysozyme as Examples, *Faraday Discuss* 99, 263 (1994).

T.A. Keiderling, in *Circular Dichroism Interpretations and Applications*, ed. K. Nakanishi, N. Berova, and R. Woody (VCH, New York, 1994), p. 497.

T.A. Keiderling and P. Pancoska, in *Advances in Spectroscopy: Biomolecular Spectroscopy Part B*, ed. R.J.H. Clark and R.E. Hester (Wiley, 1993), vol. 21, p. 267.

T.A. Keiderling, in *Physical Chemistry of Food Processes*, ed. I. Baianu, H. Pessen, and T. Kumosinski (Van Norstrand Reinhold, New York, 1993), vol. 2, p. 307.

T.A. Keiderling, P. Pancoska, S.C. Yasui, M. Urbanova, and R.K. Dukor, in *Proteins: Structure, Dynamics and Design*, ed. V. Renugopalakrishnan, P.R. Carey, I.C.P. Smith, et al. (ESCOMP, Leiden, Netherlands, 1991), p. 165.

T.A. Keiderling and P. Pancoska, Protein Conformational Studies with Vibrational Circular Dichroism, *Spec. Publ. R. Soc. Chem.* 94, 121 (1991).

T.A. Keiderling, in *Practical Fourier Transform Infrared Spectroscopy*, ed. J.R. Ferraro and K. Krishnan (Academic Press, San Diego, 1990), p. 203.

T.A. Keiderling, S.C. Yasui, P. Pancoska, R.K. Dukor, and L. Yang, Biopolymer Conformational Studies with Vibrational Circular Dichroism, *Proc. SPIE* 1057, 7 (1989).

T.A. Keiderling, S.C. Yasui, P. Malon, P. Pancoska, R.K. Dukor, P.V. Croatto, and L. Yang, Techniques and Application of FTIR Vibrational Circular Dichroism, *Proc. SPIE* 1145, 57 (1989).

T.A. Keiderling, S.C. Yasui, R.K. Dukor, and L. Yang, Application of Vibrational Circular Dichroism to Synthetic Polypeptides and Polynucleic Acids, *Polym. Prepr.* 30 (2), 423 (1989).

T.A. Keiderling, Vibrational CD of Biopolymers, *Nature* 322 (6082), 851 (1986).

T.A. Keiderling, Vibrational Circular Dichroism, *Appl. Spectrosc. Rev.* 17 (2), 189 (1981).

T.A. Keiderling and P.J. Stephens, Vibrational Circular Dichroism of Spirononadiene. Fixed Partial Charge Calculations, *J. Am. Chem. Soc.* 101 (6), 1396 (1979).

T.A. Keiderling and P.J. Stephens, Vibrational Circular Dichroism of Dimethyl Tartrate. A Coupled Oscillator, *J. Am. Chem. Soc.* 99 (24), 8061 (1977).

T.A. Keiderling and P.J. Stephens, Vibrational Circular Dichroism of Overtone and Combination Bands, *Chem. Phys. Lett.* 41 (1), 46 (1976).

E.R. Kellenbach, R.K. Dukor, and L.A. Nafie, Absolute Configuration Determination of Chiral Molecules without Crystallisation by Vibrational Circular Dichroism (VCD), *Spectrosc. Eur.* 19 (4), 15 (2007).

J. Kim and T.A. Keiderling, All-Atom Molecular Dynamics Simulations of β-Hairpins Stabilized by a Tight Turn: Pronounced Heterogeneous Folding Pathways, *J. Phys. Chem.* B 114 (25), 8494 (2010).

J. Kim, J. Kapitan, A. Lakhani, P. Bour, and T.A. Keiderling, Tight β-Turns in Peptides. DFT-Based Study of Infrared Absorption and Vibrational Circular Dichroism for Various Conformers Including Solvent Effects, *Theor. Chem. Acc.* 119 (1–3), 81 (2008).

N. Kobayashi and T. Fukuda, First Observation of the Vibrational Circular Dichroism Spectra of Synthetic Chiral Porphyrazines, *Chem. Lett.* 33 (1), 32 (2004).

A. Kocak, R. Luque, and M. Diem, The Solution Structure of Small Peptides: An IR CD Study of Aqueous Solutions of $(L-Ala)_n$ [n=3,4,5,6] at Different Temperatures and Ionic Strengths, *Biopolymers* 46, 455 (1998).

M.E. Koehler and F.L. Urbach, A Microprocessor Controlled Near-Infrared Circular Dichroism Spectrophotometer, *Appl. Spectrosc.* 33 (6), 563 (1979).

V. Král, S. Pataridis, V. Setnicka, K. Záruba, M. Urbanová, and K. Volka, New Chiral Porphyrin-Brucine Gelator Characterized by Methods of Circular Dichroism, *Tetrahedron* 61 (23), 5499 (2005).

M. Krautmann, E.C. de Riscala, E. Burgueño-Tapia, Y. Mora-Pérez, C.A.N. Catalán, and P. Joseph-Nathan, C-15-Functionalized Eudesmanolides from *Mikania campanulata*, *J. Nat. Prod.* 70 (7), 1173 (2007).

K. Krohn, D. Gehle, S.K. Dey, N. Nahar, M. Mosihuzzaman, N. Sultana, M.H. Sohrab, P.J. Stephens, J.-J. Pan, and F. Sasse, Prismatomerin, a New Iridoid from Prismatomeris Tetrandra. Structure Elucidation, Determination of Absolute Configuration and Cytotoxicity, *J. Nat. Prod.* 70 (8), 1339 (2007).

A.T. Krummel and M.T. Zanni, Interpreting DNA Vibrational Circular Dichroism Spectra Using a Coupling Model from Two-Dimensional Infrared Spectroscopy, *J. Phys. Chem.* B 110 (48), 24720 (2006).

J. Kubelka and P. Bour, Simulation of Vibrational Spectra of Large Molecules by Arbitrary Time Propagation, *J. Chem. Theory Comput.* 5 (1), 200 (2009).

J. Kubelka, R. Huang, and T.A. Keiderling, Solvent Effects on IR and VCD Spectra of Helical Peptides: DFT-Based Static Spectral Simulations with Explicit Water, *J. Phys. Chem.* B 109 (16), 8231 (2005).

J. Kubelka, R.A.G.D. Silva, and T.A. Keiderling, Discrimination between Peptide 3_{10}- and α-Helices. Theoretical Analysis of the Impact of α-Methyl Substitution on Experimental Spectra, *J. Am. Chem. Soc.* 124 (19), 5325 (2002).

J. Kubelka and T.A. Keiderling, Differentiation of β-Sheet-Forming Structures: *Ab Initio*-Based Simulations of IR Absorption and Vibrational CD for Model Peptide and Protein β-Sheets, *J. Am. Chem. Soc.* 123 (48), 12048 (2001).

J. Kubelka, P. Pancoska, and T.A. Keiderling, in *Spectroscopy of Biological Molecules: New Directions*, ed. J. Greve, G.J. Puppels, and C. Otto (Kluwer, Dordrecht, Netherlands, 1999), p. 67.

J.L. Kulp III, J.C. Owrutsky, D.Y. Petrovykh, K.P. Fears, R. Lombardi, L.A. Nafie, and T.D. Clark, Vibrational Circular-Dichroism Spectroscopy of Homologous Cyclic Peptides Designed to Fold into β Helices of Opposite Chirality, *Biointerphases* 6 (1), 1 (2011).

T. Kuppens, K. Vandyck, J. van der Eycken, W. Herrebout, B. van der Veken, and P. Bultinck, A DFT Conformational Analysis and VCD Study on Methyl Tetrahydrofuran-2-Carboxylate, *Spectrochim. Acta A* 67 (2), 402 (2007).

T. Kuppens, W. Herrebout, B. Van Der Veken, D. Corens, A. De Groot, J. Doyon, G. Van Lommen, and P. Bultinck, Elucidation of the Absolute Configuration of JNJ-27553292, a CCR2 Receptor Antagonist, by Vibrational Circular Dichroism Analysis of Two Precursors, *Chirality* 18 (8), 609 (2006).

T. Kuppens, W. Herrebout, B. van der Veken, and P. Bultinck, Intermolecular Association of Tetrahydrofuran-2-Carboxylic Acid in Solution: A Vibrational Circular Dichroism Study, *J. Phys. Chem. A* 110 (34), 10191 (2006).

T. Kuppens, K. Vandyck, J. Van der Eycken, W. Herrebout, B.J. van der Veken, and P. Bultinck, Determination of the Absolute Configuration of Three as-Hydrindacene Compounds by Vibrational Circular Dichroism, *J. Org. Chem.* 70 (23), 9103 (2005).

T. Kuppens, W. Langenaeker, J.P. Tollenaere, and P. Bultinck, Determination of the Stereochemistry of 3-Hydroxymethyl-2,3-Dihydro-[1,4]Dioxino[2,3-B]Pyridine by Vibrational Circular Dichroism and the Effect of DFT Integration Grids, *J. Phys. Chem. A* 107 (4), 542 (2003).

D. Kurouski, R.K. Dukor, X. Lu, L.A. Nafie, and I.K. Lednev, Spontaneous Inter-Conversion of Insulin Fibril Chirality, *Chem. Comm.* 48 (23), 2837 (2012).

D. Kurouski, R.A. Lombardi, R.K. Dukor, I.K. Lednev, and L.A. Nafie, Direct Observation and pH Control of Reversed Supramolecular Chirality in Insulin Fibrils by Vibrational Circular Dichroism, *Chem. Comm.* 46 (38), 7154 (2010).

K. Kwac, K.-K. Lee, J.B. Han, K.-I. Oh, and M. Cho, Classical and Quantum Mechanical/Molecular Mechanical Molecular Dynamics Simulations of Alanine Dipeptide in Water: Comparisons with IR and Vibrational Circular Dichroism Spectra, *J. Chem. Phys.* 128 (10), 105106/1 (2008).

C.C. LaBrake, L. Wang, T.A. Keiderling, and L.W.-M. Fung, Fourier Transform Infrared Spectroscopic Studies of the Secondary Structure of Spectrin under Different Ionic Strengths, *Biochemistry* 32 (39), 10296 (1993).

A. Lakhani, P. Malon, and T.A. Keiderling, Comparison of Vibrational Circular Dichroism Instruments: Development of a New Dispersive VCD, *Appl. Spectrosc.* 63 (7), 775 (2009).

A. Lakhani, A. Roy, M. De Poli, M. Nakaema, F. Formaggio, C. Toniolo, and T.A. Keiderling, Experimental and Theoretical Spectroscopic Study of 310-Helical Peptides Using Isotopic Labeling to Evaluate Vibrational Coupling, *J. Phys. Chem. B* 115 (19), 6252 (2011).

B.B. Lal and L.A. Nafie, Vibrational Circular Dichroism in Amino Acids and Peptides. 7. Amide Stretching Vibrations in Polypeptides, *Biopolymers* 21 (11), 2161 (1982).

B.B. Lal, M. Diem, P.L. Polavarapu, M. Oboodi, T.B. Freedman, and L.A. Nafie, Vibrational Circular Dichroism in Amino Acids and Peptides. 5. Carbon-Hydrogen Stretching Vibrational Circular Dichroism and Fixed Partial Charge Calculations for Deuterated Isotopomers of Alanine, *J. Am. Chem. Soc.* 104 (12), 3336 (1982).

I.J. Lalov and J.N. Kotzev, Vibrational Circular Dichroism and Vibrational Optical Rotatory Dispersion in Molecular Crystals. IV. Overtones of Degenerate Vibrations, *J. Chem. Phys.* 89 (6), 3454 (1988).

I.J. Lalov, D.A. Svetogorski, and N.P. Turkedjiev, Overtone Spectra of Helical Polymers: Infrared Absorption and Vibrational Circular Dichroism, *J. Chem. Phys.* 84 (6), 3545 (1986).

I.J. Lalov, Fermi Resonance in the Infrared Absorption Spectra and Vibrational Circular Dichroism Spectra of Helical Polymers, *Bulgarian J. Phys.* 12 (6), 560 (1985).

I.J. Lalov and D.A. Svetogorski, Vibrational Circular Dichroism and Vibrational Optical Rotatory Dispersion in Molecular Crystals. III. Overtones of Nondegenerate Vibrations, *J. Chem. Phys.* 80 (3), 1083 (1984).

I.J. Lalov and J.N. Kotzev, Vibrational Circular Dichroism and Vibrational Optical Rotatory Dispersion in Molecular Crystals. II. Selection Rules, *J. Chem. Phys.* 80 (3), 1078 (1984).

I.J. Lalov, Vibrational Circular Dichroism and Vibrational Optical Rotatory Dispersion in Molecular Crystals. I. Fundamental Tones, *J. Chem. Phys.* 80 (3), 1069 (1984).

I.J. Lalov, VCD Spectroscopy of Crystals—An Interesting Possibility, *J. Mol. Struct.* 114, 75 (1984).

G. Lamanna, C. Faggi, F. Gasparrini, A. Ciogli, C. Villani, P.J. Stephens, F.J. Devlin, and S. Menichetti, Efficient Thia-Bridged Triarylamine Heterohelicenes: Synthesis, Resolution, and Absolute Configuration Determination, *Chem. Eur. J.* 14 (19), 5747 (2008).

P.R. Lassen, L. Guy, I. Karame, T. Roisnel, N. Vanthuyne, C. Roussel, X. Cao, R. Lombardi, J. Crassous, T.B. Freedman, and L.A. Nafie, Synthesis and Vibrational Circular Dichroism of Enantiopure Chiral Oxorhenium(V) Complexes Containing the Hydrotris(1-Pyrazolyl)Borate Ligand, *Inorg. Chem.* 45 (25), 10230 (2006).

P.R. Lassen, D.M. Skytte, L. Hemmingsen, S.F. Nielsen, T.B. Freedman, L.A. Nafie, and S.B. Christensen, Structure and Absolute Configuration of Nyasol and Hinokiresinol via Synthesis and Vibrational Circular Dichroism Spectroscopy, *J. Nat. Prod.* 68 (11), 1603 (2005).

L. Laux, V. Pultz, S. Abbate, H.A. Havel, J. Overend, A. Moscowitz, and D.A. Lightner, Inherently Dissymmetric Chromophores and Vibrational Circular Dichroism. The CH_2-CH_2-C*H Fragment, *J. Am. Chem. Soc.* 104 (15), 4276 (1982).

P. Lazzeretti, R. Zanasi, T. Prosperi, and A. Lapiccirella, The Nuclear Electromagnetic Shielding Approach to IR and VCD Intensities: A Theoretical Study of Ethylene Oxide and Cyclopropane, *Chem. Phys. Lett.* 150 (6), 515 (1988).

P. Lazzeretti, R. Zanasi, and P.J. Stephens, Magnetic Dipole Transition Moments and Rotational Strengths of Vibrational Transitions: An Alternative Formalism, *J. Phys. Chem.* 90, 6761 (1986).

F. Lebon, G. Longhi, S. Abbate, M. Catellani, C. Zhao, and P.L. Polavarapu, Vibrational Circular Dichroism Spectra of Chirally Substituted Polythiophenes, *Synth. Metals* 119, 75 (2001).

K.-K. Lee, E. Kim, C. Joo, J. Song, H. Han, and M. Cho, Site-Selective Intramolecular Hydrogen-Bonding Interactions in Phosphorylated Serine and Threonine Dipeptides, *J. Phys. Chem. B* 112 (51), 16782 (2008).

K.-K. Lee, K.-I. Oh, H. Lee, C. Joo, H. Han, and M. Cho, Dipeptide Structure Determination by Vibrational Circular Dichroism Combined with Quantum Chemistry Calculations, *Chem. Phys. Chem.* 8 (15), 2218 (2007).

K.K. Lee, C. Joo, S. Yang, H. Han, and M. Cho, Phosphorylation Effect on the GSSS Peptide Conformation in Water: Infrared, Vibrational Circular Dichroism, and Circular Dichroism Experiments and Comparisons with Molecular Dynamics Simulations, *J. Chem. Phys.* 126 (23), 235102/1 (2007).

K.K. Lee, S. Hahn, K.I. Oh, J.S. Choi, C. Joo, H. Lee, H. Han, and M. Cho, Structure of N-Acetylproline Amide in Liquid Water: Experimentally Measured and Numerically Simulated Infrared and Vibrational Circular Dichroism Spectra, *J. Phys. Chem. B* 110 (38), 18834 (2006).

O. Lee and M. Diem, Infrared CD in the 6 mm Spectral Region: Design of a Dispersive Infrared Dichrograph, *Anal. Instrum.* 20 (1), 23 (1992).

O. Lee, G.M. Roberts, and M. Diem, IR Vibrational CD in Alanyl Tripeptide: Indication of a Stable Solution Conformer, *Biopolymers* 28 (10), 1759 (1989).

T. Lefevre, J. Leclerc, J.-F. Rioux-Dube, T. Buffeteau, M.-C. Paquin, M.-E. Rousseau, I. Cloutier, M. Auger, S.M. Gagne, S. Boudreault, C. Cloutier, and M. Pezolet, *In Situ* Conformation of Spider Silk Proteins in the Intact Major Ampullate Gland and in Solution, *Biomacromolecules* 8 (8), 2342 (2007).

X. Li, K.H. Hopmann, J. Hudecová, W. Stensen, J. Novotná, M. Urbanová, J.-S. Svendsen, P. Bouř, and K. Ruud, Absolute Configuration of a Cyclic Dipeptide Reflected in Vibrational Optical Activity: Ab Initio and Experimental Investigation, *J. Phys. Chem. A* 116 (10), 2554 (2012).

X. Liang and J.A.D. Haseth, The Studies of Chiral Separation Mechanisms by VCD/FT-IR Spectrometry, *Proc. SPIE* 2089, 480 (1993).

E.D. Lipp and L.A. Nafie, Vibrational Circular Dichroism in Amino Acids and Peptides. 10. Fourier Transform VCD and Fourier Self-Deconvolution of the Amide I Region of Poly(3-Benzyl-L-Glutamate), *Biopolymers* 24, 799 (1985).

E.D. Lipp and L.A. Nafie, Application of Fourier Self-Deconvolution to Vibrational Circular Dichroism Spectra, *Appl. Spectrosc.* 38 (6), 774 (1984).

E.D. Lipp and L.A. Nafie, Fourier Transform Infrared Vibrational Circular Dichroism: Improvements in Methodology and Mid-Infrared Spectral Results, *Appl. Spectrosc.* 38 (1), 20 (1984).

E.D. Lipp, C.G. Zimba, L.A. Nafie, and D.W. Vidrine, Polarization Demodulation: A New Approach to the Reduction of Polarization Artifacts from Vibrational Circular Dichroism Spectra, *Appl. Spectrosc.* 36 (5), 496 (1982).

E.D. Lipp, C.G. Zimba, and L.A. Nafie, Vibrational Circular Dichroism in the Mid-Infrared Using Fourier Transform Spectroscopy, *Chem. Phys. Lett.* 90 (1), 1 (1982).

F. Long, T.B. Freedman, T.J. Tague, and L.A. Nafie, Step-Scan Fourier Transform Vibrational Circular Dichroism Measurements in the Vibrational Region above 2000 cm^{-1}, *Appl. Spectrosc.* 51 (4), 508 (1997).

F. Long, T.B. Freedman, R. Hapanowicz, and L.A. Nafie, Comparison of Step-Scan and Rapid-Scan Approaches to the Measurement of Mid-Infrared Fourier Transform Vibrational Circular Dichroism, *Appl. Spectrosc.* 51 (4), 504 (1997).

G. Longhi, S. Abbate, P. Scafato, and C. Rosini, A Vibrational Circular Dichroism Approach to the Determination of the Absolute Configuration of Flexible and Transparent Molecules: Fluorenone Ketals of 1,N-Diols, *Phys. Chem. Chem. Phys.* 12 (18), 4725 (2010).

G. Longhi, S. Abbate, R. Gangemi, E. Giorgio, and C. Rosini, Fenchone, Camphor, 2-Methylenefenchone and 2-Methylenecamphor: A Vibrational Circular Dichroism Study, *J. Phys. Chem. A* 110 (15), 4958 (2006).

G. Longhi, R. Gangemi, F. Lebon, E. Castiglioni, S. Abbate, V.M. Pultz, and D.A. Lightner, A Comparative Study of Overtone CH-Stretching Vibrational Circular Dichroism Spectra of Fenchone and Camphor, *J. Phys. Chem. A* 108 (25), 5338 (2004).

M. Losada, H. Tran, and Y. Xu, Lactic Acid in Solution: Investigations of Lactic Acid Self-Aggregation and Hydrogen Bonding Interactions with Water and Methanol Using Vibrational Absorption and Vibrational Circular Dichroism Spectroscopies, *J. Chem. Phys.* 128 (1), 014508/1 (2008).

M. Losada, P. Nguyen, and Y. Xu, Solvation of Propylene Oxide in Water: Vibrational Circular Dichroism, Optical Rotation, and Computer Simulation Studies, *J. Phys. Chem. A* 112 (25), 5621 (2008).

M. Losada and Y. Xu, Chirality Transfer through Hydrogen-Bonding: Experimental and *Ab Initio* Analyses of Vibrational Circular Dichroism Spectra of Methyl Lactate in Water, *Phys. Chem. Chem. Phys.* 9 (24), 3127 (2007).

M.A. Lowe and J.S. Alper, *Ab Initio* Calculations of Vibrational Circular Dichroism in Propylene Oxide: Geometry and Force Field Dependance, *J. Phys. Chem.* 92 (14), 4035 (1988).

M.A. Lowe, P.J. Stephens, and G.A. Segal, The Theory of Vibrational Circular Dichroism: Trans-1,2-Dideuteriocyclobutane and Propylene Oxide, *Chem. Phys. Lett* 123 (1–2), 108 (1986).

M.A. Lowe, G.A. Segal, and P.J. Stephens, The Theory of Vibrational Circular Dichroism: Trans-1,2-Dideuteriocyclopropane, *J. Am. Chem. Soc.* 108 (2), 248 (1986).

W.-G. Lu, J.-Z. Gu, L. Jiang, M.-Y. Tan, and T.-B. Lu, Achiral and Chiral Coordination Polymers Containing Helical Chains: The Chirality Transfer between Helical Chains, *Cryst. Growth Des.* 8 (1), 192 (2008).

L. Lunazzi, M. Mancinelli, and A. Mazzanti, Atropisomers of Hindered Triarylisocyanurates: Structure, Conformation, Stereodynamics, and Absolute Configuration, *J. Org. Chem.* 77 (7), 3373 (2012).

J. Luo and L. Xu, Layered Lanthanide Molybdate Pillared by Chiral [λ-Mo₂O₄EDTA]²⁻, *Inorg. Chem.* 45 (26), 11030 (2006).

Z. Luz, R. Poupko, E.J. Wachtel, H. Zheng, N. Friedman, X. Cao, T.B. Freedman, L.A. Nafie, and H. Zimmermann, Structural and Optical Isomers of Nonamethoxy Cyclotriveratrylene: Separation and Physical Characterization, *J. Phys. Chem. A* 111 (42), 10507 (2007).

S. Ma, T.B. Freedman, R.K. Dukor, and L.A. Nafie, Near-Infrared and Mid-Infrared Fourier Transform Vibrational Circular Dichroism of Proteins in Aqueous Solution, *Appl. Spectrosc.* 64 (6), 615 (2010).

S. Ma, S. Shen, H. Lee, M. Eriksson, X. Zeng, J. Xu, K. Fandrick, N. Yee, C. Senanayake, and N. Grinberg, Mechanistic Studies on the Chiral Recognition of Polysaccharide-Based Chiral Stationary Phases Using Liquid Chromatography and Vibrational Circular Dichroism, *J. Chromatogr. A* 1216 (18), 3784 (2009).

S. Ma, S. Shen, N. Haddad, W. Tang, J. Wang, H. Lee, N. Yee, C. Senanayake, and N. Grinberg, Chromatographic and Spectroscopic Studies on the Chiral Recognition of Sulfated β-Cyclodextrin as Chiral Mobile Phase Additive: Enantiomeric Separation of a Chiral Amine, *J. Chromatogr. A* 1216 (8), 1232 (2009).

S. Ma, S. Shen, H. Lee, N. Yee, C. Senanayake, L.A. Nafie, and N. Grinberg, Vibrational Circular Dichroism of Amylose Carbamate: Structure and Solvent-Induced Conformational Changes, *Tetrahedron Asymm.* 19 (18), 2111 (2008).

S. Ma, X. Cao, M. Mak, A. Sadik, C. Walkner, T.B. Freedman, I.K. Lednev, R.K. Dukor, and L.A. Nafie, Vibrational Circular Dichroism Shows Unusual Sensitivity to Protein Fibril Formation and Development in Solution, *J. Am. Chem. Soc.* 129 (41), 12364 (2007).

S. Ma, T.B. Freedman, X. Cao, and L.A. Nafie, Two-Dimensional Vibrational Circular Dichroism Correlation Spectroscopy: pH-Induced Spectral Changes in L-Alanine, *J. Mol. Struct.* 799 (1–3), 226 (2006).

Y. Ma, Z. Han, Y. He, and L. Yang, A 3d Chiral Zn(II) Coordination Polymer with Triple Zn-Oba-Zn Helical Chains (Oba = 4,4′-Oxybis(Benzoate)), *Chem. Comm.* (40), 4107 (2007).

V. Maharaj, H. Van de Sande, D. Tsankov, A. Rauk, and H. Wieser, Towards Interpreting the FT-IR and VCD Spectra of Selected Octadeoxynucleotides, *Mikrochim. Acta Suppl.* 14, 529 (1997).

V. Maharaj, A. Rauk, J.H. van de Sande, and H. Wieser, Infrared Absorption and Vibrational Circular Dichroism Spectra of Selected Deoxyoctanucleotides Complexed with Daunorubicin, *J. Mol. Struct.* 408–409, 315 (1997).

V. Maharaj, D. Tsankov, H.J. Van de Sande, and H. Wieser, FTIR-VCD Spectra of Six Octadeoxynucleotides, *J. Mol. Struct.* 349, 25 (1995).

P. Malon and T.A. Keiderling, Spinning Quarter-Wave Plate Polarization Modulator: Test of Feasibility for Vibrational Circular Dichroism Measurements, *Appl. Spectrosc.* 50 (5), 669 (1996).

P. Malon, L.J. Mickley, K.M. Sluis, C.N. Tam, T.A. Keiderling, S. Kamath, J. Uang, and J.S. Chickos, Vibrational Circular Dichroism Study of (2s, 3s)-Dideuteriobutyrolactone. Synthesis, Normal Mode Analysis, and Comparison of Experimental and Calculated Spectra, *J. Phys. Chem.* 96 (25), 10139 (1992).

P. Malon, T.A. Keiderling, J.Y. Uang, and J.S. Chickos, Vibrational Circular Dichroism Study of [3r,4r]-Dideuteriocyclobutane-1,2-Dione. Preliminary Comparison of Experiment and Calculations, *Chem. Phys. Lett.* 179 (3), 282 (1991).

P. Malon, R. Kobrinskaya, and T.A. Keiderling, Vibrational CD of Polypeptides. XII. Reevaluation of the Fourier Transform Vibrational CD of Poly (γ-Benzyl-L-Glutamate), *Biopolymers* 27 (5), 733 (1988).

P. Malon and T.A. Keiderling, A Solution to the Artifact Problem in Fourier Transform Vibrational Circular Dichroism, *Appl. Spectrosc.* 42 (1), 32 (1988).

P. Malon and T.A. Keiderling, Fourier Transform Vibrational Circular Dichroism and the Artifact Problem, *Mikrochim. Acta* II (1–6), 279 (1988).

P. Malon, C.L. Barness, M. Budesinsky, R.K. Dukor, D. Helm, T.A. Keiderling, Z. Koblicova, F. Pavlikova, M. Tichy, and K. Blaha, (1S,7S)-7-Methyl-6,9-Diazatricyclo[6,3,0,0^{1,6}] Tridecane-5,10-Dione, a Tricyclic Spirodilactam Containing Non-Planar Amide Groups. Synthesis, NMR, Crystal Structure, Absolute Configuration, Electronic and Vibrational Circular Dichroism, *Collect. Czech. Chem. Commun.* 53 (11A), 2447 (1988).

J.J. Manríquez-Torres, J.M. Torres-Valencia, M.A. Gómez-Hurtado, V. Motilva, S. García-Mauriño, J. Ávila, E. Talero, C.M. Cerda-García-Rojas, and P. Joseph-Nathan, Absolute Configuration of 7,8-*seco*-7,8-Oxacassane Diterpenoids from *Acacia Schaffneri*, *J. Nat. Prod.* 74 (9), 1946 (2011).

C. Marcott, A.E. Dowrey, and I. Noda, Instrumental Aspects of Dynamic Two-Dimensional Infrared Spectroscopy, *Appl. Spectrosc.* 47 (9), 1324 (1993).

C. Marcott, K. Scanlon, J. Overend, and A. Moscowitz, Vibrational Circular Dichroism in (R)-Cyclohexanone-3-D, (R)-3-Methylcyclohexanone, and (1S)-2-Adamantanone-4-^{13}C, *J. Am. Chem. Soc.* 103 (3), 483 (1981).

C. Marcott, H.A. Havel, B. Hedlund, J. Overend, and A. Moscowitz, in *Optical Activity and Chiral Discrimination*, ed. by S.F. Mason (Reidel, 1979), p. 289.

C. Marcott, H.A. Havel, J. Overend, and A. Moscowitz, Vibrational Circular Dichroism and Individual Chiral Centers. An Example from the Sugars, *J. Am. Chem. Soc.* 100 (22), 7088 (1978).

C. Marcott, C.C. Blackburn, T.R. Faulkner, A. Moscowitz, and J. Overend, Infrared Circular Dichroism Associated with the OH-Stretching Vibration in the Methyl Ester of Mandelic Acid, *J. Am. Chem. Soc.* 100 (17), 5262 (1978).

C. Marcott, T.R. Faulkner, A. Moscowitz, and J. Overend, Vibrational Circular Dichroism in HCBrCLF and DCBrCLF. Calculation of the Rotational Strengths Associated with the Fundamentals and the Binary Overtones and Combinations, *J. Am. Chem. Soc.* 99 (25), 8169 (1977).

T.A. Martinek, I.M. Mándity, L. Fülöp, G.K. Tóth, E. Vass, M. Hollósi, E. Forró, and F. Fülöp, Effects of the Alternating Backbone Configuration on the Secondary Structure and Self-Assembly of β-Peptides, *J. Am. Chem. Soc.* 128 (41), 13539 (2006).

B.E. Maryanoff, D.F. McComsey, R.K. Dukor, L.A. Nafie, T.B. Freedman, X. Cao, and V.W. Day, Structural Studies on McN-5652-X, a High-Affinity Ligand for the Serotonin Transporter in Mammalian Brain, *Bioorg. Med. Chem.* 11 (11), 2463 (2003).

S. Mason, Vibrational Circular Dichroism: An Incisive Tool for Stereochemical Applications, *Enantiomer* 3, 283 (1998).

S.F. Mason, in *Advances in Infrared and Raman Spectroscopy*, ed. R.J.H. Clark and R.E. Hester (Heiden, London, 1980), vol. 8, p. 283.

W. Mästle, R.K. Dukor, G. Yoder, and T.A. Keiderling, Conformational Study of Linear Alternating and Mixed D- and L-Proline Oligomers Using Electronic and Vibrational CD and Fourier Transform IR, *Biopolymers* 36 (5), 623 (1995).

F. Maurer and H. Wieser, Vibrational Circular Dichroism Spectra of α-Pinene, β-Pinene, and Nopinone, *Proc. SPIE* 1575, 410 (1991).

J.L. McCann, D. Tsankov, and H. Wieser, Extending the Spectral Range of an FT-VCD Spectrometer, *Mikrochim. Acta Suppl.* 14, 777 (1997).

J.L. McCann, B. Schulte, D. Tsankov, and H. Wieser, Vibrational Circular Dichroism Spectra of Selected Chiral Polymers, *Mikrochim. Acta Suppl.* 14, 809 (1997).

J.L. McCann, A. Rauk, and H. Wieser, Infrared Absorption and Vibrational Circular Dichroism Spectra of Poly (Vinyl Ether) Containing Diastereomeric Menthols as Pendants, *J. Mol. Struct.* 408 (409), 417 (1997).

J. McCann, A. Rauk, G.V. Shustov, H. Wieser, and D. Yang, Electronic and Vibrational Circular Dichroism of Model β -Lactams: 3-Methyl- and 4-Methylazetidin-2-One, *Appl. Spectrosc.* 50 (5), 630 (1996).

J. McCann, D. Tsankov, N. Hu, G. Liu, and H. Wieser, VCD Study of Synthetic Chiral Polymers: Stereoregularity in Poly-Menthyl Methacrylate, *J. Mol. Struct.* 349, 309 (1995).

O. McConnell, Y. He, L. Nogle, and A. Sarkahian, Application of Chiral Technology in a Pharmaceutical Company. Enantiomeric Separation and Spectroscopic Studies of Key Asymmetric Intermediates Using a Combination of Techniques. Phenylglycidols, *Chirality* 19 (9), 716 (2007).

C.A. McCoy and J.A.D. Haseth, Modified Phase Correction Algorithms in VCD/FT-IR Spectrometry, *Mikrochim. Acta* 1 (1–6), 97 (1988).

C.A. McCoy and J.A. Haseth, Phase Correction of Vibrational Circular Dichroic Features, *Appl. Spectrosc.* 42 (2), 336 (1988).

T.J. Measey and R. Schweitzer-Stenner, Aggregation of the Amphipathic Peptides $(AAKA)_N$ into Antiparallel, β-Sheets, *J. Am. Chem. Soc.* 128 (41), 13324 (2006).

B. Mennucci, C. Cappelli, R. Cammi, and J. Tomasi, Modeling Solvent Effects on Chiroptical Properties, *Chirality* 23 (9), 717 (2011).

B. Mennucci, R. Cammi, and J. Tomasi, Medium Effects on the Properties of Chemical Systems: Electric and Magnetic Response of Donor-Acceptor Systems within the Polarizable Continuum Model, *Int. J. Quantum Chem.* 75, 767 (1999).

C. Merten, T. Kowalik, and A. Hartwig, Vibrational Circular Dichroism Spectroscopy of Solid Polymer Films: Effects of Sample Orientation, *Appl. Spectrosc.* 62 (8), 901 (2008).

G. Mezo, A. Czajlik, M. Manea, A. Jakab, V. Farkas, Z. Majer, E. Vass, A. Bodor, B. Kapuvári, M. Boldizsár, B. Vincze, O. Csuka, M. Kovács, M. Przybylski, A. Perczel, and F. Hudecz, Structure, Enzymatic Stability and Antitumor Activity of Sea Lamprey GNRH-III and Its Dimer Derivatives, *Peptides* 28 (4), 806 (2007).

O. Michalski, W. Kisiel, K. Michalska, V. Setnicka, and M. Urbanova, Absolute Configuration and Conformational Analysis of Sesquiterpene Lactone Glycoside Studied by Vibrational Circular Dichroism Spectroscopy, *J. Mol. Struct.* 871 (1–3), 67 (2007).

H.M. Min, M. Aye, T. Taniguchi, N. Miura, K. Monde, K. Ohzawa, T. Nikai, M. Niwa, and Y. Takaya, A Structure and an Absolute Configuration of (+)-Alternamin, a New Coumarin from Murraya Alternans Having Antidote Activity against Snake Venom, *Tetrahedron Lett.* 48 (35), 6155 (2007).

D.J. Minick, R.D. Rutkowske, and L.A.D. Miller, Strategies for Successfully Applying Vibrational Circular Dichroism in a Pharmaceutical Research Environment, *Am. Pharm. Rev.* 10 (4), 118 (2007).

D.J. Minick, R.C.B. Copley, J.R. Szewczyk, R.D. Rutkowske, and L.A. Miller, An Investigation of the Absolute Configuration of the Potent Histamine H3 Receptor Antagonist Gt-2331 Using Vibrational Circular Dichroism, *Chirality* 19 (9), 731 (2007).

N. Miura, T. Taniguchi, K. Monde, and S.I. Nishimura, A Theoretical Study of α- and β-D-Glucopyranose Conformations by the Density Functional Theory, *Chem. Phys. Lett.* 419 (4–6), 326 (2006).

M. Miyazawa, K. Inouye, T. Hayakawa, Y. Kyogoku, and H. Sugeta, Vibrational Circular Dichroism of Proline-Containing Oligopeptides, *Appl. Spectrosc.* 50 (5), 644 (1996).

M. Miyazawa, Y. Kyogoku, and H. Sugeta, Vibrational Circular Dichroism Studies of Molecular Conformation and Association of Dipeptides, *Spectrochim. Acta* 50A (8–9), 1505 (1994).

P. Mobian, C. Nicolas, E. Francotte, T. Burgi, and J. Lacour, Synthesis, Resolution, and VCD Analysis of an Enantiopure Diazaoxatricornan Derivative, *J. Am. Chem. Soc.* 130 (20), 6507 (2008).

K. Monde, N. Miura, and T. Taniguchi, Absolute Configuration Determination Using Vibrational Circular Dichroism. Methods for Absolute Configuration Determination in the Solution State without Induction, *Kagaku Seibutsu* 45 (7), 455 (2007).

K. Monde, T. Taniguchi, N. Miura, C.S. Vairappan, and M. Suzuki, Absolute Configurations of Brominated Sesquiterpenes Determined by Vibrational Circular Dichroism, *Chirality* 18 (5), 335 (2006).

K. Monde, T. Taniguchi, N. Miura, C.S. Vairappan, and M. Suzuki, Absolute Configurations of Endoperoxides Determined by Vibrational Circular Dichroism (VCD), *Tetrahedron Lett.* 47 (26), 4389 (2006).

K. Monde, N. Miura, M. Hashimoto, T. Taniguchi, and T. Inabe, Conformational Analysis of Chiral Helical Perfluoroalkyl Chains by VCD, *J. Am. Chem. Soc.* 128 (18), 6000 (2006).

K. Monde, Vibrational Circular Dichroism: New Chiroptical Spectroscopy without UV-VIS Chromophores, *Kobunshi* 55 (7), 516 (2006).

K. Monde, T. Taniguchi, N. Miura, and S.I. Nishimura, Specific Band Observed in VCD Predicts the Anomeric Configuration of Carbohydrates, *J.Am. Chem. Soc.* 126 (31), 9496 (2004).

K. Monde, T. Taniguchi, N. Miura, S.I. Nishimura, N. Harada, R.K. Dukor, and L.A. Nafie, Preparation of Cruciferous Phytoalexin Related Metabolites, (–)-Dioxibrassinin and (–)-3-Cyanomethyl-3-Hydroxyoxindole, and Determination of Their Absolute Configurations by Vibrational Circular Dichroism (VCD), *Tetrahedron Lett.* 44 (32), 6017 (2003).

J.R.A. Moreno, F.P. Urena, and J.J.L. Gonzalez, Conformational Preference of a Chiral Terpene: Vibrational Circular Dichroism (VCD), Infrared and Raman Study of S-(–)-Limonene Oxide, *Phys. Chem. Chem. Phys.* 11 (14), 2459 (2009).

A. Moretto, F. Formaggio, B. Kaptein, Q.B. Broxterman, L. Wu, T.A. Keiderling, and C. Toniolo, First Homo-Peptides Undergoing a Reversible 3_{10}-Helix/α-Helix Transition: Critical Main-Chain Length, *Biopolymers (Peptide Sci.)* 90 (4), 567 (2008).

T. Mori, Y. Inoue, and S. Grimme, Quantum Chemical Study on the Circular Dichroism Spectra and Specific Rotation of Donor-Acceptor Cyclophanes, *J. Phys. Chem. A* 111 (32), 7995 (2007).

T. Mori, R.G. Weiss, and Y. Inoue, Mediation of Conformationally Controlled Photodecarboxylations of Chiral and Cyclic Aryl Esters by Substrate Structure, Temperature, Pressure, and Medium Constraints, *J. Am. Chem. Soc.* 126 (29), 8961 (2004).

T. Mori, H. Izumi, and Y. Inoue, Chiroptical Properties of Organic Radical Cations. The Electronic and Vibrational Circular Dichroism Spectra of α-Tocopherol Derivatives and Sterically Hindered Chiral Hydroquinone Ethers, *J. Phys. Chem. A* 108 (44), 9540 (2004).

H. Morimoto, I. Kinoshita, M. Mori, Y. Kyogoku, and H. Sugeta, Vibrational Circular Dichroism of Metal Complexes Containing Trans-1,2-Diaminocyclohexane, *Chem. Lett.* 18 (1), 73 (1989).

H.E. Morita, T.S. Kodama, and T. Tanaka, Chirality of Camphor Derivatives by Density Functional Theory, *Chirality* 18 (10), 783 (2006).

K. Morokuma and H. Sugeta, *Ab Initio* Derivative Calculation of Vibrational Circular Dichroism, *Chem. Phys. Lett.* 134 (1), 23 (1987).

M. Moskovits and A. Gohin, Vibrational Circular Dichroism: Effect of Charge Fluxes and Bond Currents, *J. Phys. Chem.* 86 (20), 3947 (1982).

O. Muehling and P. Wessig, Stereoselective Synthesis of Cyclopropanes Based on a 1,2-Chirality Transfer, *Chem. Eur. J.* 14 (26), 7951 (2008).

T. Mugishima, M. Tsuda, Y. Kasai, H. Ishiyama, E. Fukushi, J. Kawabata, M. Watanabe, K. Akao, and J. Kobayashi, Absolute Stereochemistry of Citrinadins a and B from Marine-Derived Fungus, *J. Org. Chem.* 70 (23), 9430 (2005).

P. Mukhopadhyay, P. Wipf, and D.N. Beratan, Optical Signatures of Molecular Dissymmetry: Combining Theory with Experiments to Address Stereochemical Puzzles, *Acc. Chem. Res.* 42 (6), 809 (2009).

M.A. Muñoz, O. Muñoz, and P. Joseph-Nathan, Absolute Configuration of Natural Diastereoisomers of 6 β-Hydroxyhyoscyamine by Vibrational Circular Dichroism, *J. Nat. Prod.* 69 (9), 1335 (2006).

L.A. Nafie, *Vibrational Optical Activity: Principles and Applications* (Wiley, UK) 2011.

L.A. Nafie, Vibrational Circular Dichroism: A New Tool for the Solution-State Determination of the Structure and Absolute Configuration of Chiral Natural Product Molecules, *Nat. Prod. Commun.* 3 (3), 451 (2008).

L.A. Nafie and R.K. Dukor, in *Applications of Vibrational Spectroscopy in Pharmaceutical Research and Development*, ed. D. Pivonka, J.M. Chalmers, and P.R. Griffiths (John Wiley & Sons, Chichester, 2007), p. 129.

L.A. Nafie and R.K. Dukor, in *Chiral Analysis*, ed. K.W. Busch and M.A. Busch (Elsevier, Amsterdam, 2006), p. 505.

L.A. Nafie, C. Guo, and R.K. Dukor, Reaction Monitoring of Chiral Molecules Using Fourier-Transform Infrared Vibrational Circular Dichroism Spectroscopy, *PCT Int. Appl.* 30 (2005).

L.A. Nafie, H. Buijs, A. Rilling, X. Cao, and R.K. Dukor, Dual Source Fourier Transform Polarization Modulation Spectroscopy: An Improved Method for the Measurement of Circular and Linear Dichroism, *Appl. Spectrosc.* 58 (6), 647 (2004).

L.A. Nafie, Theory of Vibrational Circular Dichroism and Infrared Absorption: Extension to Molecules with Low-Lying Excited Electronic States, *J. Phys. Chem. A* 108 (35), 7222 (2004).

L.A. Nafie, R.K. Dukor, J.R. Roy, A. Rilling, X. Cao, and H. Buijs, Observation of Fourier Transform Near-Infrared Vibrational Circular Dichroism to 6150 cm^{-1}, *Appl. Spectrosc.* 57 (10), 1245 (2003).

L.A. Nafie, Dual Polarization Modulation: A Real-Time, Spectral-Multiplex Separation of Circular Dichroism from Linear Birefringence Spectral Intensities, *Appl. Spectrosc.* 54 (11), 1634 (2000).

L.A. Nafie and T.B. Freedman, in *Circular Dichroism: Principles and Applications*, ed. N. Berova, K. Nakanishi, and R.W. Woody (Wiley, 2000), 2nd ed., p. 97.

L.A. Nafie and T.B. Freedman, Vibrational Circular Dichroism: An Incisive Tool for Stereochemical Applications, *Enantiomer* 3, 283 (1998).

L.A. Nafie, Electron Transition Current Density in Molecules. 1. Non-Born-Oppenheimer Theory of Vibronic and Vibrational Transitions, *J. Phys. Chem. A* 101 (42), 7826 (1997).

L.A. Nafie, X. Qu, F. Long, and T.B. Freedman, Optical Design and Sampling Methods for the Measurement of Vibrational Circular Dichroism with a Nicolet Magna FT-IR Spectrometer, *Mikrochim. Acta Suppl.* 14, 803 (1997).

L.A. Nafie, X. Qu, E. Lee, G.-S. Yu, and T.B. Freedman, Comparison of Fourier-Transform Vibrational Circular Dichroism and Multichannel-Detected Raman Optical Activity, *Mikrochim. Acta Suppl.* 14, 807 (1997).

L.A. Nafie, Infrared and Raman Vibrational Optical Activity: Theoretical and Experimental Aspects, *Annu. Rev. Phys. Chem.* 48, 357 (1997).

L.A. Nafie, Vibrational Optical Activity, *Appl. Spectrosc.* 50 (5), 14A (1996).

L.A. Nafie, Circular Polarization Spectroscopy of Chiral Molecules, *J. Mol. Struct.* 347, 83 (1995).

L.A. Nafie, G.-S. Yu, X. Qu, and T.B. Freedman, Comparison of IR and Raman Forms of Vibrational Optical Activity, *Faraday Discuss* 99, 13 (1994).

L.A. Nafie, M. Citra, N. Ragunathan, G.-S. Yu, and D. Che, Instrumental Methods of Infrared and Raman Vibrational Optical Activity, *Tech. Instrum. Anal. Chem.* 14, 53 (1994).

L.A. Nafie, Instrumental Techniques for Infrared and Raman Vibrational Optical Activity, *Proc. SPIE* 1681, 29 (1992).

L.A. Nafie, Velocity-Gauge Formalism in the Theory of Vibrational Circular Dichroism and Infrared Absorption, *J. Chem. Phys.* 96 (8), 5687 (1992).

L.A. Nafie and T.B. Freedman, Electronic Current Models of Vibrational Circular Dichroism, *J. Mol. Struct.* 224, 121 (1990).

L.A. Nafie, N.-S. Lee, G. Paterlini, and T.B. Freedman, Polarization Modulation Fourier Transform Infrared Spectroscopy, *Mikrochim. Acta* 3, 93 (1988).

L.A. Nafie, in *Advances in Applied Fourier Transform Infrared Spectroscopy*, ed. M.W. Mackenzie (Wiley, New York, 1988), p. 67.

L.A. Nafie, E.D. Lipp, A. Chernovitz, and G. Paterlini, Fourier Transform Infrared Vibrational Circular Dichroism in the Carbonyl Stretching Region of Polypeptides and Urethane Amino Acid Derivatives, *Polym. Sci. Technol.* 36, 81 (1987).

L.A. Nafie and T.B. Freedman, Vibrational Circular Dichroism Theory: Formulation Defining Magnetic Dipole Atomic Polar Tensors and Vibrational Nuclear Magnetic Shielding Tensors, *Chem. Phys. Lett.* 134 (3), 225 (1987).

L.A. Nafie and T.B. Freedman, Vibrational Circular Dichroism Spectroscopy, *Spectroscopy* 2 (12), 24 (1987).

L.A. Nafie and T.B. Freedman, Ring Current Mechanism of Vibrational Circular Dichroism, *J. Phys. Chem.* 90 (5), 763 (1986).

L.A. Nafie, E.D. Lipp, A. Farrell, and G. Paterlini, Fourier Transform Vibrational Circular Dichroism in the Carbonyl Region of Peptides and Polypeptides, *Polym. Prepr.* 25 (2), 145 (1984).

L.A. Nafie, in *Advances in Infrared and Raman Spectroscopy*, ed. R.J.H. Clark and R.E. Hester (Wiley, Heyden, 1984), vol. 11, p. 49.

L.A. Nafie, Adiabatic Molecular Properties beyond the Born Oppenheimer Approximation. Complete Adiabatic Wavefunctions and Vibrationally Induced Electronic Current Density, *J. Chem. Phys.* 79 (10), 4950 (1983).

L.A. Nafie, M.R. Oboodi, and T.B. Freedman, Vibrational Circular Dichroism in Amino Acids and Peptides. 8. A Chirality Rule for Methine C^*_α-H Stretching Modes, *J. Am. Chem. Soc.* 105 (25), 7449 (1983).

L.A. Nafie and T.B. Freedman, Vibronic Coupling Theory of Infrared Vibrational Transitions, *J. Chem. Phys.* 78 (12), 7108 (1983).

L.A. Nafie and D.W. Vidrine, in *Fourier Transform Infrared Spectroscopy*, ed. J.R. Ferraro and L.J. Basile (Academic Press, New York, 1982), vol. 3, p. 83.

L.A. Nafie, The Emergence and Exploration of Vibrational Optical Activity, *Appl. Spectrosc.* 36 (5), 489 (1982).

L.A. Nafie and T.B. Freedman, A Unified Approach to the Determination of Infrared and Raman Vibrational Optical Activity Intensities Using Localized Molecular Orbitals, *J. Chem. Phys.* 75 (10), 4847 (1981).

L.A. Nafie and P.L. Polavarapu, Localized Molecular Orbital Calculations of Vibrational Circular Dichroism. I. General Theoretical Formalism and CNDO Results for the Carbon-Deuterium Stretching Vibration in Neopentyl-1-D-Chloride, *J. Chem. Phys.* 75 (6), 2935 (1981).

L.A. Nafie, E.D. Lipp, and C.G. Zimba, Fourier Transform Infrared Circular Dichroism: A Double Modulation Approach, *Proc. SPIE* 289, 457 (1981).

L.A. Nafie, in *Vibrational Spectra and Structure*, ed. J.R. Durig (Elsevier, Amsterdam, 1981), vol. 10, p. 153.

L.A. Nafie, P.L. Polavarapu, and M. Diem, Vibrational Optical Activity in Perturbed Degenerate Modes: Concepts and Model Calculations in 1-Substituted Haloethanes, *J. Chem. Phys.* 73 (8), 3530 (1980).

L.A. Nafie and M. Diem, Theory of High Frequency Differential Interferometry: Application to the Measurement of Infrared Circular and Linear Dichroism via Fourier Transform Spectroscopy, *Appl. Spectrosc.* 33 (2), 130 (1979).

L.A. Nafie and D.W. Vidrine, Differential Absorption at High Modulation Frequencies Using a Fourier Transform Infrared Spectrometer, *Proc. SPIE* 191, 56 (1979).

L.A. Nafie, P.L. Prasad, H. Khouri, and E. Doorly, Vibrational Circular Dichroism in Peptides, *Polym. Prepr.* 20 (2), 85 (1979).

L.A. Nafie, M. Diem, and D.W. Vidrine, Fourier Transform Infrared Vibrational Circular Dichroism, *J. Am. Chem. Soc.* 101 (2), 496 (1979).

L.A. Nafie and M. Diem, Optical Activity in Vibrational Transitions: Vibrational Circular Dichroism and Raman Optical Activity, *Acc. Chem. Res.* 12 (8), 296 (1979).

L.A. Nafie and T.H. Walnut, Vibrational Circular Dichroism Theory: A Localized Molecular Orbital Model, *Chem. Phys. Lett.* 49 (3), 441 (1977).

L.A. Nafie, T.A. Keiderling, and P.J. Stephens, Vibrational Circular Dichroism, *J. Am. Chem. Soc.* 98 (10), 2715 (1976).

L.A. Nafie, J.C. Cheng, and P.J. Stephens, Vibrational Circular Dichroism of 2,2,2-Trifluoro-1-Phenylethanol, *J. Am. Chem. Soc.* 97 (13), 3842 (1975).

A. Nakahashi, N. Miura, K. Monde, and S. Tsukamoto, Stereochemical Studies of Hexylitaconic Acid, an Inhibitor of P53-HDM2 Interaction, *Bioorg. Med. Chem. Lett.* 19 (11), 3027 (2009).

A. Nakahashi, T. Taniguchi, N. Miura, and K. Monde, Stereochemical Studies of Sialic Acid Derivatives by Vibrational Circular Dichroism, *Org. Lett.* 9 (23), 4741 (2007).

K. Nakao, Y. Kyogoku, and H. Sugeta, Vibrational Circular Dichroism of the OH-Stretching Vibration in 2-2′-Dihydroxy-1,1′-Binaphthyl, *Faraday Discuss* 99, 77 (1994).

Y. Nakao, Y. Kyogoku, and H. Sugeta, Vibrational Circular Dichroism in Hydrogen Bond Systems. Part II. Vibrational Circular Dichroism of the OH Stretching Vibration in 2,2-Dimethyl-1,3-Dioxolane-4-Methanol, *J. Mol. Struct.* 146, 85 (1986).

Y. Nakao, H. Sugeta, and Y. Kyogoku, Vibrational Circular Dichroism of the OH Stretching Vibrations in Methyl 3-Hydroxybutyrate and Methyl Lactate, *Chem. Lett.* 13 (4), 623 (1984).

U. Narayanan, T.A. Keiderling, C.J. Elsevier, P. Vermeer, and W. Runge, Vibrational Circular Dichroism of Optically Active Allenes. Experimental Results, *J. Am. Chem. Soc.* 110 (13), 4133 (1988).

U. Narayanan and T.A. Keiderling, Vibrational Circular Dichroism of Optically Active Substituted Allenes. Calculational Results, *J. Am. Chem. Soc.* 110 (13), 4139 (1988).

U. Narayanan, T.A. Keiderling, G.M. Bonora, and C. Toniolo, Vibrational Circular Dichroism of Polypeptides. 7. Film and Solution Studies of β-Sheet-Forming Homooligopeptides, *J. Am. Chem. Soc.* 108 (9), 2431 (1986).

U. Narayanan, T.A. Keiderling, G.M. Bonora, and C. Toniolo, Vibrational Circular Dichroism of Polypeptides. IV. Film Studies of L-Alanine Homo-Oligopeptides, *Biopolymers* 24 (7), 1257 (1985).

U. Narayanan and T.A. Keiderling, Coupled Oscillator Interpretation of the Vibrational Circular Dichroism of Several Dicarbonyl-Containing Steroids, *J. Am. Chem. Soc.* 105 (21), 6406 (1983).

J.V. Naubron, L. Giordano, F. Fotiadu, T. Bürgi, N. Vanthuyne, C. Roussel, and G. Buono, Chromatographic Resolution, Solution and Crystal Phase Conformations, and Absolute Configuration of Tert-Butyl(Dimethylamino)Phenylphosphine-Borane Complex, *J. Org. Chem.* 71 (15), 5586 (2006).

F. Neese, Prediction of Molecular Properties and Molecular Spectroscopy with Density Functional Theory: From Fundamental Theory to Exchange-Coupling, *Coord. Chem. Rev.* 253 (5–6), 526 (2009).

V.P. Nicu and E.J. Baerends, On the Origin Dependence of the Angle Made by the Electric and Magnetic Vibrational Transition Dipole Moment Vectors, *Phys. Chem. Chem. Phys.* 13 (36), 16126 (2011).

V.P. Nicu, M. Heshmat, and E.J. Baerends, Signatures of Counter-Ion Association and Hydrogen Bonding in Vibrational Circular Dichroism Spectra, *Phys. Chem. Chem. Phys.* 13 (19), 8811 (2011).

V.P. Nicu and E.J. Baerends, Robust Normal Modes in Vibrational Circular Dichroism Spectra, *Phys. Chem. Chem. Phys.* 11 (29), 6107 (2009).

V.P. Nicu, E. Debie, W. Herrebout, B. Van Der Veken, P. Bultinck, and E.J. Baerends, A VCD Robust Mode Analysis of Induced Chirality: The Case of Pulegone in Chloroform, *Chirality* 21 (1E), E287 (2009).

V.P. Nicu, J. Autschbach, and E.J. Baerends, Enhancement of IR and VCD Intensities Due to Charge Transfer, *Phys. Chem. Chem. Phys.* 11 (10), 1526 (2009).

V.P. Nicu, J. Neugebauer, S.K. Wolff, and E.J. Baerends, A Vibrational Circular Dichroism Implementation within a Slater-Type-Orbital Based Density Functional Framework and Its Application to Hexa- and Hepta-Helicenes, *Theor. Chem. Acc.* 119 (1–3), 245 (2008).

V.P. Nicu, J. Neugebauer, and E.J. Baerends, Effects of Complex Formation on Vibrational Circular Dichroism Spectra, *J. Phys. Chem. A* 112 (30), 6978 (2008).

J.A. Nieman, B.A. Keay, M. Kubicki, D. Yang, A. Rauk, D. Tsankov, and H. Wieser, Determining Absolute Configuration by Vibrational Circular Dichroism: (+)-(1S, 5S, 6S)- and (–)-(1R,5R,6R)-Spiro[4.4]Nonane-1,6-Diol, *J. Org. Chem.* 60 (7), 1918 (1995).

M. Niemeyer, G.G. Hoffmann, and B. Schrader, New Application of the Step-Scan Lock-In Technique to Vibrational Circular Dichroism, *J. Mol. Struct.* 349, 451 (1995).

N. Norani, H. Rahemi, S.F. Tayyari, and M.J. Riley, Conformational Stabilities, Infrared, and Vibrational Dichroism Spectroscopy Studies of Tris(Ethylenediamine) Zinc (II) Chloride, *J. Mol. Model.* 15 (1), 25 (2009).

J. Nový, S. Böhm, J. Králová, V. Král, and M. Urbanová, Formation and Temperature Stability of G-Quadruplex Structures Studied by Electronic and Vibrational Circular Dichroism Spectroscopy Combined with *Ab Initio* Calculations, *Biopolymers* 89 (2), 144 (2008).

J. Nový, M. Urbanová, and K. Volka, Electronic and Vibrational Circular Dichroism Spectroscopic Study of Non-Covalent Interactions of Meso-5,10,15,20-Tetrakis (1-Methylpyridinium-4-Yl)Porphyrin with $(dG-dC)_{10}$ and $(dA-dT)_{10}$, *Vibrational Spectrosc.* 43 (1), 71 (2007).

J. Nový and M. Urbanová, Vibrational and Electronic Circular Dichroism Study of the Interactions of Cationic Porphyrins with $(dG-dC)_{10}$ and $(dA-dT)_{10}$, *Biopolymers* 85 (4), 349 (2007).

J. Nový, M. Urbanová, and K. Volka, Vibrational and Electronic Circular Dichroism and Absorption Spectral Study of the DNA-5,10,15,20-Tetrakis(1-Methylpyridinium-4-Yl) Porphyrin Interaction, *J. Mol. Struct.* 748 (1–3), 17 (2005).

J. Nový, M. Urbanová, and K. Volka, The DNA-Porphyrin Interactions Studied by Vibrational and Electronic Circular Dichroism Spectroscopy, *Collection Czechoslovak Chem. Commun.* 70 (11), 1799 (2005).

M.R. Oboodi, B.B. Lal, D.A. Young, T.B. Freedman, and L.A. Nafie, Vibrational Circular Dichroism in Amino Acids and Peptides. 9. Carbon-Hydrogen Stretching Spectra of the Amino Acids and Selected Transition-Metal Complexes, *J. Am. Chem. Soc.* 107 (6), 1547 (1985).

K.I. Oh, J. Han, K.K. Lee, S. Hahn, H. Han, and M. Cho, Site-Specific Hydrogen-Bonding Interaction between N-Acetylproline Amide and Protic Solvent Molecules: Comparisons of IR and VCD Measurements with MD Simulations, *J. Phys. Chem. A* 110 (50), 13355 (2006).

S. Ostrowski, M.H. Jamróz, J.E. Rode, and J.C. Dobrowolski, On Stability, Chirality Measures, and Theoretical VCD Spectra of the Chiral $C_{58}X_2$ Fullerenes (X = N, B), *J. Phys. Chem. A* 116 (1), 631 (2012).

L. Palivec, M. Urbanová, and K. Volka, Circular Dichroism Spectroscopic Study of Non-Covalent Interactions of Poly-L-Glutamic Acid with a Porphyrin Derivative in Aqueous Solutions, *J. Peptide Sci.* 11 (9), 536 (2005).

P. Pancoska, V. Janota, and T.A. Keiderling, Novel Matrix Descriptor for Secondary Structure Segments in Proteins: Demonstration of Predictability from Circular Dichroism Spectra, *Anal. Biochem.* 267, 72 (1999).

P. Pancoska, V. Janota, and T.A. Keiderling, Interconvertibility of Electronic and Vibrational Circular Dichroism Spectra of Proteins: A Test of Principle Using Neural Network Mapping, *Appl. Spectrosc.* 50 (5), 658 (1996).

P. Pancoska, H. Fabian, G. Yoder, V. Baumruk, and T.A. Keiderling, Protein Structural Segments and Their Interconnections Derived from Optical Spectra. Thermal Unfolding of Ribonuclease T_1 as an Example, *Biochemistry* 35 (40), 13094 (1996).

P. Pancoska, E. Bitto, V. Janota, M. Urbanova, V.P. Gupta, and T.A. Keiderling, Comparison of and Limits of Accuracy for Statistical Analyses of Vibrational and Electronic Circular Dichroism Spectra in Terms of Correlations to and Predictions of Protein Secondary Structure, *Protein Sci.* 4 (7), 1384 (1995).

P. Pancoska, L. Wang, and T.A. Keiderling, Frequency Analysis of Infrared Absorption and Vibrational Circular Dichroism of Proteins in D_2O Solution, *Protein Sci.* 2 (3), 411 (1993).

P. Pancoska, S.C. Yasui, and T.A. Keiderling, Statistical Analyses of the Vibrational Circular Dichroism of Selected Proteins and Relationship to Secondary Structures, *Biochemistry* 30 (20), 5089 (1991).

P. Pancoska and T.A. Keiderling, Systematic Comparison of Statistical Analyses of Electronic and Vibrational Circular Dichroism for Secondary Structure Prediction of Selected Proteins, *Biochemistry* 30 (28), 6885 (1991).

P. Pancoska, S.C. Yasui, and T.A. Keiderling, Enhanced Sensitivity to Conformation in Various Proteins. Vibrational Circular Dichroism Results, *Biochemistry* 28 (14), 5917 (1989).

P. Pancoska, S.C. Yasui, and T.A. Keiderling, Vibrational Circular Dichroism of Proteins, *Proc. SPIE* 1145, 154 (1989).

A.H. Pandith, N. Islam, Z.F. Syed, S. Rehman, S. Bandaru, and A. Anoop, Density Functional Theory Prediction of Geometry and Vibrational Circular Dichroism of Bridged Triarylamine Helicenes, *Chem. Phys. Lett.* 516 (4–6), 199 (2011).

A. Pandyra, D. Tsankov, V. Andrushchenko, J.H. Van de Sande, and H. Wieser, Intercalation of Daunomycin into $d(CG)_4$ Oligomer Duplex Containing G·T Mismatches by Vibrational Circular Dichroism and Infrared Absorption Spectroscopy, *Biopolymers* 82 (3), 189 (2006).

A.A. Pandyra, A.P. Yamniuk, V.V. Andrushchenko, H. Wieser, and H.J. Vogel, Isotope-Labeled Vibrational Circular Dichroism Studies of Calmodulin and Its Interactions with Ligands, *Biopolymers* 79 (5), 231 (2005).

F. Partal Urena, J.R.A. Moreno, and J.J. Lopez Gonzalez, Conformational Study of (R)-(+)-Limonene in the Liquid Phase Using Vibrational Spectroscopy (IR, Raman, and VCD) and DFT Calculations, *Tetrahedron Asymm.* 20 (1), 89 (2009).

F. Partal Urena, J.-R. Aviles Moreno, and J.J. Lopez Gonzalez, Conformational Flexibility in Terpenes: Vibrational Circular Dichroism (VCD), Infrared and Raman Study of S-(–)-Perillaldehyde, *J. Phys. Chem. A* 112 (34), 7887 (2008).

M.G. Paterlini, T.B. Freedman, L.A. Nafie, Y. Tor, and A. Shanzer, Vibrational CD Studies of Interchain Hydrogen-Bonded Tripodal Peptides, *Biopolymers* 32 (7), 765 (1992).

M.G. Paterlini, T.B. Freedman, L.A. Nafie, Y. Tor, and A. Shanzer, Vibrational Circular Dichroism Studies of Interchain Hydrogen Bonding in Tripodal Peptide Molecules, *Proc. SPIE* 1145, 150 (1989).

M.G. Paterlini, T.B. Freedman, and L.A. Nafie, Ring Current Enhanced Vibrational Circular Dichroism in the CH-Stretching Motions of Sugars, *J. Am. Chem. Soc.* 108, 1389 (1986).

M.G. Paterlini, T.B. Freedman, and L.A. Nafie, Vibrational Circular Dichroism Spectra of Three Conformationally Distinct States and an Unordered State of Poly(L-Lysine) in Deuterated Aqueous Solution, *Biopolymers* 25 (9), 1751 (1986).

A.G. Petrovic, P.L. Polavarapu, R. Mahalakshimi, and P. Balaram, Characterization of Folded Conformations in a Tetrapeptide Containing Two Tryptophan Residues by Vibrational Circular Dichroism, *Chirality* 21 (1E), E76 (2009).

A.G. Petrovic, S.E. Vick, and P.L. Polavarapu, Determination of the Absolute Stereochemistry of Chiral Biphenanthryls in Solution Phase Using Chiroptical Spectroscopic Methods: 2,2′-Diphenyl-[3,3′-Biphenanthrene]-4,4′-Diol, *Chirality* 20 (3/4), 501 (2008).

A.G. Petrovic, P.L. Polavarapu, J. Drabowicz, P. Lyzwa, M. Mikolajczyk, W. Wieczorek, and A. Balinska, Diastereomers of N-α-Phenylethyl-T-Butylsulfinamide: Absolute Configurations and Predominant Conformations, *J. Org. Chem.* 73 (8), 3120 (2008).

A.G. Petrovic and P.L. Polavarapu, The Quadruplex-Duplex Structural Transition of Polyriboguanylic Acid, *J. Phys. Chem. B* 112 (7), 2245 (2008).

A.G. Petrovic and P.L. Polavarapu, Quadruplex Structure of Polyriboinosinic Acid: Dependence on Alkali Metal Ion Concentration, pH and Temperature, *J. Phys. Chem. B* 112 (7), 2255 (2008).

A.G. Petrovic and P.L. Polavarapu, Chiroptical Spectroscopic Determination of Molecular Structures of Chiral Sulfinamides: T-Butanesulfinamide, *J. Phys. Chem. A* 111 (43), 10938 (2007).

A.G. Petrovic and P.L. Polavarapu, Structural Transitions in Polyribocytidylic Acid Induced by Changes in pH and Temperature: Vibrational Circular Dichroism Study in Solution and Film States, *J. Phys. Chem. B* 110 (45), 22826 (2006).

A.G. Petrovic, P.L. Polavarapu, J. Drabowicz, Y. Zhang, O.J. McConnell, and H. Duddeck, Absolute Configuration of C_2-Symmetric Spiroselenurane: 3,3,3′,3′-Tetramethyl-1,1′-Spirobi[3h,2,1]Benzoxaselenole, *Chem. Eur. J.* 11 (14), 4257 (2005).

A.G. Petrovic, J. He, P.L. Polavarapu, L.S. Xiao, and D.W. Armstrong, Absolute Configuration and Predominant Conformations of 1,1-Dimethyl-2-Phenylethyl Phenyl Sulfoxide, *Org. Biomol. Chem.* 3 (10), 1977 (2005).

A.G. Petrovic, P.K. Bose, and P.L. Polavarapu, Vibrational Circular Dichroism of Carbohydrate Films Formed from Aqueous Solutions, *Carbohyd.Res.* 339 (16), 2713 (2004).

H. Pham-Tuan, C. Larsson, F. Hoffmann, A. Bergman, M. Froeba, and H. Huhnerfuss, Enantioselective Semipreparative HPLC Separation of PCB Metabolites and Their Absolute Structure Elucidation Using Electronic and Vibrational Circular Dichroism, *Chirality* 17 (5), 266 (2005).

S.T. Pickard, H.E. Smith, P.L. Polavarapu, T.M. Black, A. Rauk, and D. Yang, Synthesis, Experimental and *Ab Initio* Theoretical Vibrational Circular Dichroism and Absolute Configurations of Substituted Oxiranes, *J. Am. Chem. Soc.* 114 (17), 6850 (1992).

G. Pohl, A. Perczel, E. Vass, G. Magyarfalvi, and G. Tarczay, A Matrix Isolation Study on Ac-L-Pro-Nh₂: A Frequent Structural Element of β- and γ-Turns of Peptides and Proteins, *Tetrahedron* 64 (9), 2126 (2008).

G. Pohl, A. Perczel, E. Vass, G. Magyarfalvi, and G. Tarczay, A Matrix Isolation Study on Ac-Gly-NHMe and Ac-L-Ala-NHMe, the Simplest Chiral and Achiral Building Blocks of Peptides and Proteins, *Phys. Chem. Chem. Phys.* 9 (33), 4698 (2007).

P.L. Polavarapu, E.A. Donahue, G. Shanmugam, G. Scalmani, E.K. Hawkins, C. Rizzo, I. Ibnusaud, G. Thomas, D. Habel, and D. Sebastian, A Single Chiroptical Spectroscopic Method May Not Be Able to Establish the Absolute Configurations of Diastereomers: Dimethylesters of Hibiscus and Garcinia Acids, *J. Phys. Chem. A* 115 (22), 5665 (2011).

P.L. Polavarapu, J. Frelek, and M. Woźnica, Determination of the Absolute Configurations Using Electronic and Vibrational Circular Dichroism Measurements and Quantum Chemical Calculations, *Tet. Asymm.* 22 (18–19), 1720 (2011).

P.L. Polavarapu, G. Scalmani, E.K. Hawkins, C. Rizzo, N. Jeirath, I. Ibnusaud, D. Habel, D.S. Nair, and S. Haeema, Importance of Solvation in Understanding the Chiroptical Spectra of Natural Products in Solution Phase: Garcinia Acid Dimethyl Ester, *J. Nat. Prod.* 74 (3), 321 (2011).

P.L. Polavarapu and G. Shanmugam, Comparison of Mid-Infrared Fourier Transform Vibrational Circular Dichroism Measurements with Single and Dual Polarization Modulations, *Chirality* 23 (9), 801 (2011).

P.L. Polavarapu, N. Jeirath, T. Kurtán, G. Pescitelli, and K. Krohn, Determination of the Absolute Configurations at Stereogenic Centers in the Presence of Axial Chirality, *Chirality* 21 (1E), E202 (2009).

P.L. Polavarapu, N. Jeirath, and S. Walia, Conformational Sensitivity of Chiroptical Spectroscopic Methods: 6,6'-Dibromo-1,1'-Bi-2-Naphthol, *J. Phys. Chem. A* 113 (18), 5423 (2009).

P.L. Polavarapu, Why Is It Important to Simultaneously Use More Than One Chiroptical Spectroscopic Method for Determining the Structures of Chiral Molecules? *Chirality* 20 (5), 664 (2008).

P.L. Polavarapu, Renaissance in Chiroptical Spectroscopic Methods for Molecular Structure Determination, *Chem. Rec.* 7 (2), 125 (2007).

P.L. Polavarapu, Quantum Mechanical Predictions of Chiroptical Vibrational Properties, *Int. J. Quantum Chem.* 106 (8), 1809 (2006).

P.L. Polavarapu and J. He, Chiral Analysis Using Mid-IR Vibrational CD Spectroscopy, *Anal. Chem.* 76 (3), 61A (2004).

P.L. Polavarapu and C. Zhao, Vibrational Circular Dichroism: A New Spectroscopic Tool for Biomolecular Structural Determination, *J. Anal. Chem.* 366 (6–7), 727 (2000).

P.L. Polavarapu, C. Zhao, and K. Ramig, Vibrational Circular Dichroism, Absolute Configuration and Predominant Conformations of Volatile Anesthetics: 1,2,2,2-Tetrafluoroethyl Methyl Ether, *Tetrahedron Asymm.* 10 (6), 1099 (1999).

P.L. Polavarapu, C. Zhao, A.L. Cholli, and G.G. Vernice, Vibrational Circular Dichroism, Absolute Configuration, and Predominant Conformations of Volatile Anesthetics: Desflurane, *J. Phys. Chem. B* 103, 6127 (1999).

P.L. Polavarapu and Z. Deng, Measurement of Vibrational Circular Dichroism below ~600 cm⁻¹: Progress Towards Meeting the Challenge, *Appl. Spectrosc.* 50 (5), 686 (1996).

P.L. Polavarapu, Z. Deng, and G.-C. Chen, Polarization-Division Interferometry: Time-Resolved Infrared Vibrational Dichroism Spectroscopy, *Appl. Spectrosc.* 49 (2), 229 (1995).

P.L. Polavarapu, Z. Deng, and C.S. Ewig, Vibrational Properties of the Peptide Group: Achiral and Chiral Conformers of N-Methylacetamide, *J. Phys. Chem.* 98 (39), 9919 (1994).

P.L. Polavarapu, G.C. Chen, and S. Weibel, Development, Justification, and Applications of a Mid-Infrared Polarization-Division Interferometer, *Appl. Spectrosc.* 48 (10), 1224 (1994).

P.L. Polavarapu and G.C. Chen, Polarization-Division Interferometry: Far-Infrared Dichroism, *Appl. Spectrosc.* 48 (11), 1410 (1994).

P.L. Polavarapu, New Spectroscopic Tool: Absolute Configuration Determination of Pharmaceutical Compounds by Vibrational Circular Dichroism, *Spectroscopy* 9 (9), 48 (1994).

P.L. Polavarapu, S.T. Pickard, H.E. Smith, and R.S. Pandurangi, Determination of Absolute Configurations from Vibrational Circular Dichroism: (+)-2-Methylthiirane-3,3-D₂, *Talanta* 40 (5), 687 (1993).

P.L. Polavarapu, A.L. Cholli, and G. Vernice, Determination of Absolute Configurations and Predominant Conformations of General Inhalation Anesthetics: Desflurane, *J. Pharm. Sci.* 82 (8), 791 (1993).

P.L. Polavarapu, A.L. Cholli, and G. Vernice, Absolute Configuration of Isoflurane, *J. Am.Chem. Soc.* 114 (27), 10953 (1992).

P.L. Polavarapu, S.T. Pickard, H.E. Smith, T.M. Black, A. Rauk, and D. Yang, Vibrational Circular Dichroism and Absolute Configuration of Substituted Thiiranes, *J. Am. Chem. Soc.* 113 (26), 9747 (1991).

P.L. Polavarapu, P.K. Bose, and S.T. Pickard, Vibrational Circular Dichroism in Methylthiirane: *Ab Initio* Localized Molecular Orbital Predictions and Experimental Measurements, *J. Am. Chem. Soc.* 113 (1), 43 (1991).

P.L. Polavarapu and P.K. Bose, *Ab Initio* Localized Molecular Orbital Predictions of Vibrational Circular Dichroism: Trans-1,2-Dideuteriocyclopropane, *J. Phys. Chem.* 95 (4), 1606 (1991).

P.L. Polavarapu, P.G. Quincey, and J.R. Birch, Circular Dichroism in the Far Infrared and Millimetre Wavelength Regions: Preliminary Measurements, *Infrared Phys.* 30 (2), 175 (1990).

P.L. Polavarapu and P.K. Bose, *Ab Initio* Localized Molecular Orbital Predictions of Vibrational Circular Dichroism: Trans-2,3-Dideuterooxirane, *J. Chem. Phys.* 93 (10), 7524 (1990).

P.L. Polavarapu, Achromatic Long Wavelength Polarization Modulator, *Proc. SPIE* 1166, 472 (1990).

P.L. Polavarapu, Spectroscopy with Triangular Modulation, *Proc. SPIE* 1145, 174 (1989).

P.L. Polavarapu, Rotational-Vibrational Circular Dichroism, *Chem. Phys. Lett.* 161 (6), 485 (1989).

P.L. Polavarapu, New Developments in Fourier Transform Infrared Vibrational Circular Dichroism Measurements, *Appl. Spectrosc.* 43 (8), 1295 (1989).

P.L. Polavarapu, Far Infrared Circular Dichroism Measurements with Martin-Puplett Interferometer: Methods and Analysis, *Infrared Phys.* 28 (2), 109 (1988).

P.L. Polavarapu, B.A. Hess Jr., L.J. Schaad, D.O. Henderson, L.P. Fontana, H.E. Smith, L.A. Nafie, T.B. Freedman, and W.M. Zuk, Vibrational Spectra of Methylthiirane, *J. Chem. Phys.* 86 (3), 1140 (1987).

P.L. Polavarapu, C.S. Ewig, and T. Chandramouly, Conformations of Tartaric Acid and Its Esters, *J. Am. Chem. Soc.* 109 (24), 7382 (1987).

P.L. Polavarapu, Absorption and Circular Dichroism Due to Bending Vibrations of A_2B_2 Molecules with C_2 Symmetry, *J. Chem. Phys.* 87 (8), 4419 (1987).

P.L. Polavarapu, L.P. Fontana, and H.E. Smith, Vibrational Circular Dichroism of Phenylcarbinols. A Configurational Correlation, *J. Am. Chem. Soc.* 108 (1), 94 (1986).

P.L. Polavarapu, Vibrational Circular Dichroism of A_2B_2 Molecules, *J. Chem. Phys.* 85 (10), 6245 (1986).

P.L. Polavarapu, A Sum Rule for Vibrational Circular Dichroism Intensities of Fundamental Transitions, *J. Chem. Phys.* 84 (1), 542 (1986).

P.L. Polavarapu, B.A. Hess, and L.J. Schaad, Vibrational Spectra of Epoxypropane, *J. Chem. Phys.* 82 (4), 1705 (1985).

P.L. Polavarapu, in *Fourier Transform Infrared Spectroscopy*, ed. J.R. Ferraro and L.J. Basile (Academic Press, Orlando, FL, 1985), vol. 4, p. 61.

P.L. Polavarapu, D.F. Michalska, J.R. Neergaard, and H.E. Smith, Vibrational Circular Dichroism of 3-Phenylethylamine, *J. Am. Chem. Soc.* 106 (11), 3378 (1984).

P.L. Polavarapu, D.F. Michalska, and D.M. Back, On the Measurement of Vibrational Circular Dichroism of a Single Enantiomer on Fourier Transform Infrared Spectrometers, *Appl. Spectrosc.* 38 (3), 438 (1984).

P.L. Polavarapu and D.F. Michalska, Mid-Infrared Vibrational Circular Dichroism in (S)-(−)-Epoxypropane. Bond Moment Model Predictions and Comparison to the Experimental Results, *Mol. Phys.* 52 (5), 1225 (1984).

P.L. Polavarapu, in *Vibrational Spectra Struct.* (1984), vol. 13, p. 103.

P.L. Polavarapu, Circular Dichroism Measurements Down to 600 cm⁻¹, *Appl. Spectrosc.* 38 (1), 26 (1984).

P.L. Polavarapu and D.F. Michalska, Vibrational Circular Dichroism in (S)-(–)-Epoxypropane. Measurement in Vapor Phase and Verification of the Perturbed Degenerate Mode Theory, *J. Am. Chem. Soc.* 105 (19), 6190 (1983).

P.L. Polavarapu, A Comparison of Bond Moment and Charge Flow Models for Vibrational Circular Dichroism Intensities, *Mol. Phys.* 49 (4), 645 (1983).

P.L. Polavarapu, Electric Field Perturbative Calculations of Vibrational Intensities. II. Orbital Polarizability Model for Raman Optical Activity and Simultaneous Evaluation of Raman Optical Activity, Raman Scattering, Infrared Circular Dichroism, and Infrared Absorption Intensities, *J. Chem. Phys.* 77 (5), 2273 (1982).

P.L. Polavarapu, L.A. Nafie, S.A. Benner, and T.H. Morton, Optical Activity Due to Isotopic Substitution. Vibrational Circular Dichroism and the Absolute Configurations of α-deuterated Cyclohexanones, *J. Am. Chem. Soc.* 103 (18), 5349 (1981).

P.L. Polavarapu and L.A. Nafie, Localized Molecular Orbital Calculations of Vibrational Circular Dichroism. II. CNDO Formulation and Results for the Hydrogen Stretching Vibrations of (+)-(3r)-Methylcyclohexanone, *J. Chem. Phys.* 75 (6), 2945 (1981).

P.L. Polavarapu and J. Chandrasekhar, An Efficient Molecular Orbital Approach for a Simultaneous Evaluation of the Vibrational Circular Dichroism and Absorption Intensities, *Chem. Phys. Lett.* 84 (3), 587 (1981).

P.L. Polavarapu and L.A. Nafie, Vibrational Optical Activity: Comparison of Theoretical and Experimental Results for (+)-(3r)-Methylcyclohexanone, *J. Chem. Phys.* 73 (4), 1567 (1980).

P.L. Polavarapu, M. Diem, and L.A. Nafie, Vibrational Optical Activity in Para-Substituted 1-Methylcyclohex-1-Enes, *J. Am. Chem. Soc.* 102 (17), 5449 (1980).

A.M. Polyanichko, E.V. Chikhirzhina, V.V. Andrushchenko, V.I. Vorob'ev, and H. Wieser, The Effect of Manganese(II) on the Structure of DNA/HMGB1/H1 Complexes: Electronic and Vibrational Circular Dichroism Studies, *Biopolymers* 83 (2), 182 (2006).

A. Polyanichko and H. Wieser, Fourier Transform Infrared/Vibrational Circular Dichroism Spectroscopy as an Informative Tool for the Investigation of Large Supramolecular Complexes of Biological Macromolecules, *Biopolymers* 78 (6), 329 (2005).

A.M. Polyanichko, V.V. Andrushchenko, E.V. Chikhirzhina, V.I. Vorob'ev, and H. Wieser, The Effect of Manganese (II) on DNA Structure: Electronic and Vibrational Circular Dichroism Studies, *Nucleic Acids Res.* 32 (3), 989 (2004).

V. Pultz, S. Abbate, L. Laux, H.A. Havel, J. Overend, A. Moscowitz, and H.S. Mosher, Vibrational Circular Dichroism of (R)-(–)-Neopentyl-1-D Chloride and (R)-(–)-Neopentyl-1-D Bromide, *J. Phys. Chem.* 88 (3), 505 (1984).

X. Qu, E. Lee, G.-S. Yu, T.B. Freedman, and L.A. Nafie, Quantitative Comparison of Experimental Infrared and Raman Optical Activity Spectra, *Appl. Spectrosc.* 50 (5), 649 (1996).

N. Ragunathan, N.-S. Lee, T.B. Freedman, L.A. Nafie, C. Tripp, and H. Buijs, Measurement of Vibrational Circular Dichroism Using a Polarizing Michelson Interferometer, *Appl. Spectrosc.* 44 (1), 5 (1990).

N. Ragunathan, L.A. Nafie, and T.B. Freedman, Circular Dichroism Measurement Using Fourier Transform Interferometry, *Proc. SPIE* 1145, 148 (1989).

K. Ramnarayan, H.G. Bohr, and K.J. Jalkanen, Classification of Protein Fold Classes by Knot Theory and Prediction of Folds by Neural Networks: A Combined Theoretical and Experimental Approach, *Theor. Chem. Acc.* 119 (1–3), 265 (2008).

C. Rank, R.K. Phipps, P. Harris, P. Fristrup, T.O. Larsen, and C.H. Gotfredsen, Novofumigatonin, a New Orthoester Meroterpenoid from *Aspergillus novofumigatus*, *Org. Lett.* 10 (3), 401 (2008).

A. Rauk, J.L. McCann, H. Wieser, P. Bour, Y.I. El'natanov, and R.G. Kostyanovsky, Skeletal Vibrational Circular Dichroism of a Series of Bicyclic Dilactones: The Fingerprint Region? *Can. J. Chem.* 76 (6), 717 (1998).

A. Rauk, D. Yang, D. Tsankov, H. Wieser, Y. Koltypin, A. Gedanken, and G.V. Shustov, Chiroptical Properties of 1-Azabicyclo[3.1.0]Hexane in the Vacuum-UV and IR Regions, *J. Am. Chem. Soc.* 117, 4160 (1995).

A. Rauk and T.B. Freedman, Chiroptical Techniques and Their Relationship to Biological Molecules, Big or Small, *Int. J. Quantum Chem.* 28, 315 (1994).

A. Rauk, T. Eggimann, H. Wieser, G.V. Shustov, and D. Yang, The Vibrational Circular Dichroism Spectra of 2-Methylaziridine: Dominance of the Asymmetric Center at Nitrogen, *Can. J. Chem.* 72 (3), 506 (1994).

A. Rauk and D. Yang, Vibrational Circular Dichroism and Infrared Spectra of 2-Methyloxirane and Trans-2,3-Dimethyloxirane: *Ab Initio* Vibronic Coupling Theory with the 6-31g$*^{(0.3)}$ Basis Set, *J. Phys. Chem.* 96 (1), 437 (1992).

A. Rauk, in *New Developments in Molecular Chirality*, ed. P.G. Mezey (Kluwer, Dordrecht, Netherlands, 1991), vol. 5, p. 57.

A. Rauk, R. Dutler, and D. Yang, Infrared and Vibrational Circular Dichroism Intensities of Model Systems CH_3OH, CH_3NH_2, NH_2NH_2, NH_2OH, and HOOH and the Deuterated Species, ND_2ND_2, DOOH, and DOOD: A Theoretical Study Using the Vibronic Coupling Formalism, *Can. J. Chem.* 68 (2), 258 (1990).

U.M. Reinscheid, Determination of the Absolute Configuration of Two Estrogenic Nonylphenols in Solution by Chiroptical Methods, *J. Mol. Struct.* 918 (1–3), 14 (2009).

H. Rhee, Y.-G. June, J.-S. Lee, K.-K. Lee, J.-H. Ha, Z.H. Kim, S.-J. Jeon, and M. Cho, Femtosecond Characterization of Vibrational Optical Activity of Chiral Molecules, *Nature (London, UK)* 458 (7236), 310 (2009).

H. Rhee, Y.-G. June, Z.H. Kim, S.-J. Jeon, and M. Cho, Phase Sensitive Detection of Vibrational Optical Activity Free-Induction-Decay: Vibrational CD and ORD, *J. Opt. Soc. Am. B Opt. Phys.* 26 (5), 1008 (2009).

G.M. Roberts, O. Lee, J. Calienni, and M. Diem, Infrared Vibrational Circular Dichroism in the Amide III Spectral Region of Peptides, *J. Am. Chem. Soc.* 110 (6), 1749 (1988).

J.E. Rode and J.C. Dobrowolski, On Chirality Transfer in Electron Donor-Acceptor Complexes. A Prediction for the Sulfinimine···BF_3 System, *Chirality* 24 (1), 5 (2012).

J.E. Rode and J.C. Dobrowolski, Density Functional IR, Raman and VCD Spectra of Halogen Substituted β-Lactams, *J. Mol. Struct.* 651–653, 705 (2003).

J.E. Rode and J.C. Dobrowolski, Chiral Allenes: Theoretical VCD and IR Spectra, *J. Mol. Struct. (Theochem)* 635, 151 (2003).

J.E. Rode and J.C. Dobrowolski, VCD Technique in Determining Intermolecular H-Bond Geometry: A DFT Study, *J. Mol. Struct. (Theochem)* 637, 81 (2003).

I.M. Romaine, J.E. Hempel, G. Shanmugam, H. Hori, Y. Igarashi, P.L. Polavarapu, and G.A. Sulikowski, Assignment and Stereocontrol of Hibarimicin Atropoisomers, *Org. Lett.* 13 (17), 4538 (2011).

T.R. Rudd, R.J. Nichols, and E.A. Yates, Selective Detection of Protein Secondary Structural Changes in Solution Protein-Polysaccharide Complexes Using Vibrational Circular Dichroism (VCD), *J. Am. Chem. Soc.* 130 (7), 2138 (2008).

A. Rupprecht, A Modified Sum Rule Equation for the Atomic Tensors Occurring in the Theory of Vibrational Circular Dichroism (VCD). Application to the Atomic Tensors Recently Calculated for Ammonia, *J. Mol. Struct.* 272, 35 (1992).

A. Rupprecht, Application of Intensity Sum Rules to the Atomic Tensors and the Dipole and Rotational Strengths Recently Calculated for NHDT, *Acta Chem. Scand.* 45, 439 (1991).

A. Rupprecht, Sum Rules for the Rotational Strength of Fundamental Transitions in Vibrational Circular Dichroism (VCD), *Mol. Phys.* 63 (5), 955 (1988).

A. Rupprecht, On the Atomic Polar Tensor (APT) Approach for Vibrational Circular Dichroism (VCD) and Raman Optical Activity (ROA), *Mol. Phys.* 63 (5), 951 (1988).

J. Sadlej, J.C. Dobrowolski, J.E. Rode, and M.H. Jamroz, Density Functional Theory Study on Vibrational Circular Dichroism as a Tool for Analysis of Intermolecular Systems: (1:1) Cysteine-Water Complex Conformations, *J. Phys. Chem. A* 111 (42), 10703 (2007).

J. Sadlej, J.C. Dobrowolski, J.E. Rode, and M.H. Jamroz, DFT Study of Vibrational Circular Dichroism Spectra of D-Lactic Acid-Water Complexes, *Phys. Chem. Chem. Phys.* 8 (1), 101 (2006).

K. Salazar-Salinas, L.A. Jauregui, C. Kubli-Garfias, and J.M. Seminario, Molecular Biosensor Based on a Coordinated Iron Complex, *J. Chem. Phys.* 130 (10), 105101/1 (2009).

W.R. Salzman, Calculation of the Molecular G Tensor from Data for the Calculation of Rotational Strengths in Vibrational Circular Dichroism, *J. Phys. Chem.* 93 (21), 7351 (1989).

H. Sato, F. Sato, M. Taniguchi, and A. Yamagishi, Chirality Effects on Core-Periphery Connection in a Star-Burst Type Tetranuclear Ru(III) Complex: Application of Vibrational Circular Dichroism Spectroscopy, *Dalton Trans.* 41 (6), 1709 (2012).

H. Sato, Y. Mori, Y. Fukuda, and A. Yamagishi, Syntheses and Vibrational Circular Dichroism Spectra of the Complete Series of [Ru((–)- or (+)-tfac)$_n$(Acac)$_{3-n}$] (n = 0~3, Tfac = 3-Trifluoroacetylcamphorato and acac = Acetylacetonato), *Inorg. Chem.* 48 (10), 4354 (2009).

H. Sato, K. Hori, T. Sakurai, and A. Yamagishi, Long Distance Chiral Transfer in a Gel: Experimental and *Ab Initio* Analyses of Vibrational Circular Dichroism Spectra of R- and S-12-Hydroxyoctadecanoic Acid Gels, *Chem. Phys. Lett.* 467 (1–3), 140 (2008).

H. Sato, T. Taniguchi, A. Nakahashi, K. Monde, and A. Yamagishi, Effects of Central Metal Ions on Vibrational Circular Dichroism Spectra of Tris-(β-diketonato)Metal(III) Complexes, *Inorg. Chem.* 46 (16), 6755 (2007).

H. Sato, T. Taniguchi, K. Monde, S.I. Nishimura, and A. Yamagishi, Dramatic Effects of D-Electron Configurations on Vibrational Circular Dichroism Spectra of Tris(Acetylacetonato)Metal(III), *Chem. Lett.* 35 (4), 364 (2006).

D.W. Schlosser, F. Devlin, K. Jalkanen, and P.J. Stephens, Vibrational Circular Dichroism of Matrix-Isolated Molecules, *Chem. Phys. Lett.* 88 (3), 286 (1982).

R. Schweitzer-Stenner, Simulated IR, Isotropic and Anisotropic Raman, and Vibrational Circular Dichroism Amide I Band Profiles of Stacked β-Sheets, *J. Phys. Chem. B* 116 (14), 4141 (2012).

R. Schweitzer-Stenner, Distribution of Conformations Sampled by the Central Amino Acid Residue in Tripeptides Inferred from Amide I Band Profiles and NMR Scalar Coupling Constants, *J. Phys. Chem. B* 113 (9), 2922 (2009).

R. Schweitzer-Stenner and T.J. Measey, The Alanine-Rich Xao Peptide Adopts a Heterogeneous Population, Including Turn-Like and Polyproline II Conformations, *Proc. Natl. Acad. Sci. USA* 104 (16), 6649 (2007).

R. Schweitzer-Stenner, T. Measey, L. Kakalis, F. Jordan, S. Pizzanelli, C. Forte, and K. Griebenow, Conformations of Alanine-Based Peptides in Water Probed by FTIR, Raman, Vibrational Circular Dichroism, Electronic Circular Dichroism, and NMR Spectroscopy, *Biochemistry* 46 (6), 1587 (2007).

R. Schweitzer-Stenner, W. Gonzales, G.T. Bourne, J.A. Feng, and G.R. Marshall, Conformational Manifold of Alpha-Aminoisobutyric Acid (Aib) Containing Alanine-Based Tripeptides in Aqueous Solution Explored by Vibrational Spectroscopy, Electronic Circular Dichroism Spectroscopy, and Molecular Dynamics Simulations, *J. Am. Chem. Soc.* 129 (43), 13095 (2007).

R. Schweitzer-Stenner, T. Measey, A. Hagarman, F. Eker, and K. Griebenow, Salmon Calcitonin and Amyloid β: Two Peptides with Amyloidogenic Capacity Adopt Different Conformational Manifolds in Their Unfolded States, *Biochemistry* 45 (9), 2810 (2006).

R. Schweitzer-Stenner, F. Eker, K. Griebenow, X. Cao, and L.A. Nafie, The Conformation of Tetraalanine in Water Determined by Polarized Raman, FT-IR, and VCD Spectroscopy, *J. Am. Chem. Soc.* 126 (9), 2768 (2004).

R. Schweitzer-Stenner, Secondary Structure Analysis of Polypeptides Based on an Excitonic Coupling Model to Describe the Band Profile of Amide 1' of IR, Raman, and Vibrational Circular Dichroism Spectra, *J. Phys. Chem. B* 108 (43), 16965 (2004).

R. Schweitzer-Stenner, F. Eker, A. Perez, K. Griebenow, X. Cao, and L.A. Nafie, The Structure of Tri-Proline in Water Probed by Polarized Raman, Fourier Transform Infrared, Vibrational Circular Dichroism, and Electric Ultraviolet Circular Dichroism Spectroscopy, *Biopolymers (Peptide Sci.)* 71 (5), 558 (2003).

B.D. Self and D.S. Moore, Nucleic Acid Vibrational Circular Dichroism, Absorption and Linear Dichroism Spectra. II. A Devoe Theory Approach, *Biophys. J.* 74 (5), 2249 (1998).

B.D. Self and D.S. Moore, Nucleic Acid Vibrational Circular Dichroism, Absorption and Linear Dichroism Spectra. I. A Devoe Theory Approach, *Biophys. J.* 73 (1), 339 (1997).

A.C. Sen and T.A. Keiderling, Vibrational Circular Dichroism of Polypeptides III. Film Studies of Several α-Helical and β-Sheet Polypeptides, *Biopolymers* 23 (8), 1533 (1984).

A.C. Sen and T.A. Keiderling, Vibrational Circular Dichroism of Polypeptides. II. Solution Amide II and Deuteration Results, *Biopolymers* 23 (8), 1519 (1984).

V. Setnicka, J. Novy, S. Bohm, N. Sreenivasachary, M. Urbanova, and K. Volka, Molecular Structure of Guanine-Quartet Supramolecular Assemblies in a Gel-State Based on a DFT Calculation of Infrared and Vibrational Circular Dichroism Spectra, *Langmuir* 24 (14), 7520 (2008).

V. Setnicka, J. Hlavacek, and M. Urbanova, Oligopeptide-Porphyrin Interactions Studied by Circular Dichroism Spectroscopy: The Effect of Metalloporphyrin Axial Ligands on Peptide Matrix Conformation, *J. Porphyrins Phthalocyanines* 12 (12), 1270 (2008).

V. Setnicka, M. Urbanová, K. Volka, S. Nampally, and J.M. Lehn, Investigation of Guanosine-Quartet Assemblies by Vibrational and Electronic Circular Dichroism Spectroscopy, a Novel Approach for Studying Supramolecular Entities, *Chem. Eur. J.* 12 (34), 8735 (2006).

V. Setnicka, M. Urbanova, S. Pataridis, V. Kral, and K. Volka, Sol-Gel Phase Transition of Brucine-Appended Porphyrin Gelator: A Study by Vibrational Circular Dichroism Spectroscopy, *Tetrahedron Asymm.* 13 (24), 2661 (2002).

V. Setnicka, M. Urbanova, V. Kral, and K. Volka, Interactions of Cyclodextrins with Aromatic Compounds Studied by Vibrational Circular Dichroism Spectroscopy, *Spectrochim. Acta A Mol. Biomol. Spectrosc.* 58 (13), 2983 (2002).

V. Setnicka, M. Urbanová, P. Bour, V. Král, and K. Volka, Vibrational Circular Dichroism of 1,1'-Binaphthyl Derivatives: Experimental and Theoretical Study, *J. Phys. Chem. A* 105 (39), 8931 (2001).

R.D. Shah and L.A. Nafie, Spectroscopic Methods for Determining Enantiomeric Purity and Absolute Configuration in Chiral Pharmaceutical Molecules, *Curr. Opin. Drug Discov. Dev.* 4 (6), 764 (2001).

G. Shanmugam, P.L. Polavarapu, E. Láng, and Z. Majer, Conformational Analysis of Amyloid Precursor Protein Fragment Containing Amino Acids 667-676, and the Effect of D-Asp and iso-Asp Substitution at Asp_{672} Residue, *J. Struct. Biol.* 177 (3), 621 (2012).

G. Shanmugam, N. Phambu, and P.L. Polavarapu, Unusual Structural Transition of Antimicrobial VP1 Peptide, *Biophys. Chem.* 155 (2–3), 104 (2011).

G. Shanmugam and P.L. Polavarapu, Isotope-Assisted Vibrational Circular Dichroism Investigations of Amyloid β Peptide Fragment, Aβ (16–22), *J. Struct. Biol.* 176 (2), 212 (2011).

G. Shanmugam and P.L. Polavarapu, Concentration- and Dehydration-Dependent Structural Transitions in Poly-L-Lysine, *J. Mol. Struct.* 890 (1–3), 144 (2008).

G. Shanmugam and P.L. Polavarapu, Structural Transition during Thermal Denaturation of Collagen in the Solution and Film States, *Chirality* 21 (1), 152 (2008).

G. Shanmugam and P.L. Polavarapu, Structures of Intact Glycoproteins from Vibrational Circular Dichroism, *Proteins Struct. Funct. Bioinformatics* 63 (4), 768 (2006).

G. Shanmugam, P.L. Polavarapu, A. Kendall, and G. Stubbs, Structures of Plant Viruses from Vibrational Circular Dichroism, *J. Gen. Virol.* 86 (8), 2371 (2005).

G. Shanmugam, P.L. Polavarapu, B. Hallgas, and Z. Majer, Effect of D-Amino Acids at Asp[23] and Ser[26] Residues on the Conformational Preference of $A\beta_{20-29}$ Peptides, *Biochem. Biophys. Res. Commun.* 335 (3), 712 (2005).

G. Shanmugam, P.L. Polavarapu, D. Gopinath, and R. Jayakumar, The Structure of Antimicrobial Pexiganan Peptide in Solution Probed by Fourier Transform Infrared Absorption, Vibrational Circular Dichroism, and Electronic Circular Dichroism Spectroscopy, *Biopolymers* 80 (5), 636 (2005).

G. Shanmugam and P.L. Polavarapu, Film Techniques for Vibrational Circular Dichroism Measurements, *Appl. Spectrosc.* 59 (5), 673 (2005).

G. Shanmugam and P.L. Polavarapu, Structure of a $A\beta$(25-35) Peptide in Different Environments, *Biophys. J.* 87 (1), 622 (2004).

G. Shanmugam and P.L. Polavarapu, Vibrational Circular Dichoism of Protein Films, *J. Am. Chem. Soc.* 126 (33), 10292 (2004).

G. Shanmugam and P.L. Polavarapu, Vibrational Circular Dichroism Spectra of Protein Films: Thermal Denaturation of Bovine Serum Albumin, *Biophys. Chem.* 111 (1), 73 (2004).

R.A. Shaw, N. Ibrahim, and H. Wieser, The VCD Spectrum of 2-Methylthietane: Measured (800–1500 cm^{-1}) and Computed with Semiempirical Models, *Can. J. Chem.* 69 (2), 345 (1991).

R.A. Shaw, H. Wieser, R. Dutler, and A. Rauk, Vibrational Optical Activity of (S)-1-D-Ethanol, *J. Am. Chem. Soc.* 112 (14), 5401 (1990).

R.A. Shaw, N. Ibrahim, and H. Wieser, Vibrational Circular Dichroism Spectrum of 2-Methyloxetane, *J. Phys. Chem.* 94 (1), 125 (1990).

R.A. Shaw, N. Ibrahim, and H. Wieser, Determining the Absolute Configuration of Exo-7 Derivatives of 5-Methyl-6,8-Dioxabicyclo[3.2.1]Octane by VCD, *Tetrahedron Lett.* 29 (7), 745 (1988).

S. Shin, M. Nakata, and Y. Hamada, Analysis of Vibrational Circular Dichroism Spectra of (S)-(+)-2-Butanol by Rotational Strengths Expressed in Local Symmetry Coordinates, *J. Phys. Chem. A* 110 (6), 2122 (2006).

S. Shin, A.Y. Hirakawa, and Y. Hamada, Vibrational Circular Dichroism Spectrum of 1-Amino-2-Propanol, *Enantiomer (J. Stereochem.)* 7 (4), 191 (2002).

K. Shin-ya, H. Sugeta, S. Shin, Y. Hamada, Y. Katsumoto, and K. Ohno, Absolute Configuration and Conformation Analysis of 1-Phenylethanol by Matrix-Isolation Infrared and Vibrational Circular Dichroism Spectroscopy Combined with Density Functional Theory Calculation, *J. Phys. Chem. A* 111 (35), 8598 (2007).

R.A.G.D. Silva, S.C. Yasui, J. Kubelka, F. Formaggio, M. Crisma, C. Toniolo, and T.A. Keiderling, Discriminating 3_{10}- from α-Helices: Vibrational and Electronic CD and IR Absorption Study of Related Aib-Containing Oligopeptides, *Biopolymers* 65 (4), 229 (2002).

R.A.G.D. Silva, S.A. Sherman, F. Perini, E. Bedows, and T.A. Keiderling, Folding Studies on the Human Chorionic Gonadotropin β-Subunit Using Optical Spectroscopy of Peptide Fragments, *J. Am. Chem. Soc.* 122, 8623 (2000).

R.A.G.D. Silva, J. Kubelka, P. Bour, S.M. Decatur, and T.A. Keiderling, Site-Specific Conformational Determination in Thermal Unfolding Studies of Helical Peptides Using Vibrational Circular Dichroism with Isotopic Substitution, *Proc. Natl. Acad. Sci. USA* 97 (15), 8318 (2000).

R.A.G.D. Silva, S.A. Sherman, and T.A. Keiderling, β-Hairpin Stabilization in a 28-Residue Peptide Derived from the β-Subunit Sequence of Human Chorionic Gonadotropin Hormone, *Biopolymers* 50, 413 (1999).

R.D. Singh and T.A. Keiderling, Vibrational Circular Dichroism of Poly(γ-benzyl-L-glutamate), *Biopolymers* 20 (1), 237 (1981).

R.D. Singh and T.A. Keiderling, Vibrational Circular Dichroism of 3-Methylcyclohexanone. Fixed Partial Charge Calculations, *J. Chem. Phys.* 74 (10), 5347 (1981).

R.D. Singh and T.A. Keiderling, Vibrational Circular Dichroism of Six-Membered-Ring Monoterpenes. Consistent Force Field, Fixed Partial Charge Calculations, *J. Am. Chem. Soc.* 103 (9), 2387 (1981).

M.M.J. Smulders, T. Buffeteau, D. Cavagnat, M. Wolffs, A.P.H.J. Schenning, and E.W. Meijer, C3-Symmetrical Self-Assembled Structures Investigated by Vibrational Circular Dichroism, *Chirality* 20 (9), 1016 (2008).

J. Snir, R.A. Frankel, and J.A. Schellman, Optical Activity of Polypeptides in the Infrared. Predicted CD of the Amide I and Amide II Bands, *Biopolymers* 14, 173 (1975).

A. Solladié-Cavallo, C. Marsol, M. Yaakoub, K. Azyat, A. Klein, M. Roje, C. Suteu, T.B. Freedman, X. Cao, and L.A. Nafie, Erythro-1-Naphthyl-1-(2-Piperidyl)Methanol: Synthesis, Resolution, NMR Relative Configuration, and VCD Absolute Configuration, *J. Org. Chem.* 68 (19), 7308 (2003).

A. Solladié-Cavallo, C. Marsol, M. Yaakoub, K. Azyat, A. Klein, M. Roje, C. Suteu, T.B. Freedman, X. Cao, and L.A. Nafie, Erythro-1-Naphthyl-1-(2-Piperidyl)Methanol: Synthesis, Resolution, NMR Relative Configuration and VCD Absolute Configuration, *J. Org. Chem.* 68 (19), 7308 (2003).

A. Solladié-Cavallo, C. Marsol, G. Pescitelli, L.D. Bari, P. Salvadori, X. Huang, N. Fujioka, N. Berova, X. Cao, T.B. Freedman, and L.A. Nafie, (R)-(+)- and (S)-(–)-1-(9-Phenanthryl) Ethylamine: Assignment of Absolute Configuration by CD Tweezer and VCD Methods, and Difficulties Encountered with the CD Exciton Chirality Method, *Eur. J. Org. Chem.* 1788 (2002).

A. Solladié-Cavallo, O. Sedy, M. Salisova, M. Biba, C.J. Welch, L. Nafie, and T. Freedman, A Chiral 1,4-Oxazin-2-One: Asymmetric Synthesis versus Resolution, Structure, Conformation and VCD Absolute Configuration, *Tetrahedron Asymm.* 12 (19), 2703 (2001).

A. Solladié-Cavallo, M. Balaz, M. Salisova, C. Suteu, L.A. Nafie, X. Cao, and T.B. Freedman, A New Chiral Oxathiane: Synthesis, Resolution and Absolute Configuration Determination by Vibrational Circular Dichroism, *Tetrahedron Asymm.* 12 (18), 2605 (2001).

P. Soulard, P. Asselin, A. Cuisset, J.R.A. Moreno, T.R. Huet, D. Petitprez, J. Demaison, T.B. Freedman, X. Cao, L.A. Nafie, and J. Crassous, Chlorofluoroiodomethane as a Potential Candidate for Parity Violation Measurements, *Phys. Chem. Chem. Phys.* 8 (1), 79 (2006).

K.M. Spencer, S.J. Cianciosi, J.E. Baldwin, T.B. Freedman, and L.A. Nafie, Determination of Enantiomeric Excess in Deuterated Chiral Hydrocarbons by Vibrational Circular Dichroism Spectroscopy, *Appl. Spectrosc.* 44 (2), 235 (1990).

P.J. Stephens, J.J. Pan, F.J. Devlin, M. Urbanová, O. Julinek, and J. Hájíček, Determination of the Absolute Configurations of Natural Products via Density Functional Theory Calculations of Vibrational Circular Dichroism, Electronic Circular Dichroism, and Optical Rotation: The Isoschizozygane Alkaloids Isoschizogaline and Isoschizogamine, *Chirality* 20 (3–4), 454 (2008).

P.J. Stephens, F.J. Devlin, C. Villani, F. Gasparrini, and S.L. Mortera, Determination of the Absolute Configurations of Chiral Organometallic Complexes via Density Functional Theory Calculations of Their Vibrational Circular Dichroism Spectra: The Chiral Chromium Tricarbonyl Complex of N-Pivaloyl-Tetrahydroquinoline, *Inorg. Chim. Acta* 361 (4), 987 (2008).

P.J. Stephens, F.J. Devlin, S. Schürch, and J. Hulliger, Determination of the Absolute Configuration of Chiral Molecules via Density Functional Theory Calculations of Vibrational Circular Dichroism and Optical Rotation: The Chiral Alkane D_3-Anti-Trans-Anti-Trans-Anti-Trans-Perhydrotriphenylene, *Theor. Chem. Acc.* 119 (1–3), 19 (2008).

P.J. Stephens, F.J. Devlin, and J.J. Pan, The Determination of the Absolute Configurations of Chiral Molecules Using Vibrational Circular Dichroism (VCD) Spectroscopy, *Chirality* 20 (5), 643 (2008).

P.J. Stephens, J.J. Pan, and K. Krohn, Determination of the Absolute Configurations of Pharmacological Natural Products via Density Functional Theory Calculations of Vibrational Circular Dichroism: The New, Cytotoxic, Iridoid Prismatomerin, *J. Org. Chem.* 72 (20), 7641 (2007).

P.J. Stephens, J.J. Pan, F.J. Devlin, M. Urbanová, and J. Hájíček, Determination of the Absolute Configurations of Natural Products via Density Functional Theory Calculations of Vibrational Circular Dichroism, Electronic Circular Dichroism and Optical Rotation: The Schizozygane Alkaloid Schizozygine, *J. Org. Chem.* 72 (7), 2508 (2007).

P.J. Stephens, J.J. Pan, F.J. Devlin, K. Krohn, and T. Kurtán, Determination of the Absolute Configurations of Natural Products via Density Functional Theory Calculations of Vibrational Circular Dichroism, Electronic Circular Dichroism, and Optical Rotation: The Iridoids Plumericin and Isoplumericin, *J. Org. Chem.* 72 (9), 3521 (2007).

P.J. Stephens, F.J. Devlin, F. Gasparrini, A. Ciogli, D. Spinelli, and B. Cosimelli, Determination of the Absolute Configuration of a Chiral Oxadiazol-3-One Calcium Channel Blocker, Resolved Using Chiral Chromatography, via Concerted Density Functional Theory Calculations of Its Vibrational Circular Dichroism, Electronic Circular Dichroism, and Optical Rotation, *J. Org. Chem.* 72 (13), 4707 (2007).

P.J. Stephens and F.J. Devlin, in *Continuum Solvation Models in Chemical Physics: From Theory to Applications*, ed. B. Mennucci and R. Cammi (Wiley, Chichester, England, 2007), p. 180.

P.J. Stephens, D.M. McCann, F.J. Devlin, and A.B. Smith, Determination of the Absolute Configurations of Natural Products via Density Functional Theory Calculations of Optical Rotation, Electronic Circular Dichroism, and Vibrational Circular Dichroism: The Cytotoxic Sesquiterpene Natural Products Quadrone, Suberosenone, Suberosanone, and Suberosenol A Acetate, *J. Nat. Prod.* 69 (7), 1055 (2006).

P.J. Stephens, D.M. McCann, F.J. Devlin, T.C. Flood, E. Butkus, S. Stoncius, and J.R. Cheeseman, Determination of Molecular Structure Using Vibrational Circular Dichroism Spectroscopy: The Keto-Lactone Product of Baeyer-Villiger Oxidation of (+)-(1R,5S)-Bicyclo[3.3.1] Nonane-2,7-Dione, *J. Org. Chem.* 70, 3903 (2005).

P.J. Stephens, in *Computational Medicinal Chemistry for Drug Discovery*, ed. P. Bultinck, H. deWinter, W. Langenaecker, et al. (Marcel Dekker, New York, 2004), p. 699.

P.J. Stephens, A. Aamouche, F.J. Devlin, S. Superchi, M.I. Donnoli, and C. Rosini, Determination of Absolute Configuration Using Vibrational Circular Dichroism Spectroscopy: The Chiral Sulfoxide 1-(2-Methylnaphthyl) Methyl Sulfoxide, *J. Org. Chem.* 66 (11), 3671 (2001).

P.J. Stephens and F.J. Devlin, Determination of the Structure of Chiral Molecules Using *Ab Initio* Vibrational Circular Dichroism Spectroscopy, *Chirality* 12 (4), 172 (2000).

P.J. Stephens, in *Encyclopedia of Spectroscopy and Spectrometry* (Academic Press, London, 1999), p. 2415.

P.J. Stephens, F.J. Devlin, C.S. Ashvar, K.L. Bak, P.R. Taylor, and M.J. Frisch, in *Chemical Applications of Density-Functional Theory*, ed. B.B. Laird, R.B. Ross, and T. Ziegler, vol. 629, ACS Symposium Series (1996), p. 105.

P.J. Stephens, C.S. Ashvar, F.J. Devlin, J.R. Cheeseman, and M.J. Frisch, *Ab Initio* Calculation of Atomic Axial Tensors and Vibrational Rotational Strengths Using Density Functional Theory, *Mol. Phys.* 89 (2), 579 (1996).

P.J. Stephens, F.J. Devlin, C.F. Chabalowski, and M.J. Frisch, *Ab Initio* Calculation of Vibrational Absorption and Circular Dichroism Spectra Using Density Functional Force Fields, *J. Phys. Chem.* 98, 11623 (1994).

P.J. Stephens, F.J. Devlin, C.S. Ashvar, C.F. Chabalowski, and M.J. Frisch, Theoretical Calculation of Vibrational Circular Dichroism Spectra, *Faraday Discuss* 99, 103 (1994).

P.J. Stephens, C.F. Chabalowski, F.J. Devlin, and K.J. Jalkanen, *Ab Initio* Calculation of Vibrational Circular Dichroism Spectra Using Large Basis Set MP2 Force Fields, *Chem. Phys. Lett.* 225 (1–3), 247 (1994).

P.J. Stephens, K.J. Jalkanen, F.J. Devlin, and C.F. Chabalowski, *Ab Initio* Calculation of Vibrational Circular Dichroism Spectra Using Accurate Post-SCF Force Fields: Trans-2,3-Dideuterio-Oxirane, *J. Phys. Chem.* 97 (23), 6107 (1993).

P.J. Stephens, Evaluation of Theories of Vibrational Magnetic Dipole Transition Moments Using Atomic Axial Tensor Sum Rules and Paramagnetic Susceptibilities, *Chem. Phys. Lett.* 180, 472 (1991).

P.J. Stephens, K.J. Jalkanen, and R.W. Kawiecki, Theory of Vibrational Rotational Strengths: Comparison of *A Priori* Theory and Approximate Models, *J. Am. Chem. Soc.* 112, 6518 (1990).

P.J. Stephens, K.J. Jalkanen, R.D. Amos, P. Lazzeretti, and R. Zanasi, *Ab Initio* Calculations of Atomic Polar and Axial Tensors for HF, H_2O, NH_3 and CH_4, *J. Phys. Chem.* 94 (5), 1811 (1990).

P.J. Stephens, in *Vibronic Processes in Inorganic Chemistry*, ed. C.D. Flint (Kluwer, Dordrecht, Netherlands, 1989), vol. 288, p. 371.

P.J. Stephens, The *A Priori* Prediction of Vibrational Circular Dichroism Spectra: A New Approach to the Study of the Stereochemistry of Chiral Molecules, *Croat. Chem. Acta.* 62 (2B), 429 (1989).

P.J. Stephens and K.J. Jalkanen, A New Formalism for Paramagnetic Susceptibilities: CH_3F, *J. Chem. Phys.* 91, 1379 (1989).

P.J. Stephens, Gauge Dependence of Vibrational Magnetic Dipole Transition Moments and Rotational Strengths, *J. Phys. Chem.* 91, 1712 (1987).

P.J. Stephens, in *Understanding Molecular Properties*, ed. J. Avery, J.P. Dahl, and A.E. Hansen (D. Reidel, 1987), p. 333.

P.J. Stephens and M.A. Lowe, in *Annual Review of Physical Chemistry*, ed. B.S. Rabinovitch, J.M. Schurr, and H.L. Strauss (Annual Reviews, Palo Alto, CA, 1985), vol. 36, p. 213.

P.J. Stephens, Theory of Vibrational Circular Dichroism, *J. Phys. Chem.* 89 (5), 748 (1985).

P.J. Stephens and R. Clark, in *Optical Activity and Chiral Discrimination*, ed. S.F. Mason (D. Reidel, 1979), vol. C48, p. 263.

P.J. Stephens, Polarized Light in Chemistry, *Proc. SPIE* 88, 75 (1976).

C.N. Su, T.A. Keiderling, K. Misiura, and W.J. Stec, Midinfrared Vibrational Circular Dichroism Studies of Cyclophosphamide and Its Congeners, *J. Am. Chem. Soc.* 104 (25), 7343 (1982).

C.N. Su and T.A. Keiderling, Vibrational Circular Dichroism of Methyl Deformation Modes, *Chem. Phys. Lett.* 77 (3), 494 (1981).

C.N. Su and T.A. Keiderling, Conformation of Dimethyl Tartrate in Solution. Vibrational Circular Dichroism Results, *J. Am. Chem. Soc.* 102 (2), 511 (1980).

C.N. Su, V.J. Heintz, and T.A. Keiderling, Vibrational Circular Dichroism in the Mid-Infrared, *Chem. Phys. Lett.* 73 (1), 157 (1980).

H. Sugeta, Vibrational Optical Activity, *Kagaku No Ryoiki, Zokan* 115, 139 (1977).

H. Sugeta, C. Marcott, T.R. Faulkner, J. Overend, and A. Moscowitz, Infrared Circular Dichroism Associated with the C-H Stretching Vibration of Tartaric Acid, *Chem. Phys. Lett.* 40 (3), 397 (1976).

W. Sun, J. Wu, B. Zheng, Y. Zhu, and C. Liu, DFT Study of Vibrational Circular Dichroism Spectra of (S)-Glycidol-Water Complexes, *J. Mol. Struct. (Theochem)* 809 (1–3), 161 (2007).

A. Synytsya, M. Urbanová, V. Setnicka, M. Tkadlecova, J. Havlicek, I. Raich, P. Matejka, A. Synytsya, J. Copikova, and K. Volka, The Complexation of Metal Cations by D-Galacturonic Acid: A Spectroscopic Study, *Carbohyd. Res.* 339 (14), 2391 (2004).

E. Tajkhorshid, K.J. Jalkanen, and S. Suhai, Structure and Vibrational Spectra of the Zwitterion L-Alanine in the Presence of Explicit Water Molecules: A Density Functional Analysis, *J. Phys. Chem. B* 102, 5899 (1998).

C.N. Tam, P. Bour, and T.A. Keiderling, An Experimental Comparison of Vibrational Circular Dichroism and Raman Optical Activity with 1-Amino-2-Propanol and 2-Amino-1-Propanol as Model Compounds, *J. Am. Chem. Soc.* 119 (30), 7061 (1997).

C.N. Tam, P. Bour, and T.A. Keiderling, Vibrational Optical Activity of (3s,6s)-3,6-Dimethyl-1,4-Dioxane-2,5-Dione, *J. Am. Chem. Soc.* 118 (42), 10285 (1996).

T. Tanaka, M. Oelgemoller, K. Fukui, F. Aoki, T. Mori, T. Ohno, and Y. Inoue, Unusual CD Couplet Pattern Observed for the $\pi^* \leftarrow n$ Transition of Enantiopure (Z)-8-Methoxy-4-Cyclooctenone: An Experimental and Theoretical Study by Electronic and Vibrational Circular Dichroism Spectroscopy and Density Functional Theory Calculation, *Chirality* 19 (5), 415 (2007).

T. Tanaka, T.S. Kodama, H.E. Morita, and T. Ohno, Electronic and Vibrational Circular Dichroism of Aromatic Amino Acids by Density Functional Theory, *Chirality* 18 (8), 652 (2006).

T. Tanaka, K. Inoue, T. Kodama, Y. Kyogoku, T. Hayakawa, and H. Sugeta, Conformational Study on Poly[α-(α-Phenethyl)-L-Glutamate] Using Vibrational Circular Dichroism Spectroscopy, *Biopolymers (Biospectroscopy)* 62 (4), 228 (2001).

H.Z. Tang, E.R. Garland, B.M. Novak, J. He, P.L. Polavarapu, F.C. Sun, and S.S. Sheiko, Helical Polyguanidines Prepared by Helix-Sense-Selective Polymerizations of Achiral Carbodiimides Using Enantiopure Binaphthol-Based Titanium Catalysts, *Macromolecules* 40 (10), 3575 (2007).

T. Taniguchi and K. Monde, Exciton Chirality Method in Vibrational Circular Dichroism, *J. Am. Chem. Soc.* 134 (8), 3695 (2012).

T. Taniguchi, I. Tone, and K. Monde, Observation and Characterization of a Specific Vibrational Circular Dichroism Band in Phenyl Glycosides, *Chirality* 20 (3/4), 446 (2008).

T. Taniguchi, K. Monde, K. Nakanishi, and N. Berova, Chiral Sulfinates Studied by Optical Rotation, ECD and VCD: The Absolute Configuration of a Cruciferous Phytoalexin Brassicanal C, *Org. Biomol. Chem.* 6 (23), 4399 (2008).

T. Taniguchi and K. Monde, Chiroptical Analysis of Glycoconjugates by Vibrational Circular Dichroism, *Trends Glycosci. Glycotechnol.* 19 (107), 147 (2007).

T. Taniguchi and K. Monde, Spectrum-Structure Relationship in Carbohydrate Vibrational Circular Dichroism and Its Application to Glycoconjugates, *Chem. Asian J.* 2 (10), 1258 (2007).

T. Taniguchi and K. Monde, Vibrational Circular Dichroism (VCD) Studies on Disaccharides in the CH Region: Toward Discrimination of the Glycosidic Linkage Position, *Org. Biomol. Chem.* 5 (7), 1104 (2007).

T. Taniguchi, K. Monde, S.I. Nishimura, J. Yoshida, H. Sato, and A. Yamagishi, Rewinding of Helical Systems by Use of the Cr(III) Complex as a Photoresponsive Chiral Dopant, *Mol. Cryst. Liquid Cryst.* 460, 107 (2006).

T. Taniguchi, K. Monde, N. Miura, and S.I. Nishimura, A Characteristic CH Band in VCD of Methyl Glycosidic Carbohydrates, *Tetrahedron Lett.* 45 (46), 8451 (2004).

T. Taniguchi, N. Miura, S.I. Nishimura, and K. Monde, Vibrational Circular Dichroism: Chiroptical Analysis of Biomolecules, *Mol. Nutr. Food Res.* 48 (4), 246 (2004).

G. Tarczay, S. Gobi, E. Vass, and G. Magyarfalvi, Model Peptide-Water Complexes in AR Matrix: Complexation Induced Conformation Change and Chirality Transfer, *Vibrational Spectrosc.* 50 (1), 21 (2009).

G. Tarczay, G. Magyarfalvi, and E. Vass, Towards the Determination of the Absolute Configuration of Complex Molecular Systems: Matrix Isolation Vibrational Circular Dichroism Study of (R)-2-Amino-1-Propanol, *Angew. Chem. Int. Ed.* 45 (11), 1775 (2006).

J. Teraoka, F. Ibe, T. Tanaka, Y. Kyogoku, and H. Sugeta, Vibrational Circular Dichroism in Hemoprotein Ligands, *J. Inorg. Biochem.* 67, 432 (1997).

J. Teraoka, N. Yamamoto, Y. Matsumoto, Y. Kyogoku, and H. Sugeta, What Is the Crucial Factor for Vibrational Circular Dichroism in Hemoprotein Ligands? *J. Am. Chem. Soc.* 118 (37), 8875 (1996).

J. Teraoka, K. Nakamura, Y. Nakahara, Y. Kyogoku, and H. Sugeta, Extraordinarily Intense Vibrational Circular Dichroism of a Metmyoglobin Cyanide Complex, *J. Am. Chem. Soc.* 114 (23), 9211 (1992).

O.R. Thiel, M. Achmatowicz, C. Bernard, P. Wheeler, C. Savarin, T.L. Correll, A. Kasparian, A. Allgeier, M.D. Bartberger, H. Tan, and R.D. Larsen, Development of a Practical Synthesis of a P38 Map Kinase Inhibitor, *Org. Proc. Res. Dev.* 13 (2), 230 (2009).

G. Tian, G. Zhu, X. Yang, Q. Fang, M. Xue, J. Sun, Y. Wei, and S. Qiu, A Chiral Layered Co(II) Coordination Polymer with Helical Chains from Achiral Materials, *Chem. Comm.* (11), 1396 (2005).

D.M. Tigelaar, W. Lee, K.A. Bates, A. Saprigin, V.N. Prigodin, X. Cao, L.A. Nafie, M.S. Platz, and A.J. Epstein, Role of Solvent and Secondary Doping in Polyaniline Films Doped with Chiral Camphorsulfonic Acid: Preparation of a Chiral Metal, *Chem. Mater.* 14 (3), 1430 (2002).

Z. Tomankova, V. Setnicka, M. Urbanova, P. Matejka, V. Kral, K. Volka, and P. Bour, Conformational Flexibility of Corey Lactone Derivatives Indicated by Absorption and Vibrational Circular Dichroism Spectra, *J. Org. Chem.* 69 (1), 26 (2004).

J. Tomasi, R. Cammi, and B. Mennucci, Medium Effects on the Properties of Chemical Systems: An Overview of Recent Formulations in the Polarizable Continuum Model (PCM). *Int. J. Quantum Chem.* 75, 783 (1999).

C. Toniolo, F. Formaggio, S. Tognon, Q.B. Broxterman, B. Kaptein, R. Huang, V. Setnicka, T.A. Keiderling, I.H. McColl, L. Hecht, and L.D. Barron,The Complete Chirospectroscopic Signature of the Peptide 3_{10}-Helix in Aqueous Solution, *Biopolymers* 75 (1), 32 (2004).

J.M. Torres-Valencia, O.E. Chavez-Rios, C.M. Cerda-Garcia-Rojas, E. Burgueno-Tapia, and P. Joseph-Nathan, Dihydrofurochromones from Prionosciadium Thapsoides, *J. Nat. Prod.* 71 (11), 1956 (2008).

N. Toselli, D. Martin, M. Achard, A. Tenaglia, T. Bürgi, and G. Buono, Enantioselective Cobalt-Catalyzed [6+2] Cycloadditions of Cycloheptatriene with Alkynes, *Adv. Synth. Catalysis* 350 (2), 280 (2008).

C.D. Tran, V.I. Grishko, and G. Huang, Chiral Detection in High-Performance Liquid Chromatography by Vibrational Circular Dichroism, *Anal. Chem.* 66 (17), 2630 (1994).

C.D. Tran and V.I. Grishko, Universal Spectropolarimeter Based on Overtone Circular Dichroism Measurements in the Near-Infrared Region, *Anal. Chem.* 66, 3639 (1994).

D. Tsankov, V. Maharaj, J.H. van de Sande, and H. Wieser, Selected Deoxyoctanucleotides and Their Daunomycin Complexes: Infrared Absorption and Vibrational Circular Dichroism Spectra, *Vibrational Spectrosc.* 43 (1), 152 (2007).

D. Tsankov, M. Krasteva, V. Andrushchenko, J.H. van de Sande, and H. Wieser, Vibrational Circular Dichroism Signature of Hemiprotonated Intercalated Four-Stranded I-DNA, *Biophys. Chem.* 119 (1), 1 (2006).

D. Tsankov, B. Kalisch, H.V.D. Sande, and H. Wieser, Cisplatin-DNA Adducts by Vibrational Circular Dichroism Spectroscopy: Structure and Isomerization of d(CCTG*G*TCC)·d(GGACCAGG) Intrastrand Cross-Linked by Cisplatin, *J. Phys. Chem. B* 107 (26), 6479 (2003).

D. Tsankov, B. Kalisch, J.H.V.D. Sande, and H. Wieser, Cisplatin Adducts of d(CCTCTG *G*TCTCC)·d(GGAGACCAGAGG) in Aqueous Solution by Vibrational Circular Dichroism Spectroscopy, *Biopolymers (Biospectroscopy)* 72 (6), 490 (2003).

D. Tsankov, V. Dimitrov, and H. Wieser, FT-Vibrational Circular Dichroism (VCD) Spectra and *Ab Initio* 6-31g$^{*(0.3)}$ Intensities for Camphor and 2-Vinyl-Exo-Borneol, *Mikrochim. Acta Suppl.* 14, 535 (1997).

D. Tsankov, T. Eggimann, and H. Wieser, Alternative Design for Improved FT-IR/VCD Capabilities, *Appl. Spectrosc.* 49 (1), 132 (1995).

D. Tsankov, T. Eggimann, G. Liu, and H. Wieser, FT-IR Vibrational Circular Dichroism (VCD) Instrument for Studies of Polymers, *Proc. SPIE* 2089, 178 (1993).

H.P. Tuan, C. Larsson, F. Hoffmann, A. Bergman, M. Fröba, and H. Hühnerfuss, Enantioselective Semipreparative HPLC Separation of PCB Metabolites and Their Absolute Structure Elucidation Using Electronic and Vibrational Circular Dichroism, *Chirality* 17, 266 (2005).

C.M. Tummalapalli, D.M. Back, and P.L. Polavarapu, Fourier-Transform Infrared Vibrational Circular Dichroism of Simple Carbohydrates, *Faraday Trans. 1* 84 (8), 2585 (1988).

E. Tur, G. Vives, G. Rapenne, J. Crassous, N. Vanthuyne, C. Roussel, R. Lombardi, T. Freedman, and L. Nafie, HPLC Separation and VCD Spectroscopy of Chiral Pyrazoles Derived from (5r)-Dihydrocarvone, *Tetrahedron Asymm.* 18 (16), 1911 (2007).

D.R. Turner and J. Kubelka, Infrared and Vibrational CD Spectra of Partially Solvated α-Helices: DFT-Based Simulations with Explicit Solvent, *J. Phys. Chem. B* 111 (7), 1834 (2007).

A.J. Turner and R.A. Hoult, Simple Techniques for Phase Correction in VCD Measurements, *Mikrochim. Acta. Suppl.* 14, 291 (1997).

C. Uncuta, S. Ion, D. Gherase, E. Bartha, F. Teodorescu, and P. Filip, Absolute Configurational Assignment in Chiral Compounds through Vibrational Circular Dichroism (VCD) Spectroscopy, *Rev. Chim. (Bucharest, Romania)* 60 (1), 86 (2009).

M. Urbanova and P. Malon, in *Analytical Methods in Supramolecular Chemistry*, ed. C.A. Schalley (Wiley-VCH, Weinheim, 2007), p. 265.

M. Urbanova, V. Setnicka, and K. Volka, Vibrational Circular Dichroism Spectroscopy as a Novel Tool for Structure Analysis, *Chemicke Listy* 96 (5), 301 (2002).

M. Urbanova, V. Setnicka, P. Bour, H. Navratilova, and K. Volka, Vibrational Circular Dichroism Spectroscopy Study of Paroxetine and Femoxetine Precursors, *Biopolymers (Biospectroscopy)* 67 (3), 298 (2002).

M. Urbanova, V. Setnicka, V. Kral, and K. Volka, Noncovalent Interactions of Peptides with Porphyrins in Aqueous Solution: Conformational Study Using Vibrational CD Spectroscopy, *Biopolymers (Peptide Sci.)* 60 (4), 307 (2001).

M. Urbanova, V. Setnicka, and K. Volka, Measurements of Concentration Dependence and Enantiomeric Purity of Terpene Solutions as a Test of a New Commercial VCD Spectrometer, *Chirality* 12 (4), 199 (2000).

M. Urbanova, T.A. Keiderling, and P. Pancoska, Conformational Study of Some Milk Proteins. Comparison of the Results of Electronic Circular Dichroism and Vibrational Circular Dichroism, *Bioelectrochem. Bioenerg.* 41 (1), 77 (1996).

M. Urbanova, P. Pancoska, and T.A. Keiderling, Spectroscopic Study of the Temperature-Dependent Conformation of Glucoamylase, *Biochim. Biophys. Acta* 1203 (2), 290 (1993).

M. Urbanova, R.K. Dukor, P. Pancoska, and T.A. Keiderling, Conformational Study of α-Lactalbumin Using Vibrational Circular Dichroism. Evidence for a Difference in Crystal and Solution Structures, *Spec. Publ. R. Soc. Chem.* 94 (Spectrosc. Biol. Mol.), 133 (1991).

M. Urbanova, R.K. Dukor, P. Pancoska, V.P. Gupta, and T.A. Keiderling, Comparison of α-Lactalbumin and Lysozyme Using Vibrational Circular Dichroism. Evidence for a Difference in Crystal and Solution Structures, *Biochemistry* 30 (43), 10479 (1991).

M. Urbanová, V. Setnicka, F.J. Devlin, and P.J. Stephens, Determination of Molecular Structure in Solution Using Vibrational Circular Dichroism Spectroscopy: The Supramolecular Tetramer of S-2,2'-Dimethyl-Biphenyl-6,6'-Dicarboxylic Acid, *J. Am. Chem. Soc.* 127 (18), 6700 (2005).

M. Valík, J. Malina, L. Palivec, J. Foltýnová, M. Tkadlecová, M. Urbanová, V. Brabec, and V. Král, Tröger's Base Scaffold in Racemic and Chiral Fashion as a Spacer for Bisdistamycin Formation. Synthesis and DNA Binding Study, *Tetrahedron* 62 (36), 8591 (2006).

A. Vargas, F. Hoxha, N. Bonalumi, T. Mallat, and A. Baiker, Steric and Electronic Effects in Enantioselective Hydrogenation of Ketones on Platinum Modified by Cinchonidine: Directing Effect of the Trifluoromethyl Group, *J. Catalysis* 240 (2), 203 (2006).

A. Vargas, N. Bonalumi, D. Ferri, and A. Baiker, Solvent-Induced Conformational Changes of O-Phenyl-Cinchonidine: A Theoretical and VCD Spectroscopy Study, *J. Phys. Chem. A* 110 (3), 1118 (2006).

E. Vass, M. Hollósi, E. Forró, and F. Fülöp, VCD Spectroscopic Investigation of Enantiopure Cyclic β-Lactams Obtained through Lipolase-Catalyzed Enantioselective Ring-Opening Reaction, *Chirality* 18 (9), 733 (2006).

F.G. Vogt, G.P. Spoors, Q. Su, Y.W. Andemichael, H. Wang, T.C. Potter, and D.J. Minick, A Spectroscopic and Computational Study of Stereochemistry in 2-Hydroxymutilin, *J. Mol. Struct.* 797 (1–3), 5 (2006).

T.H. Walnut and L.A. Nafie, Infrared Absorption and the Born-Oppenheimer Approximation. II. Vibrational Circular Dichroism, *J. Chem. Phys.* 67 (4), 1501 (1977).

F. Wang, C. Zhao, and P.L. Polavarapu, A Study of the Conformations of Valinomycin in Solution Phase, *Biopolymers* 75 (1), 85 (2004).

F. Wang, P.L. Polavarapu, J. Drabowicz, K. Piotr, M.J. Potrzebowski, M. Mikolajczyck, M.W. Wieczorek, W.W. Majzner, and I. Lazewska, Solution and Crystal Structures of Chiral Molecules Can Be Significantly Different: Tert-Butylphenylphosphinoselenoic Acid, *J. Phys. Chem. A* 108 (11), 2072 (2004).

F. Wang and P.L. Polavarapu, Conformational Analysis of Melittin in Solution Phase: Vibrational Circular Dichroism Study, *Biopolymers (Biospectroscopy)* 70 (4), 614 (2003).

F. Wang, Y. Wang, P.L. Polavarapu, T. Li, J. Drabowicz, K.M. Pietrusiewicz, and K. Zygo, Absolute Configuration of Tert-Butyl-1-(2-Methylnaphthyl) Phosphine Oxide, *J. Org. Chem.* 67 (18), 6539 (2002).

F. Wang, P.L. Polavarapu, V. Schurig, and R. Schmidt, Absolute Configuration and Conformational Analysis of a Degradation Product of Inhalation Anesthetic Sevoflurane: A Vibrational Circular Dichroism Study, *Chirality* 14 (8), 618 (2002).

F. Wang, P.L. Polavarapu, F. Lebon, G. Longhi, S. Abbate, and M. Catellani, Conformational Analysis of (S)-(+)-1-Bromo-2-Methylbutane and the Influence of Bromine on Conformational Stability, *J. Phys. Chem. A* 106 (51), 12365 (2002).

F. Wang, P.L. Polavarapu, F. Lebon, G. Longhi, S. Abbate, and M. Catellani, Absolute Configuration and Conformational Stability of (S)-(+)-3-(2-Methylbutyl)Thiophene and (+)-3,4-Di[(S)-2-Methylbutyl)]Thiophene and Their Polymers, *J. Phys. Chem. A* 106, 5918 (2002).

F. Wang, H. Wang, P.L. Polavarapu, and C.J. Rizzo, Absolute Configuration and Conformational Stability of (+)-2,5-Dimethylthiolane and (−)-2,5 Dimethylsulfolane, *J. Org. Chem.* 66 (10), 3507 (2001).

F. Wang, P.L. Polavarapu, J. Drabowicz, M. Mikolajczyk, and P. Lyzwa, Absolute Configurations. Predominant Conformations, and Tautomeric Structures of Enantiomeric Tert-Butylphenylphosphinothioic Acid, *J. Org. Chem.* 66 (26), 9015 (2001).

F. Wang and P.L. Polavarapu, Temperature Influence on the Secondary Structure of Avidin and Avidin-Biotin Complex: A Vibrational Circular Dichroism Study, *J. Phys. Chem. B* 105 (32), 7857 (2001).

F. Wang and P.L. Polavarapu, Predominant Conformations of (2R,3R)-(–)-2,3-Butanediol, *J. Phys. Chem. A* 105 (29), 6991 (2001).

F. Wang, P.L. Polavarapu, J. Drabowicz, and M. Mikolajczyk, Absolute Configurations, Predominant Conformations and Tautomeric Structures of Enantiomeric *Tert*-Butylphenylphosphine Oxides, *J. Org. Chem.* 65, 7561 (2000).

F. Wang and P.L. Polavarapu, Vibrational Circular Dichroism, Predominant Conformations and Hydrogen Bonding in (S)-(–)-3-Butyn-2-ol, *J. Phys. Chem. A* 104 (9), 1822 (2000).

F. Wang and P.L. Polavarapu, Vibrational Circular Dichroism: Predominant Conformations and Intermolecular Interactions in (R)-(–)-2-Butanol, *J. Phys. Chem. A* 104 (46), 10683 (2000).

F. Wang and P.L. Polavarapu, Conformational Stability of (+)-Epichlorohydrin, *J. Phys. Chem. A* 104, 6189 (2000).

B. Wang, Photoelastic Modulator-Based Vibrational Circular Dichroism, *Am. Lab.* 28 (6), 36C (1996).

B. Wang and T.A. Keiderling, Observations on the Measurement of Vibrational Circular Dichroism with Rapid-Scan and Step-Scan FT-IR Techniques, *Appl. Spectrosc.* 49 (9), 1347 (1995).

L. Wang, L. Yang, and T.A. Keiderling, Vibrational Circular Dichroism of a-, B-, and Z-Form Nucleic Acids in the PO_2^- Stretching Region, *Biophys. J.* 67 (6), 2460 (1994).

L. Wang, P. Pancoska, and T.A. Keiderling, Detection and Characterization of Triple-Helical Pyrimidine-Purine-Pyrimidine Nucleic Acids with Vibrational Circular Dichroism, *Biochemistry* 33 (28), 8428 (1994).

L. Wang and T.A. Keiderling, Helical Nature of Poly (di-dC).(di-dC). Vibrational Circular Dichroism Results, *Nucleic Acids Res.* 21 (17), 4127 (1993).

L. Wang and T.A. Keiderling, Vibrational Circular Dichroism Studies of the a-to-B Conformational Transition in DNA, *Biochemistry* 31 (42), 10265 (1992).

L. Wang, L. Yang, and T.A. Keiderling, Conformational Phase Transitions (a-B and B-Z) of DNA and Models Using Vibrational Circular Dichroism, *Spec. Publ. R. Soc. Chem.* 94 (Spectrosc. Biol. Mol.), 137 (1991).

P.M. Wood, L.W.L. Woo, J.-R. Labrosse, M.N. Trusselle, S. Abbate, G. Longhi, E. Castiglioni, F. Lebon, A. Purohit, M.J. Reed, and B.V.L. Potter, Chiral Aromatase and Dual Aromatase-Steroid Sulfatase Inhibitors from the Letrozole Template: Synthesis, Absolute Configuration, and *In Vitro* Activity, *J. Med. Chem.* 51 (14), 4226 (2008).

K. Wright, R. Anddad, J.-F. Lohier, V. Steinmetz, M. Wakselman, J.-P. Mazaleyrat, F. Formaggio, C. Peggion, M. De Zotti, T.A. Keiderling, R. Huang, and C. Toniolo, Synthesis, Ion Complexation Study, and 3d-Structural Analysis of Peptides Based on Crown-Carrier, Cα-Methyl-L-Dopa Amino Acids, *Eur. J. Org. Chem.* (7), 1224 (2008).

H.R. Wyssbrod and M. Diem, IR (Vibrational) CD of Peptide β-Turns: A Theoretical and Experimental Study of Cyclo-(-Gly-Pro-Gly-D-Ala-Pro-), *Biopolymers* 32 (9), 1237 (1992).

T. Xiang, D.J. Goss, and M. Diem, Strategies for the Computation of Infrared CD and Absorption Spectra of Biological Molecules: Ribonucleic Acids, *Biophys. J.* 65, 1255 (1993).

P. Xie and M. Diem, Measurement of Dispersive Vibrational Circular Dichroism: Signal Optimization and Artifact Reduction, *Appl. Spectrosc.* 50 (5), 675 (1996).

P. Xie, Q. Zhou, and M. Diem, Conformational Studies of β-Turns in Cyclic Peptides by Vibrational CD, *J. Am. Chem. Soc.* 117 (37), 9502 (1995).

P. Xie and M. Diem, Conformational Studies of Cyclo-(-Pro-Gly-)$_3$ and Its Complexes with Cations by Vibrational Circular Dichroism, *J. Am. Chem. Soc.* 117 (1), 429 (1995).

P. Xie, Q. Zhou, and M. Diem, IR Circular Dichroism of Turns in Small Peptides, *Faraday Discuss* 99, 233 (1994).

Q. Xinhua, M. Citra, N. Ragunathan, T.B. Freedman, and L.A. Nafie, Vibrational Circular Dichroism of 1-Amino-2-Propanol and 2-Amino-1-Propanol: Experiment and Calculation, *Proc. SPIE* 2089, 142 (1993).

Q. Xu and T.A. Keiderling, Optical Spectroscopic Differentiation of Various Equilibrium Denatured States of Horse Cytochrome C, *Biopolymers* 73 (6), 716 (2004).

Y. Yaguchi, A. Nakahashi, N. Miura, D. Sugimoto, K. Monde, and M. Emura, Stereochemical Study of Chiral Tautomeric Flavorous Furanones by Vibrational Circular Dichroism, *Org. Lett.* 10 (21), 4883 (2008).

K. Yamamoto, Y. Nakao, Y. Kyogoku, and H. Sugeta, Vibrational Circular Dichroism in Hydrogen Bond Systems. Part III. Vibrational Circular Dichroism of the OH Stretching Vibrations of 1,2-Diols and β-Methoxy Alcohols, *J. Mol. Struct.* 242, 75 (1991).

S. Yamamoto, X. Li, K. Ruud, and P. Bouř, Transferability of Various Molecular Property Tensors in Vibrational Spectroscopy, *J. Chem. Theory Comput.* 8 (3), 977 (2012).

G. Yang and Y. Xu, Probing Chiral Solute-Water Hydrogen Bonding Networks by Chirality Transfer Effects: A Vibrational Circular Dichroism Study of Glycidol in Water, *J. Chem. Phys.* 130 (16), 164506/1 (2009).

G. Yang and Y. Xu, The Effects of Self-Aggregation on the Vibrational Circular Dichroism and Optical Rotation Measurements of Glycidol, *Phys. Chem. Chem. Phys.* 10 (45), 6787 (2008).

D. Yang and A. Rauk, The *A Priori* Calculation of Vibrational Circular Dichroism Intensities, *Rev. Comput. Chem.* 7, 261 (1996).

D. Yang and A. Rauk, Vibrational Circular Dichroism Intensities by *Ab Initio* Second-Order Møller-Plesset Vibronic Coupling Theory, *J. Chem. Phys.* 100 (11), 7995 (1994).

L. Yang and T.A. Keiderling, Vibrational CD Study of the Thermal Denaturation of Poly(Ra) 3Poly(Ru), *Biopolymers* 33 (2), 315 (1993).

D. Yang and A. Rauk, Sum Rules for Atomic Polar and Axial Tensors from Vibronic Coupling Theory, *Chem. Phys.* 178 (1–3), 147 (1993).

D. Yang, T. Eggimann, H. Wieser, A. Rauk, and G. Shustov, Local and Framework Stereochemical Markers in Vibrational Circular Dichroism: 1,2- and 2,3-Dimethylaziridines, *Can. J. Chem.* 71 (12), 2028 (1993).

D. Yang and A. Rauk, Vibrational Circular Dichroism Intensities: *Ab Initio* Vibronic Coupling Theory Using the Distributed Origin Gauge, *J. Chem. Phys.* 97 (9), 6517 (1992).

W.J. Yang, P.R. Griffiths, and G.J. Kemeny, Vibrational Circular Dichroism Measurements by Optical Subtraction FT-IR Spectrometry, *Appl. Spectrosc.* 38 (3), 337 (1984).

S.C. Yasui, P. Pancoska, R.K. Dukor, T.A. Keiderling, V. Renugopalakrishnan, M.J. Glimcher, and R.C. Clark, Conformational Transitions in Phosvitin with pH Variation: Vibrational Circular Dichroism Study, *J. Biol. Chem.* 265 (7), 3780 (1990).

S.C. Yasui and T.A. Keiderling, Vibrational Circular Dichroism of Polypeptides and Proteins, *Mikrochimica Acta* II (1–6), 325 (1988).

S.C. Yasui, T.A. Keiderling, and M. Sisido, Vibrational Circular Dichroism of Polypeptides. 11. Conformation of Poly (L-Z-Lysine-L-Z-Lysine-L-1-Pyrenylalanine) and Poly (L-Z-Lysine-L-Z-Lysine-L-1-Naphthylalanine) in Solution, *Macromolecules* 20 (10), 2403 (1987).

S.C. Yasui, T.A. Keiderling, and R. Katakai, Vibrational CD of Polypeptides. X. A Study of α-Helical Oligopeptides in Solution, *Biopolymers* 26 (8), 1407 (1987).

S.C. Yasui and T.A. Keiderling, Vibrational Circular Dichroism of Optically Active Cyclopropanes. 3. Trans-2-Phenylcyclopropanecarboxylic Acid Derivatives and Related Compounds, *J. Am. Chem. Soc.* 109 (8), 2311 (1987).

S.C. Yasui, T.A. Keiderling, F. Formaggio, G.M. Bonora, and C. Toniolo, Vibrational Circular Dichroism of Polypeptides. 9. A Study of Chain Length Dependence for 3$_{10}$-Helix Formation in Solution, *J. Am. Chem. Soc.* 108 (16), 4988 (1986).

S.C. Yasui, T.A. Keiderling, G.M. Bonora, and C. Toniolo, Vibrational Circular Dichroism of Polypeptides, V. A Study of 3$_{10}$-Helical-Octapeptides, *Biopolymers* 25, 79 (1986).

S.C. Yasui and T.A. Keiderling, Vibrational Circular Dichroism of Polypeptides. 8. Poly(Lysine) Conformations as a Function of pH in Aqueous Solution, *J. Am. Chem. Soc.* 108 (18), 5576 (1986).

S.C. Yasui and T.A. Keiderling, Vibrational Circular Dichroism of Polypeptides. VI. Polytyrosine α-Helical and Random-Coil Results, *Biopolymers* 25 (1), 5 (1986).

G. Yoder, A. Polese, R.A.G.D. Silva, F. Formaggio, M. Crisma, Q.B. Broxterman, J. Kamphuis, C. Toniolo, and T.A. Keiderling, Conformational Characterization of Terminally Blocked L-(α Me) Val Homopeptides Using Vibrational and Electronic Circular Dichroism. 3_{10}-Helical Stabilization by Peptide-Peptide Interaction, *J. Am. Chem. Soc.* 119 (43), 10278 (1997).

G. Yoder, P. Pancoska, and T.A. Keiderling, Characterization of Alanine-Rich Peptides, Ac-(Aakaa)$_N$-Gy-Nh$_2$ (N=1,4), Using Vibrational Circular Dichroism and Fourier Transform Infrared. Conformational Determination and Thermal Unfolding, *Biochemistry* 36 (49), 15123 (1997).

G. Yoder, T.A. Keiderling, F. Formaggio, M. Crisma, C. Toniolo, and J. Kamphuis, Helical Screw Sense of Homo-Oligopeptides of C^α-Methylated α-Amino Acids as Determined with Vibrational Circular Dichroism, *Tetrahedron Asymm.* 6 (3), 687 (1995).

G. Yoder, T.A. Keiderling, F. Formaggio, M. Crisma, and C. Toniolo, Characterization of β-Bend Ribbon Spiral Forming Peptides Using Electronic and Vibrational CD, *Biopolymers* 35 (1), 103 (1995).

D.A. Young, T.B. Freedman, and L.A. Nafie, Vibrational Circular Dichroism in Transition-Metal Complexes. 4. Solution Conformation of Λ-Mer- and Λ-Fac-Tris(β-Alaninato) Cobalt(III), *J. Am. Chem. Soc.* 109 (25), 7674 (1987).

D.A. Young, T.B. Freedman, E.D. Lipp, and L.A. Nafie, Vibrational Circular Dichroism in Transition-Metal Complexes. 2. Ion Association, Ring Conformation, and Ring Currents of Ethylenediamine Ligands, *J. Am. Chem. Soc.* 108 (23), 7255 (1986).

D.A. Young, E.D. Lipp, and L.A. Nafie, Vibrational Circular Dichroism in Bis(Acetylacetonato) (L-Alaninato) Cobalt (III). Isolated Occurrences of the Coupled Oscillator and Ring Current Intensity Mechanisms, *J. Am. Chem. Soc.* 107, 6205 (1985).

K. Zhang, K.-S. Cao, G. Wang, Y.-X. Peng, Z.-L. Chu, and W. Huang, Reversible Color Changes of Chiral Double Salts of D-(+)-Camphoric Diammonium and Transition-Metal Sulfates via Gain and Loss of Crystalline and Coordination Water Molecules, *Inorg. Chim. Acta* 378 (1), 224 (2011).

P. Zhang, P.L. Polavarapu, J. Huang, and T. Li, Spectroscopic Rationalization of the Separation Abilities of Decaproline Chiral Selector in Dichloromethane-Isopropanol Solvent Mixture, *Chirality* 19 (2), 99 (2007).

P. Zhang and P.L. Polavarapu, Spectroscopic Investigation of the Structures of Dialkyl Tartrates and Their Cyclodextrin Complexes, *J. Phys. Chem. A* 111 (5), 858 (2007).

P. Zhang and P. L. Polavarapu, Vibrational Circular Dichroism of Matrix-Assisted Amino Acid Films in the Mid-Infrared Region, *Appl. Spectrosc.* 60 (4), 378 (2006).

C. Zhao and P.L. Polavarapu, Comparative Evaluation of Vibrational Circular Dichroism and Optical Rotation for Determination of Enantiomeric Purity, *Appl. Spectrosc.* 55 (7), 913 (2001).

C. Zhao and P.L. Polavarapu, Vibrational Circular Dichroism of Gramicidin D in Vesicles and Micelles, *Biopolymers (Biospectroscopy)* 62 (6), 336 (2001).

C. Zhao, P.L. Polavarapu, H. Grosenick, and V. Schurig, Vibrational Circular Dichroism, Absolute Configuration and Predominant Conformations of Volatile Anesthetics: Enflurane, *J. Mol. Struct.* 550–551, 105 (2000).

C. Zhao, P.L. Polavarapu, C. Das, and P. Balaram, Vibrational Circular Dichroism of β-Hairpin Peptides, *J. Am. Chem. Soc.* 122 (34), 8228 (2000).

C. Zhao and P.L. Polavarapu, Vibrational Circular Dichroism of Gramicidin D in Organic Solvents, *Biospectroscopy* 5 (5), 276 (1999).

C. Zhao and P.L. Polavarapu, Vibrational Circular Dichroism Is an Incisive Structural Probe: Ion-Induced Structural Changes in Gramicidin D, *J. Am. Chem. Soc.* 121 (48), 11259 (1999).

P. Zhu, G. Yang, M.R. Poopari, Z. Bie, and Y. Xu, Conformations of Serine in Aqueous Solutions as Revealed by Vibrational Circular Dichroism, *Chem. Phys. Chem.* 13 (5), 1272 (2012).

W.M. Zuk, T.B. Freedman, and L.A. Nafie, Vibrational CD Studies of the Solution Conformation of Simple Alanyl Peptides as a Function of pH, *Biopolymers* 28 (5), 2025 (1989).

W.M. Zuk, T.B. Freedman, and L.A. Nafie, Vibrational Circular Dichroism in the CH-Stretching Region of L-α-Amino Acids as a Function of pH, *J. Phys. Chem.* 93 (5), 1771 (1989).

Index